GIS, Environmental Modeling and Engineering

SECOND EDITION

GIS, Environmental Modeling and Engineering

SECOND EDITION

Allan Brimicombe

CRC Press
Taylor & Francis Group
Boca Raton   London   New York

CRC Press is an imprint of the
Taylor & Francis Group, an **informa** business

CRC Press
Taylor & Francis Group
6000 Broken Sound Parkway NW, Suite 300
Boca Raton, FL 33487-2742

First issued in paperback 2020

© 2010 by Taylor and Francis Group, LLC
CRC Press is an imprint of Taylor & Francis Group, an Informa business

No claim to original U.S. Government works

ISBN-13: 978-0-367-57719-3 (pbk)
ISBN-13: 978-1-4398-0870-2 (hbk)

Library of Congress Cataloging-in-Publication Data

Brimicombe, Allan.
    GIS, environmental modeling and engineering / Allan Brimicombe -- 2nd ed.
        p. cm.
    Includes bibliographical references and index.
    ISBN 978-1-4398-0870-2 (hardcover : alk. paper)
    1. Geographic information systems. 2. Environmental sciences--Mathematical models. 3. Environmental engineering--Mathematical models. I. Title.

G70.212.B75 2010
628.0285--dc22   2009035961

Visit the Taylor & Francis Web site at
http://www.taylorandfrancis.com

and the CRC Press Web site at
http://www.crcpress.com

Contents

Section II

Section III

Acknowledgments

First Edition

First, a heartfelt thanks to my wife, Lily, for her unwavering support in this venture and for her hard work in preparing most of the figures.

Second, I would like to thank my colleague, Dr. Yang Li, for his assistance with some of the figures and particularly for the preparation of the coastal oil-spill modeling examples.

Third, I would like to thank Professor Li Chuan-tang for his invaluable insights into finite element methods.

Fourth, I would like to thank my sequential employers—Binnie & Partners International (now Binnie Black & Veatch, Hong Kong); Hong Kong Polytechnic University; University of East London—for providing me with the opportunities and space to do so much.

Second Edition

Again I must thank my wife, Lily, for all her effort in recapturing the figures and for reformatting and preparing the publisher's electronic copy of the first edition for me to work on.

My thanks to Irma Shagla and other staff at Taylor & Francis for supporting and seeing this project through.

The Author

Professor **Allan J. Brimicombe** is the Head of the Centre for Geo-Information Studies at University of East London, United Kingdom. He holds a BA (Hons) in Geography from Sheffield University, an MPhil in Applied Geomorphology, and a PhD in Geo-Information Systems both from the University of Hong Kong. Professor Brimicombe is a chartered geographer and is a Fellow of the Royal Geographical Society, the Geological Society, and the Royal Statistical Society. He was employed in the Far East for 19 years, first as an engineering geomorphologist with Binnie & Partners International (now Black & Veatch) including being general manager of a subsidiary company, Engineering Terrain Evaluation Ltd. In 1989, Professor Brimicombe joined the Hong Kong Polytechnic University where he founded the Department of Land Surveying and Geo-Informatics. Here he pioneered the use of geo-information systems (GIS) and environmental modeling as spatial decision support systems. In 1995, he returned to the United Kingdom as professor and head of the School of Surveying at the University of East London. His research interests include data quality issues, the use of GIS and numerical simulation modeling, spatial data mining and analysis, and location-based services (LBS).

Abbreviations

ABM: agent-based modeling
AI: artificial intelligence
ANN: artificial neural networks
API: aerial photographic interpretation
BMP: basin management plans
CA: cellular automata
CBR: case-based reasoning
CN: runoff curve number
DDE: dynamic data exchange
DEM: digital elevation model
DIME: dual independent map encoding
DSS: decision support systems
EIA: environmental impact assessment
EIS: environmental impact statement
fBm: fractional Brownian motion
FDM: finite difference method
FEM: finite element method
FoS: factor of safety
GI: geo-information
GIS: geographical information systems
GLUE: generalized likelihood uncertainty estimator
GPS: global positioning system
GPZ: Geo-ProZone, geographical proximity zones
HKDSD: Drainage Services Department, Hong Kong Government
HTML: hypertext markup language
ICS: index of cluster size
IDW: inverse distance weighted
KBS: knowledge-based systems
LBS: location-based services
LiDAR: light distancing and ranging
MAUP: modifiable areal unit problem
MC: Monte Carlo (analysis)
MCC: map cross-correlation
NEC: no effect concentration
NEPA: National Environmental Policy Act (U.S.)
NIMBY: not in my back yard
NVDI: normalized vegetation difference index
OAT: one-at-a-time
OLE: object linking and embedding
OO: object-oriented

ORDBMS: object-relational database management system
PCC: proportion correctly classified
PEC: predicted environmental concentration
PDF: probability density function
PGIS: participatory GIS
QAE: quality analysis engine
RAISON: regional analysis by intelligent systems on microcomputers
RDBMS: relational database management system
REA: representative elementary area
RS: remote sensing
SA: sensitivity analysis
SCS: Soil Conservation Service (U.S.)
SDSS: spatial decision support systems
TIN: triangular irregular networks
UA: uncertainty analysis
WWW: world wide web

Statement on Trade Names and Trademarks

In a book such as this, it is inevitable that proprietary or commercial products will be referred to. Where a name is used by a company to distinguish its product, which it may claim as a trade name or trademark, then that name appears in this book with an initial capital or all capital letters. Readers should contact the appropriate companies regarding complete information. Use of such names is to give due recognition to these products in illustrating different approaches and concepts and providing readers with practical information. Mention of proprietary or commercial products does not constitute an endorsement, or indeed, a refutation of these products.

1

Introduction

I wish to begin by explaining why this book has been written. Peter Fleming, in writing about his travels in Russia and China in 1933, put the need for such an explanation this way:

> With the possible exception of the Equator, everything begins somewhere. Too many of those who write about their travels plunge straight in *medias res*; their opening sentence informs us bluntly and dramatically that the prow (or bow) of the dhow grated on the sand, and they stepped lightly ashore. No doubt they did. But why? With what excuse? What other and anterior steps had they taken? Was it boredom, business, or a broken heart that drove them so far afield? We have a right to know.
>
> **Peter Fleming**
> *One's Company (1934)*

In 2003, I wrote in the first edition of this book: "At the time of writing this introduction, the President of the United States, George W. Bush, has already rejected the Kyoto Agreement on the control of greenhouse gas emissions; European leaders appear to be in a dither and ecowarriors alongside anti-capitalists have again clashed with riot police in the streets." A key change since then has been the Stern Review (Stern, 2006) on the economics of climate change. The likely environmental impact of climate change trajectories—rising sea levels permanently displacing millions of people, declining crop yields, more than a third of species facing extinction—had already been well rehearsed. What had not been adequately quantified and understood was the likely cost to the global economy (a 1% decline in economic output and 4% decline in consumption per head for every 1°C rise in average temperature) and that the cost of stabilizing the situation would cost about 1% of gross domestic product (GDP). It seemed not too much to pay, but attention is now firmly focused on the "credit crunch'" and the 2008 collapse of the financial sector. In the meantime, annual losses in natural capital worth from deforestation alone far exceed the losses of the current recession, severe as it is. Will it take ecological collapse to finally focus our attention on where it needs to be? This book has been written because, like most of its readers, I have a concern for the *quality of world* we live in, the urgent need for its maintenance and where necessary, its repair. In this book I set out what I believe is a key approach to problem solving and conflict resolution through the analysis and modeling of spatial phenomena. Whilst this book alone will perhaps not safeguard our world, you the reader on finishing this book will have much to contribute.

The phrase *quality of world* used above has been left intentionally broad, even ambiguous. It encompasses:

- Our natural environment—climate, soils, oceans, biological life (plants, animals, bacteria)—that can both nurture us and be hazards to us.
- The built environment that we have created to protect and house ourselves and to provide a modified infrastructure within which we can prosper.
- The economic environment that sustains our built environment and allows the organization of the means of production.
- The social, cultural, and legal environments within which we conduct ourselves and our interactions with others.

These environments are themselves diverse, continually evolving and having strong interdependence. Each of them varies spatially over the face of the globe mostly in a transition so that places nearer to each other are more likely to be similar than those farther apart. Some abrupt changes do, of course, happen, as, for example, between land and sea. They also change over time, again mostly gradually, but catastrophic events and revolutions do happen. Together they form a complex mosaic, the most direct visible manifestation being land cover and land use—our evolved cultural landscapes. Furthermore, the interaction of these different aspects of environment gives enormous complexity to the notion of "quality of life" for our transient existence on Earth. Globalization may have been a force for uniformity in business and consumerism, but even so businesses have had to learn to be spatially adaptive, so-called *glocalization*. When it comes to managing and ameliorating our world for a sustainable quality of life, there is no single goal, no single approach, no theory of it all. Let's not fight about it. Let us celebrate our differences and work toward a common language of understanding on how we (along with the rest of nature) are going to survive and thrive.

Metaphors of Nature

We often use metaphors as an aid in understanding complexity, none more so perhaps than in understanding nature and our relationship within it. These metaphors are inevitably bound up in philosophies of the environment, or knowledge of how the environment works and the technology available to us to modify/ameliorate our surrounding environment. Thus, for millennia, environmental knowledge was enshrined in folklore derived from the trial and error experiences of ancestors. Archaeology has revealed patterns of site selection that changed as we developed primitive technologies or adapted to new environments. Places for habitation had to satisfy the needs for water, food, raw materials, shelter, and safety, and humans learned to recognize those sites that offered the greatest potential for their mode of existence. Examples are numerous: caves near the feeding or watering places of animals; Neolithic cultivation of well-drained, easily worked river terraces; early fishing communities on raised beaches behind sheltered bays and so on. Undoubtedly mistakes were made and communities decimated, but those that survived learned to observe certain environmental truths or inevitabilities.

Successful early civilizations were those that had social structures that allowed them to best use or modify the landforms and processes of their physical environment. Thus, the Egyptians, Mesopotamians, and Sumerians devised irrigation systems to regulate and distribute seasonally fluctuating water supplies, while the Chinese and Japanese included widespread terracing as a means of increasing the amount of productive land. More than 2,500 years ago, the Chinese developed the Taoist doctrine of nature, in which the Earth and the sky had their own "way" or "rule" to maintaining harmony. Human beings should follow and respect nature's way or risk punishment in the form of disasters from land and sky. Thus, even at that time there were laws governing, for example, minimum mesh size on fishing nets so that fish would not be caught too young. Of course, our stewardship has not always been a continual upward journey of success. Some human civilizations have collapsed spectacularly through environmental impact and loss of natural resources (Tickell, 1993; Diamond, 2005). These disasters aside, the dominant metaphor was of "Mother Earth": a benevolent maker of life, a controlling parent that could provide for our needs, scold us when we erred, and, when necessary, put all things to right.

The industrial revolution allowed us to ratchet up the pace of development. Early warnings of the environmental consequences, such as from Marsh (1864), were largely ignored as the Victorians and their European and North American counterparts considered themselves above nature in the headlong rush to establish and exploit dominions. Our technologies have indeed allowed us to ameliorate our lifestyle and modify our environment on an unprecedented scale—on a global scale. But, from the 1960s, the cumulative effect of human impact on the environment and our increasing exposure to hazard finally crept onto the agenda and remains a central issue today. The rise of the environmental movement brought with it a new metaphor—Spaceship Earth—that was inspired by photos from the Apollo moon missions of a small blue globe rising above a desolate moonscape. We were dependant on a fragile life-support system with no escape, no prospect of rescue, if it were to irreparably break down. This coincided with the publication of seminal works, such as Rachel Carson's (1963) *Silent Spring*, which exposed the effects of indiscriminate use of chemical pesticides and insecticides;

McHarg's (1969) *Design with Nature*, which exhorted planners and designers to conform to and work within the capacity of nature rather than compete with it; and Schumacher's (1973) *Small Is Beautiful* proposed an economics that emphasized people rather than products and reduced the squandering of our "natural capital." The words *fractal, chaos, butterfly effect*, and *complexity* (Mandelbrot, 1983; Gleick, 1987; Lewin, 1993; Cohen and Stewart, 1994) have since been added to the popular environmental vocabulary to explain the underlying structure and workings of complex phenomena. Added to these is the *Gaia hypothesis* (Lovelock, 1988) in which the Earth is proposed to have a global physiology or may in fact be thought of as a superorganism capable of switching states to achieve its own goals in which we humans may well be (and probably are) dispensable organisms.

A Solution Space?

That we are capable of destroying our life support system is beyond doubt. As a species, we have already been responsible for a considerable number of environmental disasters. If I scan the chapter titles of Goudie's (1997) *The Human Impact Reader*, the list becomes long indeed, including (in no particular order): subsidence, sedimentation, salinization, soil erosion, desiccation, nutrient loss, nitrate pollution, acidification, deforestation, ozone depletion, climate change, wetland loss, habitat fragmentation, and desertification. I could go on to mention specific events, such as Exxon Valdez, Bhopal, and Chernobyl, but this book is not going to be a catalog of dire issues accompanied by finger-wagging exhortations that something must be done. Nevertheless, worrying headlines continue to appear, such as: "Just 100 months left to save the Earth" for a piece on how greenhouse gases may reach a critical level or tipping point beyond which global warming will accelerate out of control (Simms, 2008). One can be forgiven for having an air of pessimism; the environment and our ecosystems are definitely in trouble. But, we are far from empty-handed. We have a rich heritage of science and engineering, a profound knowledge of environmental processes and experience of conservation and restoration. The technologies that have allowed humankind to run out of control in its impact on the environment can surely be harnessed to allow us to live more wisely. Our ingenuity got us here and our ingenuity will have to get us out of it.

As stated above, we need a common language and, in this regard, we have some specific technologies—drawing upon science—that can facilitate this. While humankind has long striven to understand the workings of the environment, it has only been in the past 30 years or so that our data collection and data processing technologies have allowed us to reach a sufficiently detailed understanding of environmental processes so as to create *simulation*

models. I would argue that it is only when we have reached the stage of successful quantitative simulation, can our level of understanding of processes
allow us to confidently manage them. This is the importance of *environmental
modeling*. Facilitated by this in a parallel development has been *environmental
engineering*. Engineering also has a rich history, but while traditionally engineering has focused on the utilization of natural resources, environmental
engineering has recently developed into a separate discipline that focuses
on the impact and mitigation of environmental contaminants (Nazaroff and
Alvarez-Cohen, 2001). While most management strategies arising out of environmental modeling will usually require some form of engineering response
for implementation, environmental engineering provides solutions for managing water, air, and waste. Engineering in the title of this book refers to the
need to design workable solutions; such designs are often informed by computational or simulation modeling. The youngest technology I would like to
draw into this recipe for a common language is *geographic information systems*
(GIS). Because environmental issues are inherently spatial—they occur somewhere, often affecting a geographic location or area—their spatial dimension
needs to be captured if modeling and engineering are to be relevant in solving specific problems or avoiding future impacts. GIS have proved successful
in the handling, integration, and analysis of spatial data and have become an
easily accessible technology. While the link between simulation modeling and
engineering has been longstanding, the link between GIS and these technologies is quite new, offers tremendous possibilities for improved environmental
modeling and engineering solutions, and can help build these into versatile
decision support systems for managing, even saving our environment. And
that is why I have written this book.

Scope and Plan of This Book

From the early 1990s onwards, there has been an accelerating interest in the
research and applications of GIS in the field of environmental modeling.
There have been a few international conferences/workshops on the subject—
most notably the series organized by the National Center for Geographic
Information and Analysis (NCGIA), University of California, Santa Barbara
in 1991, 1993, 1996, and 2000—and have resulted in a number of edited collections of papers (Goodchild et al., 1993; 1996; Haines-Young et al., 1993; NCGIA,
1996; 2000) as well as a growing number of papers in journals, such as the
*International Journal of Geographical Information Science, Transactions in GIS,
Hydrological Processes, Computers Environment and Urban Systems, ASCE Journal
of Environmental Engineering, Photogrammetric Engineering and Remote Sensing,
Computers and Geosciences*, and so on. But, working with GIS and environmental simulation models is not just a case of buying some hardware, some

software, gathering some data, putting it all together and solving problems with the wisdom of a sage. While technology has simplified many things, there still remain many pitfalls, and users need to be able to think critically about what they are doing and the results that they get from the technology. Thus, the overall aim of this book is to provide a structured, coherent text that not only introduces the subject matter, but also guides the reader through a number of specific issues necessary for critical usage. This book is aimed at final-year undergraduates, postgraduates, and professional practitioners in a range of disciplines from the natural sciences, social sciences to engineering, at whatever stage in their lifelong learning or career they need or would like to start working with GIS and environmental models. The focus is on the use of these two areas of technology in tandem and the issues that arise in so doing. This book is less concerned with the practicalities of software development and the writing of code (e.g., Payne, 1982; Kirkby et al., 1987; Hardisty et al., 1993; Deaton and Winebrake, 2000; Wood, 2002). Nor does it consider in detail data collection technologies, such as remote sensing, GPS, data loggers, and so on, as there are numerous texts that already cover this ground (e.g., Anderson and Mikhail, 1998; Skidmore, 2002).

The overall thrust of this book can be summarized in the mapping:

$$f: \Omega \to \Re \qquad (1.1)$$

where Ω = set of domain inputs, $\Re$ = set of real decisions. In other words, all decisions (including the decision not to make a decision) should be adequately evidenced using appropriate sources of information. This is perhaps stating the obvious, but how often, in fact, is there insufficient information, a hunch, or a gut feeling? GIS, environmental modeling, and engineering are an approach to generating robust information upon which to make decisions about complex spatial issues.

The subject matter is laid out in three sections. Section I concentrates uniquely on GIS: what they are, how data are structured, what are the most common types of functionality. GIS will be viewed from the perspective of a technology, the evolution of its scientific basis, and, latterly, its synergies with other technologies within a geocomputational paradigm. This is not intended to be an exhaustive introduction as there are now many textbooks that do this (e.g., Chrisman, 1997; Burrough and McDonnell, 1998; Longley et al., 2005; Heywood et al., 2006) as well as edited handbooks (e.g., Wilson and Fotheringham, 2008). Rather, its purpose is to lay a sufficient foundation of GIS for an understanding of the substantive issues raised in Section III. Section II similarly focuses on modeling both from a neutral scientific perspective of its role in simulating and understanding phenomena and from a more specific perspective of environmental science and engineering. Section III is by far the largest. It looks at how GIS and simulation modeling are brought together, each adding strength to the other. There are examples of case studies and chapters covering specific issues, such as interoperability, data quality, model validity, space-time dynamics, and decision-support systems. Those readers who already have a substantial knowledge of GIS or have completed undergraduate studies in GIS may wish to skip much of Section I and move quickly to Sections II and III. Those readers from a simulation modeling background in environmental science or engineering should read Section I, skim through Section II, and proceed to Section III. In a book such as this, it is always possible to write more about any one topic; there are always additional topics that a reader might consider should be added. There are, for example, as many environmental models as there are aspects of the environment. GIS, environmental modeling, and engineering are quite endless and are themselves evolving. Also, I have tried not to focus on any one application of simulation modeling. Given its popularity, there is a temptation to focus on GIS and hydrology, but that would detract from the overall purpose of this book, which is to focus on generic issues of using GIS and external simulation models to solve real problems. Presented in the following chapters is what I consider to be a necessary understanding for critical thinking in the usage of such systems and their analytical outputs. Enjoy.

Section I

2

From GIS to Geocomputation

The cosmological event of the Big Bang created the universe and in so doing space–time emerged (some would say "switched on") as an integral aspect of gravitational fields. Space and time are closely interwoven and should more properly be thought of as a four-dimensional (4D) continuum in which time and space, over short durations, are interchangeable. Nevertheless, we conventionally think of separate one-dimensional (1D) time and three-dimensional (3D) space. The terrestrial space on which we live, the Earth, is at least 4.5 billion years old and has been around for about 40% of the time since time began. Since our earliest prehistory, we have grappled with the problems of accurately measuring time and space. Crude measures of time probably came first given the influences of the regular cycles of the day, tides, the moon, and seasons on our lives as we evolved from forager to agriculturist. With technology, we have produced the atomic clock and the quartz watch. Measuring position, distances, and area were less obvious in the absence of the type of benchmark that the natural cycles provided for time. Early measurements used a range of arbitrary devices—the pace, the pole, the chain— and longer distances tended to be equated with the time it took to get to destinations. Much later, the development of accurate clocks was the key to solving the problem of determining longitudinal position when coupled with observations of the sun. Measurement requires numerical systems, and 1D time requires either a linear accumulation (e.g., age) or a cyclical looping (e.g., time of day). Measurement of 3D space requires the development of higher order numerical systems to include geometry and trigonometry. Let us not forget that at the root of algebra and the use of algorithms was the need for precise partitioning of space (land) prescribed by Islamic law on inheritance. Calculus was developed with regard to the changing position (in time) of objects in space as a consequence of the forces acting upon them.

Three fundamental aspects of determining position are: a datum, a coordinate system (both incorporating units of measurement), and an adequate representation of the curved (or somewhat crumpled) surface of the Earth in the two dimensions of a map, plan, or screen. The establishment of a datum and coordinate system is rooted in geodetic surveying, which aims to precisely determine the shape and area of the Earth or a portion of it through the establishment of wide-area triangular networks by which unknown locations can be tied into known locations. Cartographers aim to represent geographic features and their relationships on a plane. This involves both the *art* of reduction, interpretation, and communication of geographic features

11

and the *science* of transforming coordinates from the spherical to a plane through the construction and utilization of map projections. The production of quality spatial data used to be a time-consuming, expensive task and for much of the twentieth century there was a spatial data "bottleneck" that held back the wider use of such data. Technology has provided solutions in the form of the global positioning system (GPS), electronic total stations, remote sensing (RS), digital photogrammetry, and geographic information systems (GIS). GPS, RS, and GIS are now accessible to every citizen through inexpensive devices and the Internet. Determining where is no longer difficult and, through mobile devices such as GPS-enabled smartphones, determining one's geographic position and location has become no more difficult than telling the time.

This chapter will chart the rise of the GIS as a *technology*, consider its main paradigms for representing the features of the Earth and structuring data about them. The basic functionality of GIS will be described with examples. A "systems" view of GIS will then be developed bringing us to the point where GIS can be formally defined. The limitations of modern GIS will be discussed leading us to consider the rise of geocomputation as a new paradigm and the role of GIS within it.

In the Beginning …

It would be nice to point to a date, a place, an individual and say, "That's where it all started, that's the father of GIS." But no. As Coppock and Rhind put it in their article on the History of GIS (1991), "unhappily, we scarcely know." In the beginning, of course, there were no GIS "experts" and nobody specifically set out to develop a new body of technology nor a new scientific discipline for that matter. In the mid-1960s, there were professionals from a range of disciplines, not many and mostly in North America, who were excited by the prospect of handling spatial data digitally. There were three main focal points: the Harvard Graduate School of Design, the Canada Land Inventory, and the U.S. Census Bureau. In each of these organizations were small groups of pioneers who made important contributions toward laying the foundations for today's GIS industry.

The significance of the Harvard Graduate School of Design lies in its Laboratory for Computer Graphics and Spatial Analysis, a mapping package called SYMAP (1964), two prototype GIS, called GRID (1967), and ODYSSEY (c. 1978), and a group of talented individuals within the laboratory and the wider graduate school: N. Chrisman, J. Dangermond, H. Fisher, C. Steinitz, D. Sinton, T. Peucker, and W. Warntz, to name a few. The creator of SYMAP was Howard Fisher, an architect. His use of line printers to produce three types of map—isoline, choropleth, and proximal—was a

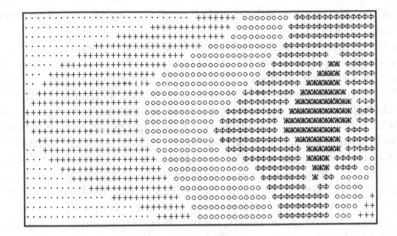

FIGURE 2.1
Sample of a SYMAP-type line printer contour map showing emphasis on similarities. The contour lines are perceived only through the "gap" between the areas of printed symbols.

way of visualizing or recognizing spatial similarities or groupings in human and physical phenomena (McHaffie, 2000). The other leap was a recognition (rightly or wrongly) that just about any such phenomenon, no matter how ephemeral or whether described quantitatively or qualitatively could be represented as a map of surfaces or regions. The printing of these maps using equally spaced characters or symbols, line by line, naturally resulted in a "blocky," cell-based map representation (Figure 2.1). David Sinton, a landscape architect, took cell-based (raster) mapping forward with GRID, which allowed analyses to include several thematic data sets (layers) for a given area. Furthermore, by 1971 a rewrite of GRID allowed users to define their own logical analyses rather than being restricted to a limited set of prepackaged procedures. Thus, a flexible user interface had been developed. By the late 1970s, ODYSSEY, a line-based (vector) GIS prototype had been written capable of polygon overlay. In this way, it can be seen that the overlay or co-analysis of several thematic layers occupied the heart of early GIS software strategies (Chrisman, 1997).

In 1966, the Canada Geographic Information System (CGIS) was initiated to serve the needs of the Canada Land Inventory to map current land uses and the capability of these areas for agriculture, forestry, wildlife, and recreation (Tomlinson, 1984). Tomlinson had recognized some years earlier that the manual map analysis tasks necessary for such an inventory over such a large area would be prohibitively expensive and that a technological solution was necessary. Within this solution came a number of key developments: optical scanning of maps, raster to vector conversion, a spatial database management system, and a seamless coverage that was nevertheless spatially partitioned into "tiles." The system was not fully operational until 1971, but has subsequently grown to become a digital archive of some 10,000 maps (Coppock and Rhind, 1991).

The significance of the U.S. Bureau of Census in developing its Dual Independent Map Encoding (DIME) scheme in the late 1960s is an early example of inserting additional information on spatial relationships into data files through the use of topological encoding. Early digital mapping data sets had been unstructured collections of lines that simply needed to be plotted with the correct symbology for a comprehensible map to emerge. But the demands for analysis of map layers in GIS required a structuring that would allow the encoding of area features (polygons) from lines and their points of intersection, ease identification of neighboring features, and facilitate the checking of internal consistency. Thus, DIME was a method of describing urban structure, for the purposes of census, by encoding the topological relationships of streets, their intersection points at junctions and the street blocks and census tracts that the streets define as area features. The data structure also provided an automated method of checking the consistency and completeness of the street block features (U.S. Bureau of Census, 1970). This laid the foundation of applying topology or graph theory now common in vector GIS.

Technological Facilitation

The rise of GIS cannot be separated from the developments in information and communication technology that have occurred since the 1960s. A timeline illustrating developments in GIS in relation to background formative events in technology and other context is given in Table 2.1. Most students and working professionals today are familiar at least with the PC or Mac. I am writing the second edition of this book in 2008/09 on a notebook PC (1.2 GHz CPU, 1 GB RAM, 100 GB disk, wireless and Bluetooth connectivity) no bigger or thicker than an A4 pad of paper. My GIS and environmental modeling workhorse is an IBM M Pro Intellistation (dual CPU 3.4 GHz each, 3.25 GB RAM, 100 GB disk). They both run the same software with a high degree of interoperability, and they both have the same look and feel with toolbars, icons, and pull-down menus. Everything is at a click of a mouse. I can easily transfer files from one to the other (also share them with colleagues) and I can look up just about anything on the Internet. Even my junk mail has been arriving on CD and DVD, so cheap and ubiquitous has this medium become, and USB data sticks are routinely given away at conferences and exhibitions. It all takes very little training and most of the basic functions have become intuitive. I'm tempted to flex my muscles (well, perhaps just exercise my index finger) for just a few minutes on the GIS in this laptop ... and have indeed produced Figure 2.2—a stark contrast to Figure 2.1.

TABLE 2.1

Timeline of Developments in GIS in Relation to Background Formative Events in Technology and Other Context

Year	GIS	Context
1962		Carson's *Silent Spring*
1963	Canadian Geographic Information System	
1964	Harvard Lab for Computer Graphics & Spatial Analysis	GPS specification
1966	SYMAP	WGS-66
1967	U.S. Bureau of Census DIME	
1968		Relational database defined by Codd
1969	ESRI, Intergraph, Laser-Scan founded	Man on the noon; NEPA; McHarg's *Design with Nature*
1970	Acronym GIS born at IGU/UNESCO conference	Integrated circuit
1971		ERTS/Landsat 1 launched
1973	U.K. Ordnance Survey starts digitizing	
1974	AutoCarto conference series; *Computers & Geosciences*	UNIX
1975		C++; SQL
1978	ERDAS founded	First GPS satellite launched
1980	FEMA integrates USGS 1:2 m mapping into seamless database	
1981	*Computers, Environment & Urban Systems*; Arc/Info launched	8088 chip; IBM PC
1983		Mandelbrot's *The Fractal Geometry of Nature*
1984	1st Spatial Data Handling Symposium	80286 chip, RISC chip; WGS-84
1985		GPS operational
1986	Burrough's *Principles of Geographical Information Systems for Land Resources Assessment*; MapInfo founded	SPOT 1 launched
1987	*International Journal of Geographical Information Systems*; GIS/LIS conference series; "Chorley" Report	80386 chip
1988	NCGIA; *GIS World*, U.K. RRL initiative	Berlin Wall comes down
1989	U.K. Association for Geographic Information	
1990		Berners–Lees launches WWW
1991	USGS digital topo series complete 1st International Symposium on Integrating GIS and Environmental Modeling	Dissolution of Soviet Union
1992		Rio Earth Summit – Agenda 21
1993	GIS Research U.K. conference series	Pentium chip; full GPS constellation
1994	Open GIS Consortium	HTML

Internet; mobile phones

Continued

TABLE 2.1 (*Continued*)

Timeline of Developments in GIS in Relation to Background Formative Events in Technology and Other Context

Year	GIS	Context
1995	OS finished digitizing 230,000 maps	Java
1996	1st International Conference on GeoComputation; *Transactions in GIS*	
1997	*IJGIS* changes "*Systems*" to "*Science*"; last AutoCarto; *Geographical and Environmental Modeling*	Kyoto Agreement on CO_2 reduction
1998	*Journal of Geographical Systems*; last GIS/LIS	GPS selective availability off
2000		"Millennium Bug"
2003	1st ed.: *GIS, Environmental Modeling & Engineering*	
2005	Google Maps; Google Earth	
2006		*Stern Review: The economics of climate change*
2008	Google Street View	

To fully comprehend the technological gulf we have crossed, let me briefly review a late 1970s GIS-based land capability study in South Dakota (Schlesinger et al., 1979). The project was carried out on an IBM 370/145 mainframe computer using 10 standalone program modules written in FORTRAN IV and IBM Assembler. A digitizing tablet and graphics terminal were available, but all hardcopy maps were produced using a line printer. Maps wider than a 132-character strip had to be printed and glued together. The study area covered 115 km²; size of cell was standardized at one acre (~0.4 ha). With the objective to identify land use potential, four base data layers were digitized: 1969 and 1976 land use from aerial photographic interpretation (API), soils, and underlying geology from published map sheets. Through a process

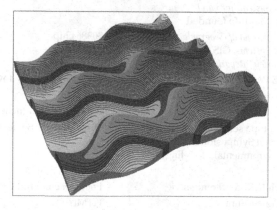

FIGURE 2.2
Laptop GIS of today: 3-D topographic perspective of a landscape.

TABLE 2.2

Multiple Layer Production from Three Source Data Sets

Base Maps → ↓ Factor Maps	1969 Land Use	1976 Land Use	Soils	Geology
Slope			✓	
Flood hazards			✓	
Potential for building sites			✓	
Potential for woodland wildlife habitat			✓	
Potential for rangeland habitat			✓	
Potential for open land habitat			✓	
Limitations to road and street construction			✓	
Limitations for septic tank absorption fields			✓	
Soils of statewide importance for farmland			✓	
Sliding hazards				✓
Groundwater recharge areas				✓
Land use change	✓	✓		
Limitations to sewage lagoons			✓	✓
Important farmland		✓	✓	
Important farmland lost to urban development	✓	✓	✓	
Limitations to urban development			✓	✓
Land suitable for urban development, but not important agricultural land			✓	✓
Limitations for septic tanks	✓	✓	✓	
Limitations for new urban development	✓	✓	✓	✓

Source: Based on Schlesinger, J., Ripple, W., and Loveland, T.R. (1979) *Harvard Library of Computer Graphics* 4: 105–114.

of either reclassification of single layers or a logical combination (overlay) of two or more layers with reclassification, a total of 19 new factor maps were created (Table 2.2) to answer a range of spatial questions where certain characteristics are concerning land suitability for development. Typical of the many pioneering efforts of the time, this study achieved its goals and was well received in the community despite the rudimentary hardware and software tools available.

Some of the changes are obvious. Over the intervening 30 years, the action of Moore's Law, by which the hardware price to performance ratio is expected to double every 18 months, means that the laptop I'm writing on far outstrips the IBM mainframe of that time in terms of power, performance, and storage by several orders of magnitude at a fraction of the cost in real terms. Instead of using a collection of software modules that may need to be modified and recompiled to satisfy the needs of the individual project, we have a choice of off-the-shelf packages (e.g., MapInfo, ArcGIS) that combine a wide range of functionality with mouse- and icon/menu-driven interfaces. For project-specific needs, most of these packages have object-oriented scripting languages

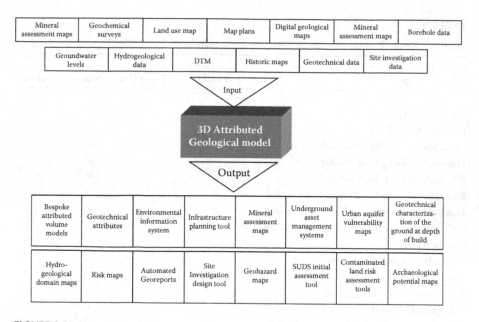

Mineral assessment maps	Geochemical surveys	Land use map	Map plans	Digital geological maps	Mineral assessment maps	Borehole data

Groundwater levels	Hydrogeological data	DTM	Historic maps	Geotechnical data	Site investigation data

Input

3D Attributed Geological model

Output

Bespoke attributed volume models	Geotechnical attributes	Environmental information system	Infrastructure planning tool	Mineral assessment maps	Underground asset management systems	Urban aquifer vulnerability maps	Geotechnical characterization of the ground at depth of build

Hydro-geological domain maps	Risk maps	Automated Georeports	Site Investigation design tool	Geohazard maps	SUDS initial assessment tool	Contaminated land risk assessment tools	Archaeological potential maps

FIGURE 2.3

A contemporary geological application using spatial modeling tools. (Adapted from Royse, K.R., Rutter, H.K., and Entwisle, D.C. (2009) *Bulletin of Engineering Geology and the Environment* 68: 1–16.)

that facilitate customization and the addition of new functionality with many such scripts available over the Internet. Moreover, analysis can now be vastly extended to include external computational models that communicate either through the scripting or use of common data storage formats. Although the availability of digital map data is uneven across the world, particularly when it comes to large-scale mapping, off-the-shelf digital data ready for use in GIS are much more common today to the point where, certainly for projects in North America and Europe, there is hardly the need anymore to manually digitize. As mentioned above, the bottleneck in the production of digital spatial data has been burst not only by technologies, such as GPS, RS, and digital photogrammetry, but through palm-top data loggers, high-speed scanners, digital data transfer standards, and, above all, the computer capacity to cost-effectively store, index, and deliver huge data sets. In contrast to Table 2.2 in which only four data sources were used, Figure 2.3 summarizes the many input sources and output derivative data sets designed by the British Geological Survey in a recent project to build an integrate 3D geological and hydrogeological model. This model is to support development in the Thames Gateway, U.K., which at the time of writing is Europe's largest regeneration program. Nevertheless, despite the technological advancement that has made spatial tools and particular GIS more widespread, sophisticated, and easier to use, many of the underlying principles have remained largely the same.

FIGURE 2.4
A view of a sample landscape. (Photo courtesy of the author.)

Representing Spatial Phenomena in GIS

The dominant paradigm in the way GIS data are structured comes from the idea that studies of landscape (both human and physical) and the solution to problems concerning the appropriate use of land can be achieved by describing the landscape as a series of relevant factor maps or layers that can then be overlaid to find those areas having particular combinations of factors that would identify them as most suited to a particular activity. The methodology in its modern GIS context derives from the seminal work of McHarg (1969) as well as the conventional cartographic tradition of representing spatial phenomena. Although the use of manual overlay of factor maps considerably predates McHarg (Steinitz et al., 1976), he provided a compelling case for the methodology as a means of organizing, analyzing, and visualizing multiple landscape factors within a problem-solving framework. Consider the landscape shown in Figure 2.4.

This landscape can be viewed both holistically as a piece of scenery and as a series of constituent elements, such as its topography, geology, hydrology, slope processes, flora, fauna, climate, and manmade (anthropomorphic) features, to name but a number that could be separated out. At any place within this land-scape there are several or all constituents to be considered: stand on any point and it has its topography, geology, hydrology, microclimate, and so on. Any comprehensive map of all these constituents would quickly become cluttered and complex—almost impossible to work with. So, consider then the mapped constituents of a very similar landscape in Figure 2.5(a–i).

Although this particular landscape has been artificially created to demon-strate a number of issues throughout this book, it illustrates well a number of aspects of the layer or coverage paradigm and the graphic primitives used in any one layer. First, in order for a selection of layers to be used together, superimposed and viewed as a composite, *they must all conform to the same coordinate system and map projection*. This is critically important, otherwise the layers will be distorted and wrongly positioned in relation to one another. Individual layers, however, need not necessarily cover exactly the same area of the landscape in their extent as may happen, for example, if they have been derived from different surveys or source documents. Each layer can neverthe-less be clipped to a specific study area as has happened in Figure 2.5. Second, some of the layers are given to represent *discrete objects* in the landscape (e.g., landslides, streams, land cover parcels) while others represent a *continuous field* (e.g., topography, gradient, rainfall), which varies in its value across the landscape. What aspects of the landscape should be treated as continuous or discrete and how they should be presented cartographically is an old, but significant problem, which can still be debated today (Robinson and Sale, 1969; Peuquet, 1984; Goodchild, 1992a; Burrough, 1992; Burrough and Frank 1996; Spiekermann and Wegener, 2000; Goodchild et al., 2007). To a consider-able extent, it is a matter of data resolution, scale of representation, conven-tion, and convenience. For example, landslides can be quickly mapped at a regional level as individual points representing each scar in the terrain (as in Figures 2.5(h) and 2.6(a)). Another approach would be to represent each landslide as a line starting at the scarp and tracing the down slope extent of the debris to the toe (Figure 2.6(b)). Clearly any laterally extensive landslide in Figure 2.5(h) would represent a methodological problem for which a sin-gle point or a line would be an oversimplification. So, yet another approach would be to represent either the whole landslide or its morphological ele-ments according to a consistent scheme (e.g., source, transport, deposition) as polygons (Figure 2.6(c)). This latter approach, while providing more informa-tion, is more time consuming and expensive to produce. Finally, these land-slides could be represented as a field of varying numbers of landslides within a tessellation of cells (Figure 2.6(d)), or as densities (Figure 5.11(a)).

To pursue this issue just a bit further, topography is a continuous field, but is conventionally represented by contours that in geometric terms are nested polygons. Gradient on the other hand is also a continuous field, but would generally be confusing to interpret if drawn as contours and, thus, is usually represented by a tessellation of cells, each having its own gradient value. Soils are conventionally classified into types and each type is represented

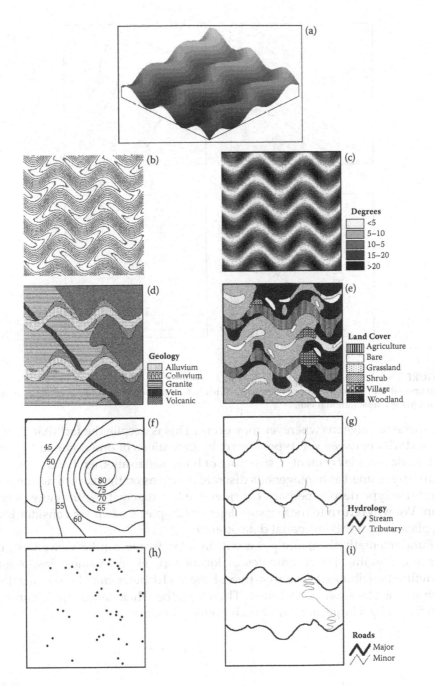

FIGURE 2.5
Mapped constituents of an example landscape in eight layers (coverages): (a) oblique view of topography, (b) contours, (c) slope gradient, (d) geology, (e) land cover, (f) rainfall isohyets from a storm event, (g) drainage network, (h) landslide scars, (i) transport.

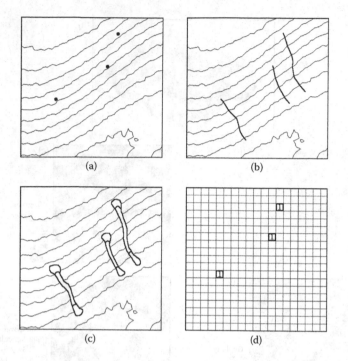

FIGURE 2.6
Four possible methods of representing landslides in GIS: (a) as points, (b) as lines, (c) as polygons, (d) as a tessellation (raster).

by discrete polygons wherever they occur. This is despite the fact that many boundaries between soil types are really gradations of one dominant characteristic (say, clay content or structure of horizons) to another. Land uses are similarly defined as homogenous discrete polygons on the basis of dominant land-use type despite perhaps considerable heterogeneity within any polygon. We will return to these issues later in Chapter 8 when we consider the implications of this on spatial data quality.

Fundamentally then, any point within a landscape can be viewed as an array containing the coordinates of location $\{x, y\}$ and values/classes for n defined attributes a. The first two of these attributes may be specifically defined as elevation z and time t. Therefore, the whole landscape $\mathbf{L}$ can be described by a large number of such points p in a matrix:

$$\mathbf{L} = \begin{pmatrix} x_1 & y_1 & z_1 & t_1 & a_{13} & a_{14} & a_{15} & \bullet & \bullet & \bullet & a_{1n} \\ & & & & & \bullet & & & & & \\ & & & & & \bullet & & & & & \\ & & & & & \bullet & & & & & \\ & & & & & \bullet & & & & & \\ x_p & y_p & z_p & t_p & a_{p3} & a_{p4} & a_{ps} & \bullet & \bullet & \bullet & a_{pn} \end{pmatrix} \qquad (2.1)$$

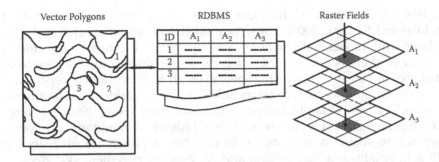

FIGURE 2.7
Basic organization of geometry and attributes in layered GIS: vector and raster.

In practical terms, time *t* is often fixed and the matrix is taken to be a single snapshot of the landscape. Also, because the number of points used to describe the landscape is usually only a tiny proportion of all possible points, **L** is considered to be a sample of one. Elevation *z* is taken to be an attribute of a location and, therefore, is not really a third dimension in the traditional sense of an {*x, y, z*} tuple. GIS are commonly referred to as 2½D rather than 3D. The points themselves can be organized into a series of *points, lines,* or *polygons,* that is, discrete objects of 0, 1, and 2 dimensions, respectively, to form *vector layer(s).* Usually, objects that are points, lines, and polygons are not mixed within a layer, but are kept separate. This describes the planar geometry and disposition of the objects within the landscape. The attributes of each object are stored in a database (either as flat files or in a relational database management system (RDBMS)) and are linked to the graphics via a unique identifier (Figure 2.7). The other approach to **L** is for the landscape to be tessellated, that is, split into a space-filling pattern of *cells* and for each cell to take an attribute value according to the distribution of points to form a *raster layer.* Thus, there may be *n* layers, one for each attribute. Although the objective in both vector and raster approaches is to achieve spatially seamless layers that cover an entire area of interest; it may be that for large areas the data volume in each layer becomes too large and cumbersome to handle conveniently (e.g., response times in display and analysis). When this occurs, layers are usually split into a series of nonoverlapping *tiles,* which when used give the impression of seamless layers.

Thus far, I have described the mainstream approach to representing spatial phenomena in GIS. Since the early 1990s, an alternative has emerged—the object-oriented (OO) view of spatial features, which should not be confused with the above object-based approach of vector representation. Spatial objects as discernible features of a landscape are still the focus, but rather than splitting their various aspects or attributes into layers (the geology, soils, vegetation, hydrology, etc., of a parcel of land), an object is taken as a whole with its properties, graphical representation, and behavior in relation to other spatial objects embedded within the definition of the object itself (Worboys et al., 1990; Milne et al., 1993; Brimicombe and Yeung, 1995; Wachowicz, 1999; Shekhar and Vatsavai, 2008). Thus, the modeling of "what" is separated from "where" and, in fact, both "where" and whether to use raster or vector (or both, or neither) as a means of graphical representation can be viewed as attributes of "what." This then allows even abstract spatial concepts, such as sociocultural constructs to be included in GIS alongside more traditional physical features of a landscape (see Brimicombe and Yeung, 1995). Although from a personal perspective the OO view provides a superior, more robust approach to spatial representation in GIS, the market share for truly OO GIS (e.g., Smallworld, Laser-Scan) and database management systems (e.g., ObjectStore) has remained comparatively small. Instead, hybrid object-relational database management systems (ORDBMS, e.g., Oracle Spatial) have emerged to combine the best of both approaches to database management and spatial query.

Putting the Real World onto Media

Having introduced the representation of geographic phenomena in GIS from a practical "what you see on the screen" perspective, it is now necessary to do so from a computer science "what technically underpins it" perspective. Essentially, we want to achieve a representation of a landscape that can be stored digitally on a machine in such a way that the representation is convenient to handle and analyze using that machine. Ultimately, the intended purpose of the representation, the nature of software tools available and the types of analyses we wish to undertake will strongly influence the form of representation that is deemed appropriate.

A machine representation of a landscape as a digital stream of binary zeros and ones on a hard disk or diskette necessitates a considerable amount of abstraction, to say the least. The process of abstraction and translation into zeros and ones needs to be a formally controlled process if the results are going to be of any use. This process is known as *data modeling* and is discussed at some length by Peuquet (1984) and Molenaar (1998). Two diagrammatic views of the data modeling process are given in Figure 2.8.

In general, four levels can be recognized within data modeling:

1. The first of these is *reality* itself, which is the range of phenomena we wish to model as they actually exist or are perceived to exist in all their complexity.

2. The second level is the *conceptual model*, which is the first stage abstraction and incorporates only those parts of reality considered to be relevant to the particular application. A cartographic map is a good metaphor for the conceptual model as a map only contains

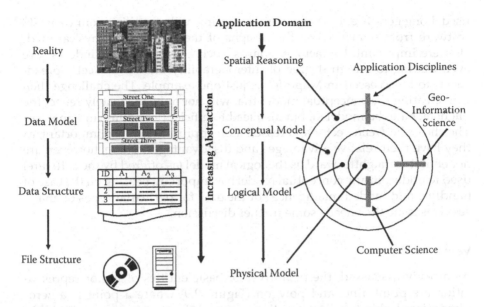

FIGURE 2.8
Stages in the data modeling process. (Partly based on Molenaar, M. (1998) *An introduction to the theory of spatial object modeling*. Taylor & Francis, London.)

those features that the cartographer has chosen to represent and all other aspects of reality are omitted. This provides an immediate simplification, though a sense of the reality can still be readily interpreted or reconstituted from it. Just as a cartographer must decide in creating a map what symbologies should be used for the various features, so it is at the conceptual modeling stage that decisions are generally made as to whether to use raster or vector and what the theme for each layer is going to be. The conceptual model is often referred to as the *data model*, which in a data modeling process can give rise to confusion.

3. The third level is the *logical model*, often called the *data structure*. This is a further abstraction of the conceptual model into lists, arrays, and matrices that represent how the features of the conceptual model are going to be entered and viewed in the database, handled within the code of the software, and prepared for storage. The logical model can generally be interpreted as reality only with the assistance of software, such as by creating a display.

4. The fourth level is the *physical model* or *file structure*. This is the final abstraction and represents the way in which the data are physically stored on the hardware or media as bits and bytes.

The third and fourth levels, the logical and physical models, are usually taken care of in practical terms by the GIS software and hardware being used. Long gone are the days of programming and compiling your own GIS software from scratch when the designs of the logical and physical models were important. De facto standards, such as Microsoft® Windows® are even leading to a high degree of interoperability allowing Excel® spreadsheets to be accessed in MapInfo, as just one example. The challenge then is in creating the conceptual model that will not only adequately reflect the phenomena to be modeled, but also lead to efficient handling and analysis. The choice between vector and tessellation approaches can be important, as they have their relative advantages and disadvantages. These, however, are not entirely straightforward as the logical model (as offered by the software) used to underpin any conceptual model has important bearing on the ease of handling and "added intelligence" of the data for particular types of analyses. This issue then needs some further discussion.

Vector

As already discussed, the primitives or basic entities of vector representation are point, line, and polygon (Figure 2.9) where a point is a zero-dimensional object, a line is a linear connection between two points in one-dimension, and a polygon is one or more lines where the end point of the line or chain of lines coincides with the start point to form a closed two-dimensional (2D) object. A line need not be straight, but can take on any weird shape as long as there are no loops. Any nonstraight line, from a digital perspective, is in fact made up of a series of segments and each segment will, of course, begin and end at a point. In order to avoid confusion then, points at the beginning and end of a line or connecting two or more lines are referred to as *nodes*. Lines connected at their nodes into a series can form a *network*. Polygons (also known as *area features*) when adjacent to one another will share one or more lines. Because all lines have orientation from their start node to their end node, they have a direction and on the basis of this have a left and right side. Thus, within a logical model that records *topology*, which is explicitly recording connectivity (as in a network) or adjacency (as for polygons), the polygon to the left and right of a line can be explicitly recorded in the database (Figure 2.10). In this way, a fully topological database has additional intelligence so that locating neighboring lines and polygons becomes straightforward. Some desktop GIS do not go so far, leaving each feature to be recorded separately without reference to possible neighbors. These are commonly referred to as *shape-files*. Finally, by providing a unique identifier to each point, line, and polygon (usually done automatically by the software), a join can be made to a database containing relevant attributes for each object (see Figure 2.7). Thus, by selecting specific map features in a vector-based GIS, their attributes can be displayed from the database. Conversely, by selecting specific attributes from the database, their spatial representation on the map can be highlighted.

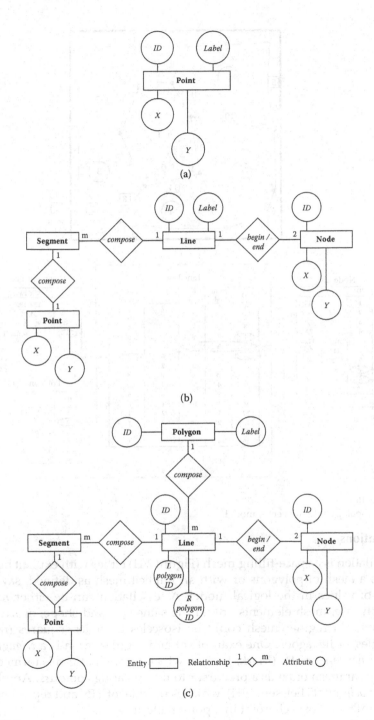

FIGURE 2.9
Entities of the vector model: (a) point, (b) line, (c) polygon.

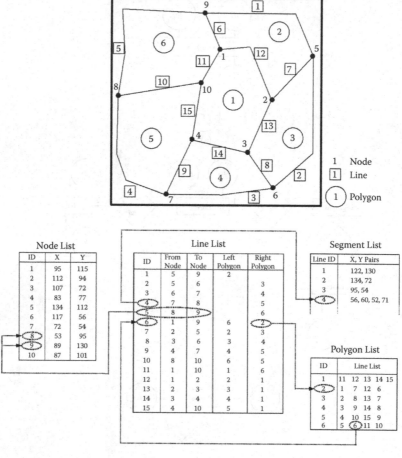

Node List

ID	X	Y
1	95	115
2	112	94
3	107	72
4	83	77
5	134	112
6	117	56
7	72	54
8	53	95
9	89	130
10	87	101

Line List

ID	From Node	To Node	Left Polygon	Right Polygon
1	5	9	2	
2	5	6		3
3	6	7		4
4	7	8		5
5	8	9		6
6	1	9	6	2
7	2	5	2	3
8	3	6	3	4
9	4	7	4	5
10	8	10	6	5
11	1	10	1	6
12	1	2	2	1
13	2	3	3	1
14	3	4	4	1
15	4	10	5	1

Segment List

Line ID	X, Y Pairs
1	122, 130
2	134, 72
3	95, 54
4	56, 60, 52, 71

Polygon List

ID	Line List
1	11 12 13 14 15
2	1 7 12 6
3	2 8 13 7
4	3 9 14 8
5	4 10 15 9
6	5 6 11 10

FIGURE 2.10
Building topology into the vector model.

Tessellations

A tessellation is a space-filling mesh (Figure 2.11) either with explicit boundaries as a mesh of polygons or with an implicit mesh as defined, say, by a matrix of values in the logical model. A tessellation can be either *regular*, in which case, mesh elements are all the same size and shape, or *irregular*. Elements of a regular mesh could be isosceles triangles, squares (raster), rectangles, or hexagons. One example of an irregular mesh is a *triangulated irregular network* or TIN (Mark, 1975) in which a point pattern is formed into a triangular mesh often as a precursor to interpolating contours. Another is *Theissen polygons* (Theissen, 1911), which is the dual of TIN and represents the area of influence of each point in a point pattern.

Tessellations can also be *recursive*, that is, the basic mesh shape can be progressively split into a finer mesh in order to represent higher resolution

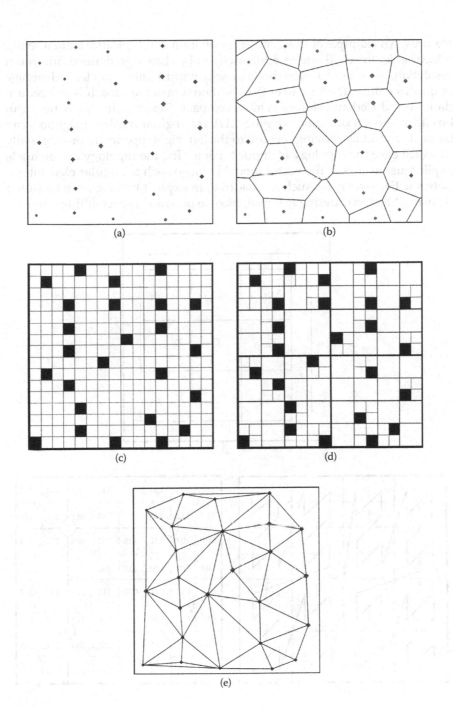

FIGURE 2.11
Examples of mesh types within the tessellation model: (a) point data set from which the tessellations are derived, (b) Theissen polygons, (c) raster, (d) quadtree, (e) TIN.

features. An example of this type of tessellation is the *quadtree* (Samet, 1984), which seeks to subdivide in a hierarchy, subject to a predefined minimum resolution, in order to achieve homogeneity within cells. One clear advantage of quadtree data structure over the traditional raster approach is that redundancy is reduced and storage is more compact. Topology in tessellations can be either implicit or explicit (Figure 2.12). For regular meshes, neighbors can be easily found by moving one cell to the left, right, up, down, or diagonally in which case the topology is implicit. For a TIN, the topology can be made explicit just as it is in the vector model because each triangular element is a polygon. For structures such as quadtree, an explicit topology can be stored by use of Morton ordering (Morton, 1966) to produce a space-filling curve (in

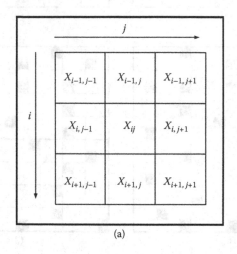

(a)

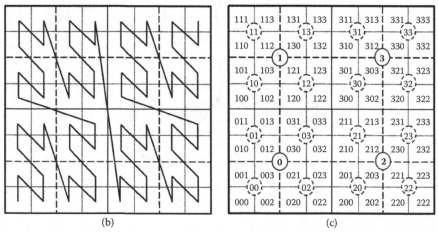

(b)   (c)

FIGURE 2.12
Examples of implicit and explicit topology in the tessellation model: (a) implicit neighbors, (b) Peano scan, (c) Morton ordering.

TABLE 2.3

Relative Advantages and Disadvantages of Vector and Tessellation Models

	Vector	Tessellations
POSITIVE	Good portrayal of individual object geometry: versatility of point, line, polygon primitives Portrayal of networks Explicit topology Multiplicity of object attributes in RDBMS Topological (polygon) overlay	Good portrayal of spatially continuous phenomena (fields) Relatively simple data structures Map algebra (on raster) Better conformance with remote sensing imagery
NEGATIVE	Relatively complex data structure, requires conflation of common object boundaries between layers and edge matching of tiles Poor representation of natural variation	Implicit and explicit topology only at cell, not feature level Single attribute layers only Blocky cartographic appearance

this case an N-shaped Peano scan), which reflects position of a cell within a hierarchical decomposition. The generally accepted relative advantages and disadvantages of vector and tessellation approaches are given in Table 2.3. Even so, as will be discussed below, most GIS provide adequate functionality for transforming vector to raster and vice versa and for transforming point patterns to area features (Theissen polygons), areas to points (centroids), lines to areas, points to TIN, and so on. More often than not, choice of an initial conceptual model is by no means a straightjacket.

Object-Oriented

Object-oriented (OO) analysis seeks to decompose a phenomenon into identifiable, relevant classes of objects and to explicitly relate them into a structured theme (Coad and Yourdan, 1991). A *class* represents a group of objects having similar or shared characteristics. These are made explicit in the *attributes* and *services* of a class, where attributes that describe or characterize the class and the services (or methods) are computer coded for handling that class (e.g., transformation, visualization). Thus, the class PUB includes all objects that can be called a pub; attributes would include general characteristics shared by all pubs (opening hours, license); services might include the code for plotting a symbol of appropriate size on a map or screen. A specific pub, say the George & Dragon, would be an *instance* of a class and would *inherit* the attributes and services of that class as well as having some attributes and services specific to itself. The way classes are structured in a theme is shown explicitly by the links between them and which determine the form of association and, in turn, the form that inheritance takes. IS _ A denotes generalization–specification structures while PART _ OF denotes whole-to-part structures; other forms of association, such as POSSESS, START, STOP, and so on are possible. An example of an OO analysis is given in Figure 2.13 for some classes that constitute a landscape. The top most class (or super class) is LANDSCAPE, which, for the sake of simplicity has two parts: COMMUNITY and TOPOGRAPHY. The class COMMUNITY can be further partitioned into classes, one of which is VILLAGE, which in turn can be further partitioned into classes, one of which is BUILDING. Class PUB IS _ A BUILDING is a specific class of building with the George & Dragon being an instance of PUB. Within this structure there are mixtures of classes that can be physical objects (VILLAGE, TOPOGRAPHY) or those that are social constructs (COMMUNITY). Clearly it is very difficult to map a "community" and while it may physically consist, in this example, of a village and its surrounding hamlets and farms, it will have other dimensions that are neither easily quantified nor easily portrayed in map form (e.g., degree of cohesion, social structure, political outlook). In a traditional vector or raster GIS, it is not possible to include abstract, conceptual features that are not distinctly spatial objects no matter how important they might be to planning and environmental decision making. In OO, it is possible to include such classes of features, and while they may not have distinct geographic boundaries they can be included in the data structure and analyzed alongside those classes of features that are geographically distinct. For a more detailed example of this, see Brimicombe and Yeung (1995). So far in our LANDSCAPE example, we haven't touched on the issue of geometry. Whether a class is portrayed by its services in a vector or tessellation representation (or both, or even as 3D virtual reality) will depend on the attributes and services that are encapsulated within a class or instance of a class. Thus, in Figure 2.13, the FARM, HAMLET, VILLAGE, and its components may well be all represented by vector geometry while TOPOGRAPHY may be represented both as tessellations (raster, TIN) and vector (contours). Overall, while OO provides for much greater versatility, it is not so straightforward to implement as a traditional vector and raster GIS.

Data Characteristics

Data sources for GIS are broadly classified as *primary* or *secondary*. Primary data are those collected through first-hand surveys and can be termed raw data if they are unprocessed observations. Secondary data are those collected by others, perhaps even for a different purpose, or have been derived from published/marketed sources. All data used in connection with GIS that have dimensionality can be categorized by measurement type and have characteristics of scale and resolution. Furthermore, the data may be an exhaustive compilation (e.g., census) or it may be a sample. With data so central to GIS, it is important to have an understanding of these issues.

A GIS layer of data has a locational, temporal, and thematic dimension or component, usually represented as a cube, whereby one component is always

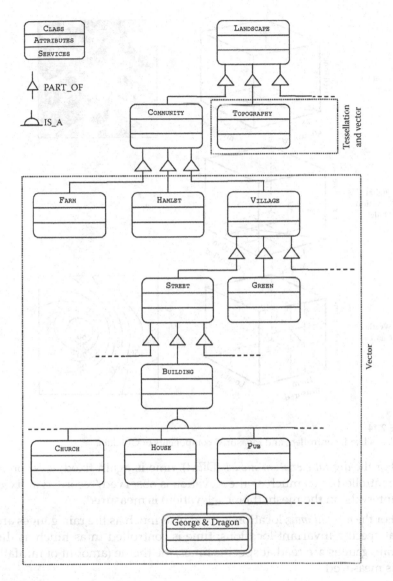

FIGURE 2.13
Object-oriented modeling of geographic features.

fixed, another is allowed to vary in a controlled manner, and the third is measured (Sinton, 1978). Some examples are given in Figure 2.14:

- For the land *cover layer*, time is fixed as a snapshot; the theme is controlled through defining a fixed number of land cover categories; location is measured in as much as the land cover is observed/recorded at all places.

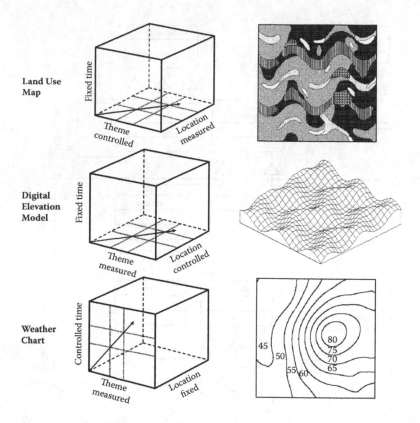

FIGURE 2.14
Examples of fixed, controlled, and measured components of GIS data.

- For the *digital elevation model* (DEM), time is again fixed; location is controlled in as much as the elevation is observed/recorded at fixed intervals on the mesh; theme (elevation) is measured.

- For the *rainfall map,* location is fixed in as much as the rain gauges are at specific invariant locations; time is controlled in as much as the rain gauges are read at specific times; the theme (amount of rainfall) is measured.

The dimensions of the data cube can take a range of values according to the scale of measurement used. Since the use of the term *scale* here can easily be confused with the cartographic scale of a map, I prefer to use *measurement type* to denote the system of measurement in use. It has become conventional in the social sciences to classify measurement into four types (Stevens, 1946):

- *Nominal* where objects are classified into named groups (e.g., land cover classes).

- *Ordinal* where objects are ranked in some order (e.g., from smallest to largest, lowest to highest).
- *Interval* where objects are measured according to a scale that has both an arbitrary zero point and an arbitrary interval (e.g., measurements of temperature: 0°C is arbitrary defined as the freezing point of water and 20°C is not twice as hot as 10°C, just 10°C hotter).
- *Ratio* where objects are measured against an absolute zero and where relative ratios are preserved (e.g., velocity: 0 kph is absolute and 60 kph is twice as fast as 30 kph, which in turn is twice as fast as 15 kph).

This typology goes farther in being prescriptive about appropriate statistical procedures. For example, the mean should not be calculated for nominal and interval data and instead the mode and median should be used respectively. Stevens' typology has been criticized for not being sufficiently inclusive and for being too formal for good data analysis (Velleman and Wilkinson, 1993; Chrisman, 1995; 1997). Data types frequently used in GIS and yet not sitting conformably with Stevens' typology are *counts* (non-negative integers), *probabilities* (where the whole range is absolute between zero and one), *direction* (where the measurement is circular), *fuzzy sets* (where membership of a nominal class can be graded), and *reference systems* (where at least two scales are required simultaneously for measurement, e.g., $\{x, y\}$ coordinates). Despite these difficulties, Stevens' typology remains popular. Users of GIS should make themselves fully aware of the measurement type in use for each data layer and consequently how to handle each layer both individually (e.g., calculating summary statistics) and in conjunction with other layers. Some examples of GIS data layers classified according to Stevens' typology are given in Figure 2.15.

The issues of scale and data resolution are fundamental to GIS and yet remain problematic (Lam and Quattrochi, 1992; Goodchild and Proctor, 1997; Atkinson and Tate, 2000; Goodchild, 2001). *Scale* as a noun has at least 10 different meanings in common use. From a spatial perspective, it can refer to a measurement type (above), the extent of some area (e.g., large-scale process acting over an extensive area) and the representative fraction of a map (e.g., 1:10,000 scale). The fact that large-scale processes are likely to be shown on small-scale maps just adds to the confusion. *Resolution*, that is the smallest discernible feature, has been traditionally linked in broad terms with map scale in as much as the larger the representative fraction the smaller the objects it can feature. Thus, at 1:100,000 scale, a village may be portrayed as a dot whereas at 1:1,000 scale each house, shed, and garage in the village can be shown separately. It is widely accepted that environmental processes are *scale-dependent* (Davis et al., 1991), that is, have both temporal and spatial resolutions at which they can be observed/measured. For example, plate tectonics occur at continental scales over millions of years

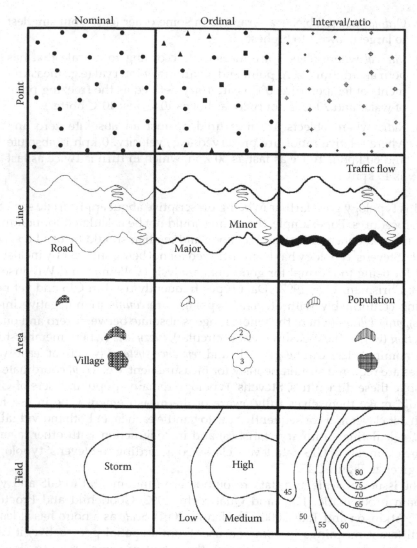

FIGURE 2.15
Examples of GIS layers classified according to Stevens' typology.

while landslides tend to affect just a few hundred meters and happen in seconds. This leads to the notion of a *characteristic scale*: the spatiotemporal scale (or narrow range of scales) at which a process (or the landform imprint of a process) best manifests itself for observation/measurement. Mapping scale and frequency of observation, therefore, should be tuned to the phenomenon under investigation. However, the relevance of using a representative fraction as a measure of scale in a digital database has been questioned (Goodchild and Proctor, 1997; Goodchild, 2001). Goodchild suggests the use of a dimensionless ratio:

$$\log_{10}(L/S) \tag{2.2}$$

where L is length measure equal to the square root of spatial extent and S is the spatial resolution in terms of the shortest distance over which change is recorded. If $\log_{10}(L/S)$ is kept to within a value of 3 to 4, then for any resolution it will determine the upper spatial extent of a tile (in terms of data volume) that is manageable and vice versa for any spatial extent to determine a workable resolution.

As commented above, a digital realization of a landscape is considered to be a sample of one because any number of slightly different realizations are possible. Nevertheless, data compilation can have as its objective the complete enumeration at a given resolution or it can set out to achieve a representative sample. For example, at the resolution of the individual person, a census intends to enumerate everybody and achieve exact counts of age groups, employment, ethnicity, and so on. Such an exercise might be considered too expensive and instead the same data might be estimated by sampling. Often within a GIS database there is a mixture of both types. So, referring back to Figure 2.5, the layers representing landslides and land cover are intended to be exhaustive surveys in as much as every landslide above a minimum size is intended to be mapped and all areas are mapped in terms of their land cover. On the other hand, topography and rainfall are mapped on the basis of sampling. Rainfall is sampled only where there are rain gauges while topography is usually captured according to a sampling scheme (e.g., regular profile, regular grid, progressive grid) and then interpolated into contours or a gridded DEM. Certain types of data, such as census, crime, and health are more often than not referenced by administrative units. Although the data at its highest resolution may be *geocoded* (adding of $\{x, y\}$ coordinates) as individual points based on addresses, it is often deemed for reasons of confidentiality not to release the point data, but present them as *aggregated data* by administrative unit. Since the boundaries of administrative units are frequently arbitrarily defined in relation to natural and social phenomena, the shape and size of the units themselves can significantly affect the aggregation and hence the interpretation of the data. We will return to this issue in Chapter 9.

Data Collection Technologies

Systems for the consistent measurement of length, area, weight, and time are fundamental to any organized society. As with most areas of science and technology, the microchip has revolutionized the way measurements of the Earth's surface and its cover are collected, stored, and processed to form usable inputs to GIS. For the spatial sciences, the 1990s were a transition from data-poverty to data-richness. This state change in the availability of digital spatial data was facilitated by:

- Improved technology and wider use of GPS, RS, and digital photogrammetry for data collection.
- The introduction of new technologies, such as LiDAR and radar interferometry.
- The operation of Moore's Law resulting in increased computing power to process raw data coupled with the falling cost of data storage.
- The advent of data warehousing technologies.
- Increasingly efficient ways of accessing and delivering data online, particularly through portals.

While it is still possible to discern particular approaches to measurement, such as land surveying and remote sensing, the various technologies are being increasingly integrated into digital mapping systems. These systems are increasingly aimed toward automated data collection for the construction and visualization of 3D models (Grejner-Brzezinska et al., 2004). In this section, I will describe some key components that go to make up state-of-the-art systems, but the reader must understand that this is a rapidly evolving area.

GPS and Inertial Navigation Systems

Central to nearly all forms of measurement and mapping is the Global Positioning System (GPS). Initiated as a program in 1973 by the United States with the first satellites launched in 1978, it became fully operational by 1993. GPS is now indispensable for geographical positioning and navigation. The Russians have developed their own system (GLONASS), the European Union (EU) is rushing ahead with its own (Galileo), and other states, such as China, are also either implementing new or supplementing existing systems with their own satellites. For simplicity, I will refer to all satellite-based positioning systems as GPS.

The U.S. GPS is based on a constellation of 24 satellites that orbit the Earth at an altitude of approximately 20,000 km. Radio signals are emitted by the satellites over a number of frequencies that can be picked up by receivers regardless of weather conditions, time of day, or position on the Earth. The only criterion is that receivers must have an unimpeded "view" of the sky. Generally the signals from three satellites are required for a dependable 2D (x,y) position fix and from four satellites for a 3D (x,y,z) position fix. Positioning accuracy depends on a number of factors: number and configuration of satellites in view, type of receiver, length of time allowed for a fix, whether a differential mode is used, and the amount of postprocessing carried out on the stored signals. In differential mode GPS (DGPS), a reference receiver at a known stationary position of accurately surveyed coordinates

(x,y,z) is used to supply correction data either in real time by wireless or for use in postprocessing. DGPS on receivers designed for geodetic surveying can provide subcentimeter accuracy. Handheld receivers typically have an operational accuracy of ±3 m to ±15 m depending on the configuration of the satellites in view. Most manufacturers provide a range of equipment and accuracies (see http://www.trimble.com; http://www.garmin.com).

Inertial navigation systems (INS) are quite different from GPS, although they still allow the user to track position from some known point. INS use three sets of gyroscopes and accelerometers carefully calibrated inside a vehicle (car, helicopter, aircraft) and aligned to the orthogonal axes of the 3D coordinate system in use (northing, easting, elevation). The vehicle can then travel in 3D space and have its position tracked by measuring the forces applied in acceleration and changing its position. By coupling INS with GPS allows INS to have ongoing calibration and for both the position and orientation (pitch, roll, yaw) of the vehicle to be tracked with accuracies down to centimeters for position and tens of arc-seconds for orientation. This has implications for remote sensing (discussed below) in that the position and orientation of an imaging or measurement sensor (that is being flown in a helicopter or aircraft) can be known at all times and thus can provide for an automated means to rectify and transform RS data into the desired ground coordinate system.

Remote Sensing

Remote sensing can be defined as the acquisition of data about objects using a sensing devise that does not require direct contact with the objects themselves. Thus, the use of a camera to obtain data about objects (as opposed to taking souvenir snaps) would constitute remote sensing. Use of cameras in this way from balloons and aircraft goes back at least a century and had certainly become routine by World War II. Aerial photographs have very high resolution, down to the manhole cover in the street, and can either be interpreted for the features contained within the images (API), such as the nature of the geology or vegetation, or used for measurement (photogrammetry), such as in the derivation of topographic contours or the mapping of buildings, land parcels, and roads (see http://www.getmapping.co.uk). For both these uses, it is usual to use partially overlapping images, which when viewed together permit a 3D stereoscopic visualization of the landscape. As with nearly all technologies, aerial photography and photogrammetry have moved into the digital age with digitally acquired or scanned photographs being rectified and measured in a semiautomated fashion by software for the fast production of maps.

Satellite imaging for civilian purposes started in the 1960s with meteorological satellites, but was quickly followed in the early 1970s by satellites with imaging systems designed to observe the Earth's surface rather than its atmosphere. While some traditional film-based cameras were used from space by the Americans and the Russians, unmanned satellite imaging devices are wholly digital systems so that data are transmitted back to receiving stations and compiled into images. More than 200 Earth observation satellites have been launched (though not all are still in operation) with 19 launched in 2008 alone. Up-to-date information on these satellites and their sensors can be found at http://www.tbs-satellite.com/tse/online/mis_telediction_res.html. One set of imaging systems generally works by scanning a strip or swathe orthogonal to the direction of orbit with successive scans used to construct the images. These were also designed to be multispectral, that is, each swathe being split into different bands of the electromagnetic spectrum, most commonly blue, green, red, and infrared, and thus extending the imaging beyond the visible spectrum. Just as digital cameras are sold today labeled according to the number of megapixels with which an image is captured, satellite imaging is most usefully classified according to its resolution, i.e., the ground area covered by one pixel of the image. This can range from 1 square kilometer pixels down to the higher resolution imaging systems, such as GeoEye's Ikonos (http://www.satimagingcorp.com), which has a pixel size of one square meter (for panchromatic) and 4 m (for multispectral), sufficient then to discern large vehicles. Swathe width also varies with higher resolution having narrower swathe width. For Ikonos, the swathe width is 11 km. In other words, to provide coverage for, say, Greater London, would require six passes of the Ikonos satellite given that there is lateral overlap of the swathes taken on successive orbits. Furthermore, the Ikonos imaging system can be tilted ±30° in any direction allowing the acquisition of stereoscopic imagery, which can then be used for 3D visualization and photogrammetric purposes.

Described thus far has been *passive* remote sensing, in other words, imagery that passively records reflected light from the Earth's surface or off objects. Such systems are limited to daytime operation and, if the Earth's surface is to be imaged, the weather must be cloud- and haze-free. *Active* systems are those that provide their own source of energy and then record the strength of the reflected signal. Aircraft- and satellite-borne radar has been used since the 1960s with higher resolutions being deployed in the 1990s onwards. These have a resolution of up to 10 square meters, and while this is a generally lower resolution than the passive systems described above, nevertheless has two important attributes: (1) they can operate day and night (since they generate their own energy source) and (2) in all weathers (since radar wavelengths can be cloud penetrating). This provides an invaluable capability, for example, to detect and map flooding and other hazards/disasters that occur during poor weather or poor light conditions. LiDAR (light detection and ranging) is a system for laser-based remote sensing. Usually mounted on an aircraft or helicopter with GPS and INS for positioning, LiDAR emits vertically downward pulses of light and measures the properties of the return signal to determine very accurate measurements of height. Because the pulses are spaced every few centimeters, a very dense data set is collected that then can be filtered and used to visualize the height and shape of buildings, vegetation, and all manner of street furniture as well as the slightest change in topography (see http://www.earthdata.com).

Ground Survey

Land surveying is the art and science of measuring distance (horizontal and vertical) between objects, measuring the direction of line between objects and the angles between lines. It has been the time-honored approach to mapping features and boundaries, to calculating areas and subdividing areas and to drawing up cross-sections for land management, construction, and a host of other applications. It used to be that all maps were compiled using land-surveying techniques, but for the past 50 years has been supplemented first by aerial photography and then by satellite remote sensing to increase the speed and cost-effectiveness with which maps and now digital coverages can be compiled. Nevertheless, land surveying remains the most accurate. Again the microchip, together with GPS and the ability to generate and measure the return signal from beams of infrared and laser light, land surveying has been revolutionized almost beyond recognition. The equipment has become considerably automated while software is used to carry out the calculations. Land surveying has also become digitally integrated with GPS and remote sensing to form automated systems for data integration and production (Grejner-Brzezinska et al., 2004).

While land surveying, GPS, and remote sensing provide geometric data, field surveys are carried out to sample check (ground truth) automated mapping methods, to collect more detailed attribute data, and as a means of monitoring changes to attribute data. While some attributes can be collected during a land survey or from remote sensing, many attributes tend to be collected separately and from a range of sources. Key to field surveys these days are mobile GIS deployed using personal digital assistants (PDA) or tablet PCs. This has come about due to the increasing power of PDA, their wireless connectivity, the availability of add-on GPS, and the increasing sophistication of the GIS software that can be installed. The position of the field operative can be displayed in relation to base mapping, thematic overlays, and remote sensing imagery; attributes can be entered through a series of customized data entry forms that do preliminary onsite checking of consistency so that gross errors (or blunders) can be rectified before moving on. This digital approach allows new data to be checked and integrated into the main database more quickly (Wagtendonk and de Jeu, 2007).

Another form of ground survey that has risen in popularity in recent years is the drive-by survey (see, for example, Google Street View). This is a field survey technique in which a vehicle is equipped with high accuracy differential GPS, laptop(s), pen tablets (for digital note taking), voice recording (also for annotation), roof-top digital cameras or video providing 360° view, and wireless communication. The vehicle is then driven along road networks and even off-road while an operative supervises the recording of new data and changes to the existing database.

Nontraditional Approaches to Data Collection

The ever-growing need for automated and cost-effective methods of spatial and attribute data collection with improved granularity is fostering the development of nontraditional techniques. Closed-circuit television (CCTV), although technically a form of remote sensing, is used for recording traffic and people flow rates. However, a range of roadside and inroad devices using radar, acoustic sensors, and infrared detectors are being increasingly deployed not only to count traffic and pedestrians round-the-clock, but also to classify the traffic into vehicle types and to determine levels of congestion so as to provide near-instant warnings of events that are happening on our roads. Laurini et al. (2001) have classified this type of spatial data collection where remote sensors telemeter data across fixed-line or wireless links to an operations center that is monitoring some spatial phenomenon (e.g., weather, river flow, traffic flow, transport of hazardous materials) as TeleGeoInformation.

Basic Functionality of GIS

From one perspective, all the basic functionality of GIS packages can be viewed as *data transformations*. From the initial loading of the data through its analysis to visualization as a thematic map merely requires GIS functions to appropriately transform data from one form to another. While this view of a GIS is a useful perspective in thinking about what GIS packages really do (coordinate geometry, matrices, computer graphics), it is less useful in conceptualizing strategies for using GIS functionality for achieving specific analytical goals. We will still apply the term though to a subset of GIS functionality. Another broad view is that of *cartographic modeling* (Tomlin, 1990) involving the manipulation of representations of maps as a high-level computational language. While in theory this can be applied to both raster and vector data, in practice it was limited to raster data and while versatile does not, to my mind, capture the essence of modern GIS. There have also been attempts to specify a set of universal functionality we might reasonably expect from a GIS package (e.g., Albrecht, 1996a). Again, this is problematic in as much as it does not fully recognize the role of algorithm in GIS. One can specify a function, such as "calculate gradient," as being part of the universal set, but as we shall see in Chapter 9, there are a number of different formulae one can adopt, each likely to give you a slightly different answer.

The range of functionality commonly associated with GIS as a technology is as follows:

Data entry and editing: This includes not only the import of purchased data sets, but also digitizing from secondary sources (with copyright permission), though the amount of digitizing that needs to be carried out on projects these days has dwindled because of the greater availability of data in the market. Where additional attributes come from internal sources, there is normally functionality that facilitates the reading of database and spreadsheet files and the joining of these attributes to vector objects using a unique key (see Figure 2.7). Attributes that have no immediate graphic representation, but include a geographic locator, such as a postcode, can be *geocoded* from geographic base files that list, for example, {x, y} coordinates for all postcodes. These base files usually need to be purchased separately and loaded into GIS. A geocoded data set can then be imported as point event themes. Functionality is also available for the onscreen editing of feature geometry and individual attributes as well as the means to calculate new attributes from existing ones.

Transformation: This includes a number of processes key among which are vector-to-raster transformation and vice versa. Included here are transformations of coordinate system, map projection, and reclassification of attributes. Another important transformation is *spatial aggregation*, such as the clumping of point data into zones or the clumping of smaller zones into larger ones. This may also take the form of *dissolving* redundant lines separating adjacent polygons that have been reclassified so they fall within the same class. Finally, but not universally present, is line simplification, which together with reclassification and aggregation can be used in *generalization*.

Query: This includes search by area to extract their attributes, search by attribute to extract corresponding areas and selection of features from one layer on the basis of the features in another. Queries can include the measurement of distances and areas. Also included here would be basic statistics (count, central tendency, maximum, minimum, range) and cross-tabulations of the attributes of one layer against another.

Interpolation: This includes point to area (Theissen polygons, see Figure 2.11(b)), point to field (contours generated using TIN and other computational modeling techniques, see Chapter 4), area to point (creation of a centroid point from a polygon), area to field (by contouring from the centroid), and area-to-area (from one set of zones to another).

Cartographic processing: This covers the manipulation of vector layers and principally includes *overlay* and *buffering*. Buffering is the systematic enlargement of features whether they are point, line, or polygon features. An example of a 500 m buffer around the roads in our example landscape is given in Figure 2.16(a). Buffers can be concentric, say every 100 m. For polygons, buffers can be internally made to reduce its size or even to create doughnut shapes that follow the polygon boundary. Overlay is when two or more layers are fused to create a new layer (Figure 2.16(b)) where the topology is rebuilt so that the polygons thus created carry with them the attributes of the parent polygons in the source layers. Selections of pertinent combinations of attributes can then be made using Boolean operators as a means of query or reclassification.

Map algebra: This covers the manipulation of raster layers mostly in an algebraic equivalent way where layers can be weighted by some constant if necessary and then layers can be added, subtracted, multiplied, divided, compared for maximum or minimum values, and so on. An example of map algebra is given in Figure 5.12 where a wildfire hazard layer is calculated from a reclassification and weighted summation of gradient, land cover, and elevation. Operations can include calculating or interpolating the value of cells from their surrounding cells, as in the calculation of the gradient of a cell from the elevation of its surrounding cells.

Thematic mapping: These are operations for the production, layout, visualization, and printing of thematic maps including choice of layer combinations, class intervals, color, pattern (or texture), symbology, and symbol size (or gradations). The production of good thematic maps that communicate well is both an art and science and is by no means trivial (see Chapter 10). To assist users, software developers often include certain defaults but these can easily result in "throw away" graphics of little value.

A Systems Definition of GIS

The discussion about GIS thus far has focused on data and software. But GIS are more than just these two entities. Taking a holistic systems view, GIS should also encompass: hardware, data collection/updating processes, dissemination of the products (maps, graphics, tables, reports), and the people who work with GIS along with their organizational structures. GIS can be viewed narrowly as just a tool, but should be viewed more broadly as an *approach* or *way of working*. We have now covered sufficient aspects of GIS to have an informed consideration of some definitions. The trouble is, there is no one single accepted definition. Maguire (1991) quotes no less than 11 definitions from the literature. Perhaps the most often quoted definition is that

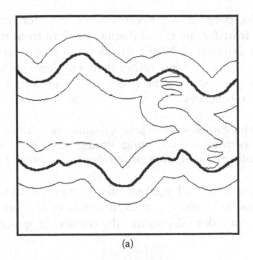

(a)

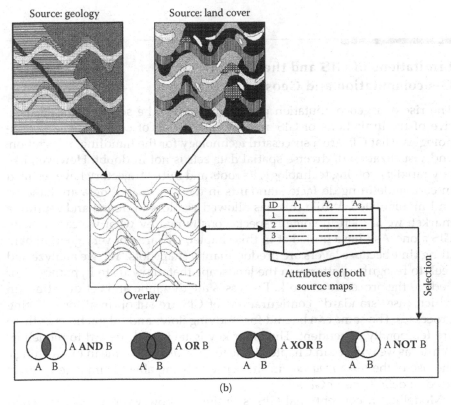

(b)

FIGURE 2.16
Examples of cartographic processing: (a) buffer and (b) overlay with Boolean selection of pertinent combinations of polygon attributes.

of Burrough (1986a, p. 6): "… a powerful set of tools for collecting, storing, retrieving at will, transforming, and displaying data from the real world for a particular set of purposes." While this definition captures much of what GIS are about, it is a toolbox view rather than a systems view. The following definition from Dueker and Kjerne (1989, p. 7), therefore, is preferred with my modifications in brackets:

> … a system of hardware, software, data, people, organizations and insti-
> tutional arrangements for collecting, storing, analyzing, [visualizing]
> and disseminating [spatial] information about areas of the Earth.

Conventionally, and not so long ago, this chapter would have ended right here. But there has been some paradigm shifts of late and we must go a stage farther and consider GIS within the context of geocomputation and geosimulation.

Limitations of GIS and the Rise of Geocomputation and Geosimulation

The rise of a geocomputation paradigm needs to be seen from a perspective of the limitations of GIS and the maturation of complementary technologies. That GIS are a successful technology for the handling, integration, and visualization of diverse spatial data sets is not in doubt. However, like any rapidly evolving technology, its roots and initial trajectory have resulted in certain defining de facto standards in the way spatial data are handled and manipulated. While this has allowed vendors to define and capture a market, we are left in some respects constrained by what we can do with GIS alone. As I have hinted at in this chapter, GIS are not very good in handling time because layers are predominantly snapshots. Yet, we analyze and seek to recognize patterns in the landscape that allow us to hypothesize or deduce the processes at work. Processes are dynamic and act over time, in which case "standard" configurations of GIS are suboptimal for studying processes. The same can be said for studying flows and interactions as these are temporally dependent. The other key issue is with regard to modeling. While, as we will see in Chapter 4, the term *modeling* can mean many things, the use of the term *modeling* in the sense of being able to simulate is rarely used in definitions of GIS.

Modeling in conventional GIS is limited to how we express real world objects as data (data modeling) and ways in which we might transform and analyze that data (e.g., cartographic processing, map algebra). Fairly simple simulations can be achieved, but not of complex environmental processes. Then there is the meaning of "analyze" in a GIS context. Many vendors would have us believe that they are supplying us with "spatial analysis" tools when what is, in fact, provided is mostly database and geometric manipulation. Statistical functionality is limited and there is almost universal absence of tools for analyzing point patterns, spatial autocorrelation, carrying out geostatistics or even good exploratory analysis (see, for example, Bailey and Gatrell, 1995; Fotheringham et al., 2000; Lee and Wong, 2001, as most of these topics are beyond the scope of this book). In these areas, conventional GIS quickly run up against the buffers. In the meantime, other tools and technologies have been developing as fast, if not faster, than GIS (e.g., statistical and spreadsheet packages, RS packages, artificial intelligence, neural networks, and so on), which can complement GIS and broaden our analytical and modeling capability.

What is touted as the "GIS revolution" was, in fact, over by the late 1990s in as much as GIS had become an accepted technology. While diffusion of GIS is far from complete, one is no longer pushing back the frontiers merely by adopting GIS. A more important distinction comes from the way GIS are being used. A paradigm shift for many areas of science in the latter part of the 1990s was the increasingly central position of *computation*—the use of computers having a pivotal role in the form of analysis—as an essential ingredient alongside observation, experimentation, and theory (Openshaw and Abrahart, 1996; Fotheringham, 1998; Longley et al., 1998; Armstrong, 2000). Thus, computers are no longer just accessories to research, they form the research environment itself. *Geocomputation* then can be defined as the use of "spatial computation tools as a means of solving *applied* problems" (Brimicombe and Tsui, 2000) and contributing to the development of theory (Macmillan, 1997). An important precondition has been the rapid improvement in the power, speed, and economics of computing in the 1990s, but equally important has been the sudden proliferation of spatial data sets during the same period, which permit "almost limitless possibilities for creating digital representations of the world" (Longley, 1998). The solution of nontrivial problems usually requires large (if not massive) nontrivial data sets that, in turn, require processing by high-performance computers. So what might be included on the menu of geocomputational tools? Certainly GIS are there, but other tools and techniques used singly or in combination with GIS would include spreadsheets, statistical packages, data mining and knowledge discovery, neural networks, artificial intelligence, heuristics, geostatistics, fuzzy computation, fractals, genetic algorithms, cellular automata, simulated annealing, and parallel computing (Couclelis, 1998; Armstrong, 2000). Such a powerful array means that geocomputation provides new opportunities to:

- Find better solutions for old problems.
- Find solutions for previously unsolved problems.
- Develop new quantitative approaches in modeling spatial phenomena.

Another recent paradigm shift involves the use of agent technologies. Agents are problem-solving software that can be embedded in a dynamic and open environment to autonomously pursue their goals (Jennings, 2000; Ferber, 2005). Agent technologies have been widely used in artificial intelligence, computer networking, software engineering, and human–computer interaction. In the area of spatial modeling, agents (as multiagent systems) have been used to simulate the behavior of individual objects (people, vehicles, animals) in order to understand what macro patterns of behavior emerge from the micro behaviors of these individuals (Batty and Torrens, 2005). Although an aspect of geocomputation has always been numerical simulation, the use of multiple agents systems alongside GIS to simulate spatial processes has led to the term *geosimulation* (Albrecht, 2005) and can be viewed as a major shift in the simulation of spatial phenomena.

The geocomputation and geosimulation paradigms offer exciting new avenues for research and application incorporating GIS, environmental modeling, and engineering. So dramatic have been the technical advances that they have changed the very problems we think about solving (Macmillan, 1998). But, before we move on to the modeling and engineering, it should be pointed out that we have treated GIS in this chapter wholly as a technology. Is this technology backed up by a science? If technology is the application of science and if there is no *geo-information science* (GIScience), then GIS are hollow and without foundation. The science behind GIS is the subject of Chapter 3.

3

GIScience and the Rise of Geo-Information Engineering

In Chapter 2, we looked at the nature of geographical information systems (GIS) from their beginnings and took a broad view of what they do and how they do it. It will not have been lost on the reader that GIS have their roots in technology and are commonly viewed as tools for handling and analyzing spatial data. We also saw how GIS have become bound up in geocomputation and how this opens up exciting new vistas for research and application. So are GIS just toolboxes for "turning the handle" on spatial data—shove some data in at one end and churn out some maps at the other? Or is there something more to it that merits the title of "Science" that creates a discipline in its own right? This is not just idle academic musing, but has implications for those numerous disciplines where GIS are applied. If it is not just a case of "turning the handle" and there are indeed substantive issues, which are important in the *way* we use GIS and spatial data, then users should be aware of them. A number of these issues provide the basis for Section III of this book. This chapter, therefore, aims to set up a framework within which these issues can be placed and their importance understood. This is not intended to be a lengthy chapter and readers wanting to follow up the debates around GIS as science, technology, and engineering should refer to: Goodchild (1992b), Wright et al. (1997a; 1997b), Pickles (1997), Burrough (2000), Frank (2000), Frank and Raubal (2001), Berry et al. (2008), and Brimicombe and Li (2009).

Technology First ...

The introduction of geographic information systems, as a technology, unleashed across a number of disciplines, an increasingly complex debate as to what it is, what it should be, should it be theirs, how and at what stage should it be taught, and what is its academic and professional standing? That such chaos should arise from simple beginnings should perhaps not be surprising given that the external technological environment has been changing faster than the subject domain. This technological forcing on GIS is of little comfort, however, to those who wish to grasp it. The early commercializing stage of GIS can be identified as being from the late 1970s until the end of the 1980s. This was a period when GIS were universally regarded as tools and a considerable body of literature developed to define the nature of the technology both conceptually (e.g., Peuquet, 1984), functionally (e.g., Dangermond, 1983), and to differentiate them from other tools (e.g., Cowen, 1988). Governments even discussed how best to reap the benefits offered by such technology (e.g., Dept. of Environment, 1987—the "Chorley Report"). GIS technology in the early stages had considerable diversity of form, not only in terms of functionality and interfacing, but also in terms of what it was supposed to be used for (e.g., cadastre, resource mapping, automated cartography). The dominance of the MSWindows®-style user interface together with the eventual market dominance of a few GIS software vendors have given the technology a more consistently recognizable form. That the technology was under constant development would suggest that there was a formidable engineering component during this period in the sense of "toolmaking" (Wright et al., 1997a). This may need however to be discounted as "engineering" if we take engineering in its more usual sense as the practical application of science, in which case, what might be called the science of GIS has largely emerged from the technology.

The technology emerged in the first place to fill a niche because the handling of spatial data was somehow more difficult than other types of data. Certainly the collection of spatial data—land and hydrographic surveying, photogrammetry—was afforded a distinct niche and was largely the province of professionals who were either formally registered or chartered. Mapping in most countries was vested in public sector agencies subject to state control. Spatial data were voluminous, which in the 1960s and 1970s caused data entry, storage, and processing problems on the computers of the day. Spatial data also seemed to need a "special kit" consisting digitizers, scanners, more than one screen, and plotters, all of which required technical expertise. Spatial data also seemed to need rather different data structures and algorithms with a distinct separation of the geometric component and the attribute database. But technological development has overtaken most of these aspects of GIS such that spatial data are no longer "difficult." Those professions that served in the collection of spatial data have seen their niche space seriously impacted by a host of disciplines, which, armed with GPS, RS imagery, digital photogrammetry, total stations, data obtained or purchased over the Internet, or even just a GPS-enabled mobile phone, can compile their own specialist base mapping. The traditional mapping authorities find themselves under competition from the private sector. Furthermore, to "do" GIS no longer requires "special kit" as the peripherals (if needed at all) have become normal, out-of-the-box, plug-and-play technologies. GIS software sits conformably in an interoperable environment alongside word processing, spreadsheets, statistics, multimedia, Web browsers, and virtual reality, and we think nothing special of it. The act of adopting GIS by an organization is no longer pushing back the frontiers. So, does GIS still have a claim to fame?

Well, yes it does in as much as integrating, handling, and analyzing spatial data is still more easily and efficiently carried out using GIS. You could try typing a page of text using a spreadsheet and would probably succeed, but it's more efficiently done—typing, editing, formatting—in a word processor. So, it is still with spatial data and GIS and will probably remain so. The other aspect of this is *what* we use GIS for. Crain and Macdonald (1984) articulated an evolutionary model of GIS facilities (Figure 3.1) in which the dominance and mix of activities changed from inventory through analysis into management. While the early years of GIS were dominated by inventory-type activities and applications, the way we tend to use GIS today places us firmly in Stage III with a predominance of analytical- and managerial-type activities and applications. We are much more into modeling, decision support, and "what if"-type analyses where GIS are coupled with other technologies,

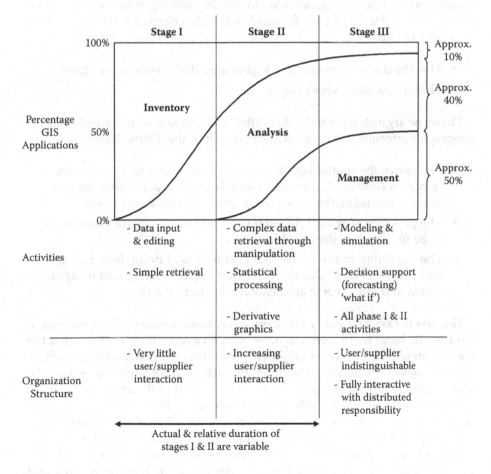

FIGURE 3.1
Operational evolution of GIS facilities. (Based on Crain, I.K., and Macdonald, C.L. (1984) *Cartographica* 21: 39–46.)

including environmental simulation models, in a geocomputational environment. New applications are emerging, such as *location-based services* (LBS), which ensure the continued relevance of GIS and are assisting its emergence from niche technology to ubiquitous technology (Li and Maguire, 2003; Brimicombe and Li, 2009).

Science to Follow ...

If, as we have seen, there is and remains a substantive technology that we refer to as GIS, then as with all other established technologies, is there a foundation in science? Goodchild, in his keynote opening of the debate (Goodchild, 1990a; 1992b), set forward two main criteria for the recognition of science in GIS:

- That the domain contained a legitimate set of scientific questions.
- That spatial data were unique.

These, he argued, were indeed fulfilled and that GIS were indeed tools for geographic information science (GIScience). Thus, the distinctiveness rests on:

- The use of the spatial key $\{x, y, a1, a2,..., an\}$, where $\{x, y\}$ define location as continuous dimensions and $\{a1, a2, ... , an\}$ define the attributes of location either as continuous or discrete dimensions.
- The presence of spatial dependence—nearer things are more likely to be similar than distant things.
- The durability of the spatial data primitives of point, line, polygon, and cell/pixel that have underwritten the technology and its application in many diverse applications (Burrough, 2000).

Yet, the notion of a coherent science has somehow remained uncertain. On the one hand there seems to have been an overarching premise that the science should be based on GIS technology (as commonly understood). A narrow view of the pertinent technologies may well limit what questions the science can effectively answer. On the other, much of the argument around GIScience has been overly influenced, I believe, by the notion that it is part of geographic science or geography. For example, Frank (2000) states that: "Geography is about how people interact with the environment. GIScience is computational geography, modeling processes in space." It's a view I have some trouble with as it implies that other disciplines that extensively use and carry out research with GIS technology, such as planning, environmental science, and surveying, somehow have less to offer GIScience. They have and will continue to make valuable contributions, after all, many of the early and significant developments in GIS did not come from geography departments. At the same time, significant branches of academic geography today remain unconvinced of the benefits in GIS for their discipline. Furthermore, there is now the insertion of computational science into the equation in order to handle both the complexity of models that are being developed and the massive data sets that are now available (Armstrong, 2000). These spatial data sets are often disaggregated, of higher resolution, and of greater temporal frequency than previously available for analysis, giving rise to new knowledge-discovery techniques that are underpinned by theoretical developments in spatial data handling (e.g., Miller and Han, 2001). The emergence of a geo-computational paradigm in which GIS are but one contributing technology serves as a marker that GIS are not the only tool of GIScience. "Even as many geographers disavow social science, geospatial science has emerged as a lusty arena marked by intellectual vigor, conceptual growth, and enhanced analytic abilities. What now is taking shape is a spatially integrated socio-environmental science that is transcending older disciplinary attachments, boundaries, and constraints" (Berry et al., 2008)

In order to fully understand the scientific plot, we need to go back to the beginning, to Goodchild's keynote. In it, he states that "we must first establish that spatial, or rather, geographical data are unique, ..." (Goodchild, 1992b). Setting aside any subtle differences between "spatial" and "geographical," the uniqueness lies in the nature of the *data* and the scientific questions that will arise out of its collection, handling, and use. Such data need not only be quantitative, but also qualitative (e.g., Brimicombe and Yeung, 1995; Brimicombe, 1997). As we have already noted above, spatial data can claim distinctiveness from other forms of data. Molenaar (1998) uses the term *geodata* and goes on to present a conceptual, linguistic model of geodata comprising three elements: syntax, semantics, and uncertainty (Figure 3.2). *Syntax* describes how we define and encode data objects, how we link their geometric representation with their thematic attributes, and how we identify and encode their spatial relationships. Geodata syntax is thus the specific data structures we employ to handle the distinctive spatial key $\{x, y, a1, a2,..., an\}$. *Semantics* represents the link between the objects in the data and file structures and the real world of landscape features as we interpret them to be. Thus, semantics is about the meaning we impart to geodata. But, there is some indeterminacy about these meanings for two important reasons. The first is that statements about the real world can have contested meanings and what we as individuals see and recognize in a landscape depends on how we interpret what we see through our various cognitive filters. For example, I tend to view landscape as a geomorphological construct, my wife sees it as predominantly a cultural construct, and we are both GIS professionals. Second, and very importantly, our data structures are necessarily simplifications of reality. In creating a data structure and populating it with data, we strip off large swathes of reality—sound, smell, unimportant features,

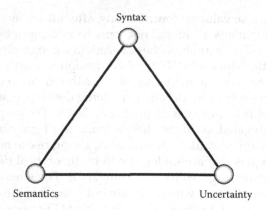

FIGURE 3.2
A conceptual model of geodata. (Based on Molenaar, M. (1998) *An introduction to the theory of spatial object modeling.* Taylor & Francis, London.)

ephemeral features, features that are too small, and so on—we inductively classify and simplify those objects that are of interest and create a static snapshot in the form $\{x, y, a1, a2, ..., an\}$. In so doing, errors can easily creep in (see Chapter 8). Not surprisingly then, when we try and reverse the process from data to reality, something can go amiss in the translation. Therefore, there is an inevitable level of *uncertainty* in both syntax and semantics and, thus, needs to be recognized as an important dimension of geodata.

I believe this geodata model can be extended to provide a conceptual model or linguistic framework with which to express the key focus of a GIScience (Figure 3.3 through Figure 3.7). To Molenaar's geodata model, I would add a fourth item: *operations*. This represents the algorithmic, software basis for the entry, storage, handling, analysis, and visualization of geodata—in other words, the basic functionality of what we would recognize to be a GIS package, but could well be some other type of package that included such functionality (e.g., an environmental simulation model with embedded spatial data handling functionality). I will refer to these four items as key *aspects* of GIScience built on the distinctive features of spatial or geographic data and which together visually form a tetrahedron (or pyramid), as illustrated in Figure 3.3.

The six edges of the tetrahedron forming the links between the aspects at the vertices can be viewed as defining certain *models* that are central to GIScience (Figure 3.4), many of which have been subject to research over the lifetime of GIS. These are

> *Data models and ontologies*: These link semantics and syntax. They represent the way in which reality is abstracted into the data structures, which, as we have seen in Chapter 2, is dominated by topological vector (object-based) and raster (field-based) representations with

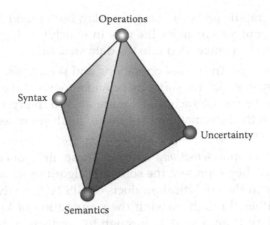

FIGURE 3.3
Key aspects of a conceptual model of GIScience.

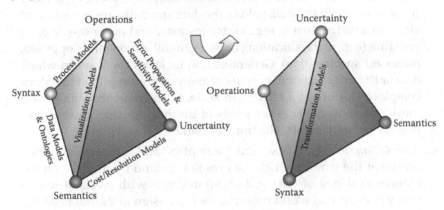

FIGURE 3.4
Models that connect the key aspects of GIScience.

object-oriented (OO) data models making their presence felt. Data models for geo-information (GI) can be extended to fully 3D or to include a temporal dimension. The term ontology originated in philosophy and in computer science is taken to refer to an explicit and formal specification of a conceptualization (Gruber, 1995) that, in practical terms, comprises an agreed list of terms used in data modeling and the relationships between them (Antoniou and van Harmelen, 2004). Thus, in GIScience, ontology has to do with how we understand terms used in data models and how they relate to reality as we might perceive it.

Process models: These link syntax and operations. They represent the way in which we handle and manipulate data resident in the data structure in order to achieve specific goals. The dominant focus here is on cartographic processing (e.g., overlay, buffer) and map algebra techniques, but also includes the way in which topology is built by the software and embedded into the data structure.

Visualization models: These link operations and semantics. They represent the way in which the software provides views of the data (maps, tables, and diagrams) such that we are able to interpret either the data itself or the informational outputs of analyses in terms of their real world meaning.

Error propagation and sensitivity models: These link operations and uncertainty. They represent the software/algorithmic way in which uncertainty in the analytical products of GIS-type analyses may be assessed either through tracking the propagation of known levels of uncertainty from base data through to products or by means of *sensitivity analysis* (SA).

Transformation models: These link syntax and uncertainty. They represent the way in which the data held in the data structure are transformed either as a restructuring (e.g., vector-to-raster and raster-to-vector), a recoding (e.g., reclassification), or as a generalization (or its opposite, increased specification). Generalization includes the ways in which data or the informational outputs of analysis are simplified to reduce complexity or effect a change in scale. Such transformations have inevitable consequences for levels of uncertainty that must appropriately be managed by the transformation.

Cost/resolution models: These link semantics and uncertainty. They represent the way in which data resolution (and by implication the manner and cost of data collection) interacts with levels of uncertainty and the real world meaning we can assign to the data and the informational products where they are used. Little of this has been researched in GIScience to date, though see Burrough et al. (1996) for a cost-quality comparison in modeling heavy-metal pollution on flood plains. See also an example in Chapter 9.

We can go farther with this conceptual model. The faces of the tetrahedron as subtended by a grouping of aspects and models, represent a number of key *issues* in GIScience (Figure 3.4), plus one *integrative issue*, which can be visualized as either being in the center of the tetrahedron (Figure 3.5) or as an outer envelope. They are

Cognitive issues: These relate both to internal representation, use of language and metaphors in the way we navigate software, databases, and construct solutions to applied problems, and to the way in which these assist us in navigating, understanding, and applying solutions to the external real world.

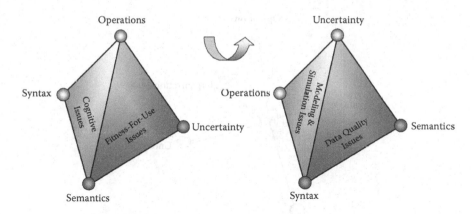

FIGURE 3.5
Issues of GIScience as selective groupings of aspects and models.

Data quality issues: These relate to how data quality is managed including the establishment and use of metadata standards in creating spatial data from reality, establishing data reliability, and longevity (shelf life) in relation to a specific application and the internal means by which quality needs to be maintained during data transformations.

Fitness-for-use issues: These relate specifically to how the fitness-for-purpose of the informational outputs and analytical products can be established. Whereas the previous issue concentrated largely on *data input*, these issues refer to *information output* as a consequence of base data quality and pertinence, the propagation of uncertainty in analyses and the visualization both of the outputs and their associated levels of uncertainty in such a way that they can be correctly interpreted in decision making.

Modeling and simulation issues: These concern the way in which analyses are carried out using standard functionality, but also including the use of embedded computational models (such as for environmental simulation), and the use of simulation approaches to establish the quality of those models and their analytical products.

Organizational issues: These concern the way in which geodata and geo-information are organized, handled, managed, and used within an organizational setting and, thus, represent an integrative issue concerning all the above aspects and models as well as the other issues (Figure 3.6).

An extension of the organizational issues is *spatial decision support systems* (SDSS). These are discussed in Chapter 10. They are a means of finding solutions to ill-structured problems where the nature of the problem itself may not be clear, let alone finding a straightforward solution. They are also a means

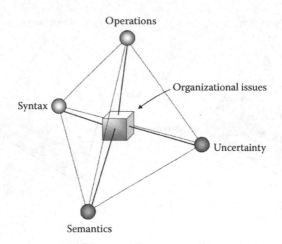

FIGURE 3.6
The organizational setting as an integrative issue of GIScience.

of reducing uncertainty in finding appropriate or reliable solutions. Such approaches these days often require an organizational arrangement of linking GIS with external models in a networked environment. However, they also go beyond just technology and must include the wider decision-making environment. SDSS as an integrative issue in GIScience is conceptualized in Figure 3.7.

All the aspects, models, and issues just discussed have been the focus of GIScience research and represent the domain of legitimate scientific questions alluded to by Goodchild above. The conceptual model described here

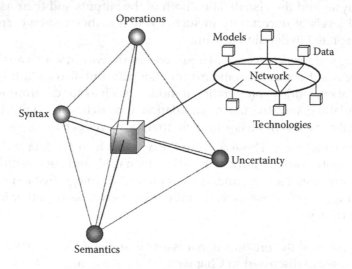

FIGURE 3.7
Spatial decision support systems as an integrative issue of GIScience.

was originally developed as an organizing framework for students to see how a curriculum in GIS and spatial analysis hangs together and that it is indeed distinguishable as a discipline. The conceptual model is not predicated on traditional discipline boundaries, nor is it predicated exclusively on GIS technology. Nevertheless, as will be evident by now, wherever spatial data are in use, then there are a set of substantive issues that need to be understood if researchers and professionals, regardless of their disciplinary roots, are to be critical users able to evaluate their own work. For this reason, Section III of this book looks at a range of issues that arise out of GIScience necessary for the application of GIS in environmental modeling and engineering.

And Now … Geo-Information Engineering

Science is about discovering new knowledge and making it available to society. Engineering is about applying that knowledge systematically using the results of scientific research to successfully and dependably solve real-world problems. For dependability and minimizing risk of failure, engineers rely on scientific laws, design standards, and codes of practice that arise not only from the results of scientific research, but also from the accumulated experience of applying those results in diverse situations. What is more, those diverse situations will necessarily be contextualized and parameterized through their own investigations. As discussed in the previous section, we have a body of endeavor that we can recognize as GIScience and which has a reasonably mature (though still evolving) technological base. Are we then beginning to see the emergence of *geo-information engineering* (GI Engineering)? I believe we are.

GI Engineering has been defined as "the design of dependably engineered solutions to society's use of geographical information" (Brimicombe and Li, 2009). Such solutions may be products, such as SatNavs, but may also be methods, such as approaches to spatial analysis that reliably support decision making. The stage is certainly set. Spatial data collection technologies have reached maturity in which resolution, precision, speed, time between resurvey, and falling unit costs have improved dramatically since the early 1990s. These technologies may further improve incrementally, but are probably reaching long-term stable states. Such data will form the raw material for GI Engineering. The basic technological infrastructures for GI Engineering have also matured—fast, low-cost PCs and handheld devices, computer networking (whether cabled or wireless), the Internet, mobile communications—and, although bandwidth will need to further increase to adequately serve GIS applications to thin client mobile devices, there are reliable platform configurations to mount GI Engineering applications. There is also a mature software environment, not just in terms of GIS packages, but in the level of

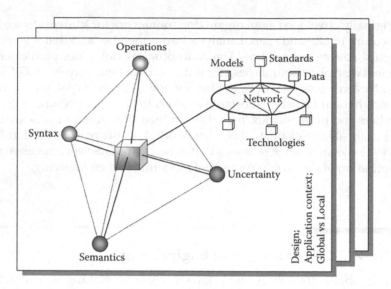

FIGURE 3.8
A conceptual model for the application of GI Engineering.

tool interoperability and ease with which effective wrappers with common interfaces can be established to bind a suite of tools. GI Engineering poses a considerable and exciting research agenda for GIScience (see, for example, Brimicombe and Li, 2009, and issues raised in Chapter 6 to Chapter 10 in this book), though there is already a substantial body of results that can be applied. Just as research in applied engineering subjects is far from complete as new problems, materials, and approaches emerge, so one would not expect to have all answers to all possible questions in GIScience before endeavoring to produce artifacts of GI Engineering. There are only two real obstacles to GI Engineering: clients (or customers) and GI engineers. The former depends on the revenue-earning potential of GI as either saleable commodity (e.g., just-in-time way finding information, real-time flood forecasting) or contributing to cost reductions (e.g., reducing the cost to society from crime and disorder, natural hazards, pollution, and so on). There is evidence that this type of client base is certainly growing both in the private sector (e.g., LBS) and in the public sector (e.g., joined-up, integrated government and services). The obstacle to achieving a sufficient body of GI engineers will depend on how GI studies are taught in universities (Frank and Raubal, 2001).

Bringing forward the conceptual model for GIScience, the conceptual model of GI Engineering is shown in Figure 3.8 where there are changing emphases on the aspects, models, and issues of GIScience within the specific application context—the professional area or discipline being served—as well as the scale for both information representation and accessibility. A key element though is the design of the solution that brings all the necessary elements together in an intuitively usable and dependable way.

Section II

Section II

4

Approaches to Modeling

Model is an everyday word. Look it up in the dictionary and you are likely to find upwards of eight definitions as a noun and four or more as a verb. Meanings range from making things out of clay to using mathematics to someone who struts down the catwalk. Leaving aside this last category, most of the definitions have one thing in common: simplification or reduction in complexity. At its easiest, a model is a simplification of reality in terms that we can easily understand. As illustrated in Figure 4.1, a map can be viewed as a model.

The types of models that we in the field of the geographical information systems (GIS), environmental modeling, and engineering would be most interested in are those that express our understanding of the way the world works with sufficient precision and accuracy to allow prediction and confident decision making. Such models may be qualitative, pictorial, or more usually quantitative and relate to the dynamics and processes of the physical, social, and economic environment. Many of these models of interest are used to explain and/or predict what happens *somewhere* and the spatial patterns that arise. Because GIS are preeminent in handling spatial data, it is natural that efforts have been made to use or link such models with GIS in order to have an enhanced management tool. Typical examples of the types of models used include:

- Regression lines predicting an output variable according to an input variable.
- Spreadsheets expressing more complex interrelationships and processes; manipulation of map layers according to a predetermined theory.
- A series of mathematical equations describing complex transport and transformation processes compiled as a stand-alone computer program.

An important issue that will have to be considered is how external models are made to work with GIS, but this will be the subject of Chapter 7. Also, since all models are a simplification of reality, they may work more or less well in certain circumstances and there will always be a level of uncertainty. This is a theme that runs through Chapter 8 to Chapter 10. In this chapter, however, we will discuss the nature and construction of models in general

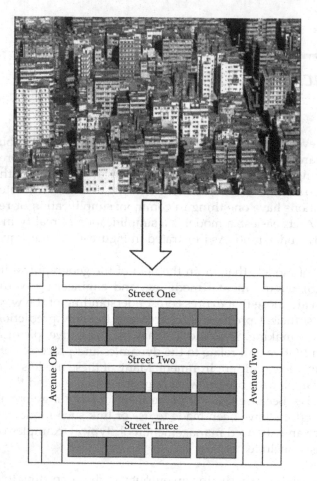

FIGURE 4.1
A map as a model, a simplification of reality.

terms, and in the following chapter look more specifically at the nature of environmental models in the context of GIS. Chapter 6 will focus on a range of examples.

Model of an x

Much has been written about the nature, construction, and use of models. Of the extensive literature, Harvey (1973) provides an in-depth treatment of models in relation to scientific explanation in geography; Taber and Timpone (1996) provide a good overview of computational modeling while Deaton and Winebrake (2000) provide a sound introduction to modeling dynamic systems. Martin and Church (2004) have reviewed the development and testing of numerical models of landscape evolution.

Models perform many different functions within a scientific framework. Chorley and Haggett (1967) identified the following uses as:

- A *psychological device* that allows complex processes or interactions to be visualized and understood.

- A *descriptive device* that allows a stylized view of the main workings of a phenomenon or process.

- A *normative device* that permits broad comparisons to be made between differing situations or processes.

- An *organizational device* for managing the collection and manipulation of data.

- A direct *exploratory device* and a *constructional device* in the search for new theories or the extension of existing theories.

These are not mutually exclusive and given this plurality, it is necessary when constructing and using models to be clear which type of device is intended and the appropriateness of any chosen model to act as such a device (Harvey, 1973). This plurality is also the reason why we cannot have a single, universal definition of a model.

Models are central to the way in which we seek to scientifically explain and develop theories about the phenomena we experience in the world around us. There are two broad routes to scientific explanation: *induction* and *deduction* (Figure 4.2). Using the inductive route, empirical observations of the real world are ordered and classified through subjective generalization. While induction offers few safeguards against jumping to the wrong conclusions, a theory derived inductively can be represented by an *a posteriori* model and tested for its ability to predict correctly (and, hence, confirm or otherwise the veracity of the theory). This route is often used in data mining in order to discover and extract knowledge from a large body of existing data. Using the deductive route, these same empirical observations are configured into *a priori* models, which, while suggesting a theory, are nevertheless open to doubt. This model of an *x* (Achinstein, 1964) allows one or more hypotheses to be deduced and tested through appropriate experimental design. While these hypotheses can never be proved in an absolute sense, they can be accepted (or rejected) with a certain degree of confidence. Thus, the relationship of models to theories may take a number of forms: discovering a theory, extending or restructuring a theory, establishing the domain of a theory, validating a theory, representing a theory, and, finally, as a means of prediction on the basis of a theory. All of these are relevant to the domain of environmental modeling and engineering. However, it is important to recognize that although models are central to explaining our world, they tend to be

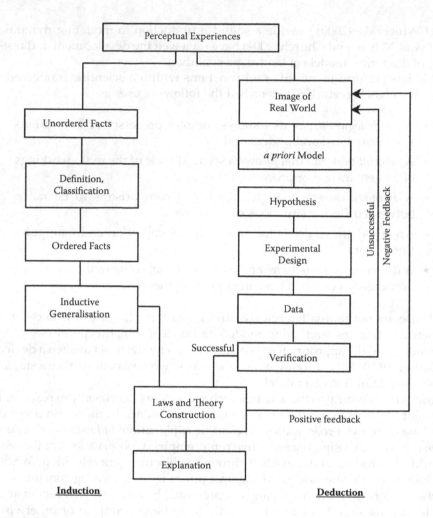

FIGURE 4.2
Inductive and deductive routes to scientific explanation.

temporary rather than permanent. Depending on the new question being investigated, the specifics of a location where a model is to be applied, or even depending on the availability of data, an existing model may need to be altered or evolved.

Typology of Models

Just as models can fulfill a range of functions, so models can be classified in a number of ways. For example, Ackoff (1964) classified models into:

- *Iconic*: Using same materials, but with a change in scale.
- *Analogue*: Having changes in materials as well as scale.
- *Symbolic*: Use of a symbolic system, such as mathematics.

While at first this typology may seem reasonable, it is, in fact, restrictive when, for example, we look at concepts of using natural language and maps. A more useful classification, from the perspective of this book, is loosely based on Chorley and Haggett (1967) and is comprised of models that are:

1. *Natural analogues*: This is the use of some past event to explain the present (*historical analog*) or use of events in one place to explain events in another (*spatial analog*). These are frequently descriptive models but, can also use diagrams and maps.

2. *Hardware*: This is where a physical miniaturization of a phenomenon is created out of the same or similar acting materials in order to study development, general behavior, changes in state, influence of variables, and so on.

3. *Mathematical*: This is where a phenomenon and its workings are described using equations, functions, or statistics. Where the model provides a unique solution, then it is labeled *deterministic*. On the other hand, where the model takes account of random behavior and provides a probabilistic answer, then it is labeled *stochastic*.

With the rise of ubiquitous computing over the past 25 years, we should now add a fourth category:

4. *Computational*: This is where a computer is used for symbolic manipulation (using code and data) in order to express phenomena and their workings. Such models may include deterministic and/or stochastic elements alongside *heuristics* (rules), logical operators (if, then, else), set operators (and, or, not), text operators, and so on. Since most mathematical models are now implemented using a computer, one could be forgiven for confusing the difference between mathematical and computational models. Mathematical models are formal, that is, are constructed in accordance with recognized forms and rules that have a truth condition (an accepted means of proof). But many real-world phenomena we would like to model are too intractable to be reduced to a formal mathematical model; specific elements may be, but not the whole thing. Computational models, on the other hand, while perhaps offering less precision and clarity, do offer greater flexibility and can increase the level of realism. From another perspective, mathematical models can be used without a computer while for computational models, computers are central to

the form of enquiry. Computational models allow complex experiments to be conducted that for ethical, logistical, and cost reasons can't be carried out in reality. Thus, studying the effects of changing CO_2 concentrations in the atmosphere is not an experiment that can be conducted live, but can be computationally simulated. It is also worth making the distinction here between monolithic and distributed component approaches to computational modeling. Under monolithic approaches, a computational model is programmed so that all the necessary components are explicitly written into the code. This has been the predominant approach in creating GIS and environmental modeling software; it results in large software packages that can only be run on sufficiently powerful computers. The currently evolving distributed component approach allows program elements to be developed independently and called for reuse by different computational models (even across a network) as and when required. This is the principle behind the use of, say, Java classes in object-oriented software development (see, for example, Faulkner, 2002) and provides for the possibility of software to run on thin clients typically through an Internet browser interface.

For each of these four classes of models, it is possible to add further descriptors depending on the role of time, the degree of specification of the model as a system, and the way in which the model is being used. Thus, for example, a computational model might be:

- *Static*: Where the elements of the model are fixed over time or where the model focuses on an equilibrium situation.
- *Dynamic*: Where variables in a model are allowed to fluctuate in time.
- *White box*: Where the internal workings of a model representing a system are fully specified.
- *Grey box*: Where the internal workings are only partially specified.
- *Black box*: Where the internal workings are not specified, for example, the details of how inputs are transformed into outputs are not known, but co-vary in some observed way.
- *Exploratory*: Where models seek to reveal the workings of some phenomenon.
- *Prescriptive*: Where models are used to provide answers, such as what will be the outputs, given certain inputs.

These are further discussed below within a dual framework of modeling landslides as one form of natural hazard, and modeling topography as continuous surfaces being typical functionality within GIS.

Building Models

The type of model one sets out to build ultimately depends on the use to which the model will be put, the extent to which the processes of the system are known, and the time and resources available to build and test the model. Though, more often than not given the abundance of models already devised, it is often a case of identifying which model(s) best fit or can be adapted to the desired modeling objectives within acceptable limits of uncertainty and for which suitable data are likely to be available or can be collected (see Beven, 2001). However, the starting point of any good project that will devise or use models is some observation, speculation, or consternation that drives one to find a satisfactory answer or solution (Lave and March, 1993). Clearly one needs a good understanding of the existing theory and domain knowledge relevant to solving the substantive question at hand and it can even be stated that without this knowledge one is unlikely to raise these questions in the first place. So understanding the relevant literature is paramount.

It is the regularities or patterns that most often catch our attention or interest. These empirical observations are at the heart of modeling—both for suggesting the possibility of models and for validating these models against reality. Patterns, either directly observed or "discovered" through exploratory analyses are often the means by which we can hypothesize or infer the presence of underlying processes (Fotheringham, 1992; Unwin, 1996; O'Sullivan and Unwin, 2003). The presence of a process that can be more or less understood allows us in turn to produce models. Models allow us to manage. The types of patterning referred to here are very broadly defined from the arrangement of objects in space (e.g., volcanoes, traffic accidents, cancer patients), morphological regularity (e.g., alluvial fans, floodplains), regularity in time (e.g., tides) and input–output response regularities (e.g., high rainfall resulting in floods). Patterns detected through spatial data analysis are usually broadly classified as random, regular (uniform), or clustered (Cliff and Ord, 1981). There is a widespread tendency to assume that spatially random data have no underlying process of interest that can be modeled, but as Phillips (1999) pointed out, such apparent randomness may be attributable to chaotic, complex, deterministic patterns. For spatial regularity to occur, a space-filling, mutual exclusion process can be hypothesized (such as competition between plants for space and light). Nevertheless, it is clustered patterns that raise the strongest hypotheses for and interest in identifying underlying processes and have an important role, for example, in such diverse areas as spatial epidemiology (Lawson, 2001), crime analysis (Goldsmith et al., 2000), knowledge discovery in large databases (Miller and Han, 2001), and statistical approaches to landslide hazard assessment (van Westen, 1993; Aleotti and Chowdhury, 1999). That said, there are times when irregularities catch our attention particularly when such irregularities do not conform with or would not seem to be predicted by the prevailing theories. Models in these instances may lead to the emergence of new theories. Whatever the spark, building a model presumes the existence of a theory or hypothesis.

It is as well here to introduce the two modeling examples for landslides and topographic surfaces as they will help in understanding the process of building, evaluating, and applying models. Two good references for more in-depth material are Turner and Schuster (1996) and Burrough and McDonnell (1998), respectively. Each example area will be introduced, the theory stated, and an explanation of how aspects of the simpler models are constructed. This is not intended to be an exhaustive technical treatment, but a means of illustrating model building.

Modeling Landslides

Landslides, stated simply, are the mass movement of soil and/or rock down a slope. They have been observed and recognized as serious hazards since time immemorial. Individual landslides can vary in size from just a few cubic meters to tens of cubic kilometers (e.g., Rossberg, central Switzerland in 1806; Bindon, south coast of England in 1839; Elm, Switzerland in 1881; Vaoint, northeastern Italy in 1963 with loss of about 1,900 lives; Armero, Columbia in 1985 with an estimated loss of 20,000 lives). The economic and social costs of landslides worldwide are getting higher with increased urban development and population densities. Total annual cost of landslides in Japan is estimated as $4 billion while in Italy, India, and the United States it is estimated as $1 billion to $2 billion each per year (Schuster, 1996). Loss of life worldwide due to landslides is estimated by the same source as having averaged well above 600 a year during the twentieth century. The initial *perception* of an explainable regularity was that landslides didn't just happen at random, they occurred almost exclusively during earthquakes (including tremors associated with volcanic eruption) or heavy rainfall and that there was also a tendency toward morphological regularity, as illustrated in Figure 4.3.

The theory underpinning landslide events can be stated verbally as: A slope will fail as a landslide when the strength of the slope materials fall below or is exceeded by the downward forces acting on the slope. Like most theories stated verbally, it seems simple and obvious—if the slope has enough strength, it will stand; if it doesn't, it will fail. However, if we are to create models that allow us to identify which slopes are in danger of collapse, we need to flesh out this theory in terms of its physical parameters that allow us to construct one or more models within the typology given above on page 67. By expressing our verbal theory in the form of an equation, we can move from a *perceptual* to a *conceptual model*:

$$F = \frac{\text{shear strength}}{\text{shear stress}} \text{ or rephrased, } F = \frac{\text{resisting moment}}{\text{disturbing moment}}$$

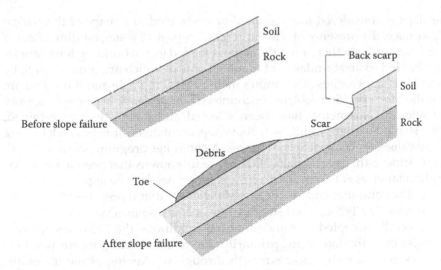

FIGURE 4.3
Generalized morphological features of a landslide.

By substituting in the relevant parameters, this can be expressed as:

$$F = \frac{c + \sigma \tan \phi}{\tau} \qquad (4.1)$$

where F = the factor of safety (FoS), c = cohesion of the soil, σ = normal stress on the slip surface, ϕ = angle of internal friction, τ = shear stress. This is for soil slopes (rock slopes are modeled more in terms of discontinuities (joints, faults) within the rock mass). If FoS ≥ 1, then the slope will stand; if FoS < 1, then it will fail. The cohesion of the soil can be roughly translated as its stickiness (e.g., clay has high cohesion, sand has low cohesion), normal stress is the amount of compression acting on the slip surface to stop movement along it, angle of internal friction can be visualized as the angle of repose permitted by the interlocking of the individual soil grains while the shear stress subsumes a number of elements, such as weight of the soil mass, angle of the slip surface, and the internal pore-water pressure. Of course, in the eyes of a geotechnical engineer, this is a gross simplification, but it serves the interest here of keeping things easy so that we can focus on modeling issues. Water, for example, affects both shear strength and shear stress: a film of water around individual grains creates surface tension and adds to cohesion (e.g., wet sand has the ability to stand at a steeper angle than dry sand) yet down slope flow of water within the soil as a consequence of rainfall not only adds weight, but exerts positive pore-water pressure that increases shear stress. Clearly there are many possible parameters, each of which needs to be measured in the field or laboratory (or simply estimated on the basis of experience and secondary data), with a different mix of parameters depending on whether the slope is considered to be drained or undrained, the shape of the critical slip surface, the presence of tension cracks, seepage of water, binding effect of tree roots, and whether or not the slope is loaded (e.g., a building foundation). This has led to an abundance of models, some of which are considered to be "standard" approaches (e.g., Janbu's method (Janbu, 1968)) and those that are modified for specific problems encountered on projects. Nevertheless, once the relevant parameters have been selected and their values ascertained, slope stability equations like (4.1) above represent deterministic models. Plug in the values and calculate the answer. A computer program is usually written to achieve this, particularly if there are unknowns that need to be solved by simultaneous equations and/or the solution needs to be approached iteratively. The computer code is the *procedural model* that represents the conceptual model. The FoS of a soil slope is, however, by no means static over time. It is generally accepted that under natural conditions, the FoS of a slope will decrease over the long term primarily due to weathering, where weathering acts to reduce the shear strength through weakening of the materials. Cutting at the base of a slope, either naturally by streams and rivers or due to excavation by man, increases the effective steepness of a slope and the sheer stress. Such action may not lead to immediate failure, but rather makes a slope susceptible to triggering events that cause a sharp decrease in the FoS to less than unity (Figure 4.4). There are generally two such events. The first is seismic shock whereby the shaking reduces the strength of the soil and temporarily increases the pore-water pressure thus leading to liquefaction. The second is heavy or prolonged rainfall, which can result in high pore-water pressures as the water moves downhill within the slope (the same effect can be achieved by rapid draining or draw down of reservoirs and canals, and by

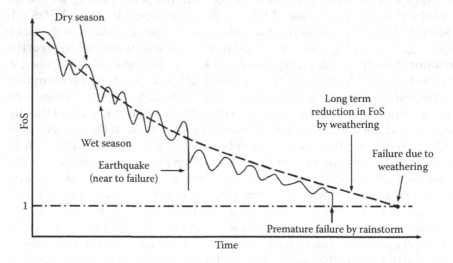

FIGURE 4.4
Reduction in soil slope factor of safety (FoS) in time.

rapid snowmelt). It is clear from Figure 4.4 that a high magnitude event may not cause failure of a slope, but may cause weakening so that a later lower magnitude event may be the trigger that causes final failure.

All of this gives increasing complexity to which can be added considerations, such as the spatial nonuniformity of weathering (and, hence, soil parameters) over quite short distances and that hidden discontinuities can lead to localized weaknesses. As Beven (2001) remarked for hydrology, "There is one fundamental problem ... most of the action takes place underground." So, too, with landslides. There are even legal issues in defining when a landslide has occurred (e.g., for insurance purposes) and what the causes of that landslide were (Griffiths, 1999). Where relevant parameters may not all be known or where they may not be accurately quantified, deterministic models are difficult to apply on complex natural slopes as opposed to many manmade slopes (cut and fill during construction, dams, and embankments). In cases where large areas need to be assessed for hazard, alternative modeling approaches can be adopted (see Aleotti and Chowdhury (1999) for a summary review).

One qualitative approach is the use of historical and spatial analog models. Here the theory is treated as a black box. The exact causes of failure may not be known nor is the FoS for each slope known. Rather, the visible presence of landslide scars in relation to the general geology, geomorphology, land use, and climate allows us to infer, on the basis of theory, the extent to which slopes can be categorized as "stable," "marginally stable," and "unstable," the likely mode of failure, and the likely trigger mechanism or mechanisms. By mapping and thus understanding one area, some broad predictions about future landslides can be made for that same area (historical analog) or for another area where similar conditions prevail (spatial analog). Bulut et al. (2000) show that by contouring percentage landslide per unit area (density of landslides) as a result of a rainstorm event in 1983 in the Findikli region of Turkey, the hazard zonation that these contours represented was found to be a reliable model in relation to the landslides that occurred over the subsequent 12 years. As will be further discussed in the next chapter, such models are eminently GIS-able. Figure 4.5 provides an example of the type of landslide mapping that has been described here.

As discussed above, Equation (4.1) looks simple, but a parameter, such as τ (shear stress), represents a bundle of parameters; additional parameters can be added to represent special conditions (e.g., seepage of water) and, if the slip surface is likely to have a complicated geometry, the entire calculation can become elaborate and cumbersome despite the use of computers. Therefore, some researchers have sought empirical equations that are easier to use for direct estimation of FoS. One solution is that of Sah et al. (1994) that represents a stochastic model in as much as it gives an answer that has the highest probability of being correct. Sah and his co-researchers took 46 cases of slope stability analyses for circular slip failure as reported in 38 publications from 1948 to 1988 from different locations around the world. Taking the model to be a linear regression equation in the form:

$$F_i = \alpha X_i + \beta Y_i \qquad (4.2)$$

where F_i = estimates of the FoS for each case $i = 1, 2, ..., n$, α, β = constants to be determined, X_i, Y_i = variables defining the characteristics of each slope such that:

$$X = \frac{c \cdot \operatorname{cosec} \psi}{\gamma \cdot H} \qquad (4.3)$$

$$Y = (1 - r) \cdot \cot \psi \cdot \tan \phi \qquad (4.4)$$

where c = cohesion of the soil, φ = angle of internal friction, r = pore-water pressure ratio, γ = unit weight of soil, ψ = slope angle, H = height of slope.

The maximum likelihood estimates of α and β are obtained by solving the normal equations:

$$\left. \begin{array}{l} \alpha \displaystyle\sum_{i=1}^{n} X_i^2 + \beta \displaystyle\sum_{i=1}^{n} X_i Y_i = \displaystyle\sum_{i=1}^{n} F_i X_i \\[2em] \alpha \displaystyle\sum_{i=1}^{n} X_i Y_i + \beta \displaystyle\sum_{i=1}^{n} Y_i^2 = \displaystyle\sum_{i=1}^{n} F_i Y_i \end{array} \right\} \qquad (4.5)$$

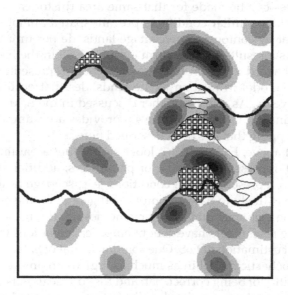

FIGURE 4.5
An example of landslide hazard mapping as an analog model—landslide density in relation to villages and roads.

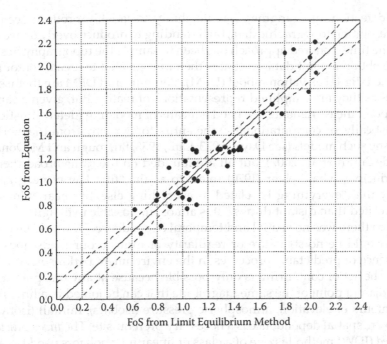

FIGURE 4.6
Regression fit and confidence intervals between FoS obtained deterministically by the limit equilibrium method and probabilistically using Equation (4.6). (Based on data in Table 3 from Sah, N.K., Sheorey, P.R., and Upadhyaya, L.N. (1994) *International Journal of Rock Mechanics, Mineral Science and Geomechanics Abstracts* 31: 47–53.)

The final equation was thus:

$$F = 2.27 \left(\frac{c \cdot \operatorname{cosec} \psi}{\gamma \cdot H} \right) + 1.54(1 - r) \cdot \cot \psi \cdot \tan \phi \qquad (4.6)$$

This, too, is GIS-able (an ArcGIS implementation is given in Sakellariou and Ferentinou, 2001). Comparing results calculated using Equation (4.6) against a full implementation of Equation (4.1) for the 46 case studies used by Sah et al. gives a close fit, as given in Figure 4.6.

Modeling Topography

After landslides, with much of the action taking place underground away from view except for the final defining event of a collapsing slope, one might well think that modeling topography, which we can all see and stand on, would not pose any difficulties. Think again. Traditionally, it has been neither feasible nor economically viable to measure every square meter let alone every square centimeter of a topography. New technology, such as LiDAR (light detection and ranging), has its ability to quickly produce accurate, high-resolution topographic data, but is tending to produce overly dense data sets. The time-honored approach has been to sample the topography in some way to obtain a point data set that is then interpolated to create contours or a gridded digital elevation model (DEM). Contours and DEM are themselves models as they are simplified representations of reality. But, given a sample set of point measurements, we need a means of modeling (deterministically, stochastically, or computationally) the entire topography from them. There are many such models (see Davis, 1973; Lam, 1983; Burrough and McDonnell, 1998; de Smith et al., 2007), but all are founded on the same broad theoretical principle, what Tobler (1970) refers to as the "first law of geography," namely that "everything is related to everything else, but near things are more related than distant things." It is almost impossible to imagine a world in which there was no spatial and temporal dependence—it would be chaos. There would be no structure or regularity of form to our landscapes and an absence of predictable processes in the environment (Atkinson and Tate, 2000). The act of contouring is only possible because at any one point there is an adjacent point of the same magnitude that can be joined in a line. Thus, for contours of elevation, atmospheric pressure (isobars), rainfall (isohyets), and so on, spatial dependence is a necessary prerequisite. The *inverse distance weighted* (IDW) method is one of a class of linear interpolators used to calculate a regular grid of points from a pattern of measured points, an entirely separate algorithm then being used to thread the contours. IDW calculates the value of an unknown point (on the regular grid) by averaging the known data points within the neighborhood. But, in the spirit that near points are more related than distant points, the known points are weighted inversely according to their distance from the point being calculated. Thus, the further away a known point is, the less influence it has on determining the value for the unknown point. The formula (a mathematically deterministic model) is given as:

$$Z_{ij} = \sum_{k=1}^{n} z_k \cdot d_k^{-r} \Big/ \sum_{k=1}^{n} d_k^{-r} \qquad (4.7)$$

where Z_{ij} = elevation of the unknown points in a grid, z_k = elevation of the known points from a survey, d_k = distance between a known point and the unknown point, r = a power weighting function.

The power function r determines the rate of distance decay of the weighting (Figure 4.7), the initial default value in practice usually being $r = 2$. As can be appreciated, this deterministic model of interpolation can be applied not just to topography, but more or less well to any spatial point measurement whether on the ground, above ground, or, indeed, underground. The formula (4.7) is easy enough to calculate, but identifying which known

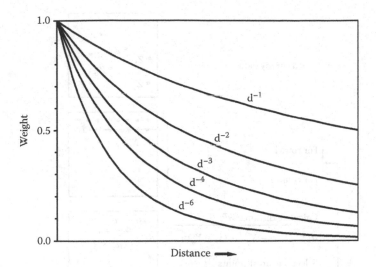

FIGURE 4.7
Power weighting and distance decay in inverse distance weighted (IDW) interpolation.

points are relevant, calculating the distances, summing the values, and then working one's way around all the unknown points becomes cumbersome and tedious to say the least. Therefore, it is usual to create a computational model that reads a data file, asks for the necessary parameter values, creates the regular grid of points to be calculated, proceeds in a loop to calculate the values, and finally stores them. The same computer program may also run straight on with a second computational model that threads and displays the contours. IDW is often a basic function of GIS packages and is commonly offered in environmental modeling packages where interpolation of point data is required. A computational model for IDW interpolation to a regular grid is illustrated in Figure 4.8. IDW can be regarded as an exact solution because for any unknown point spatially coinciding with a known point, the z_i value of the known point, will be copied to z_j without change (Lam, 1983; Burrough and McDonnell, 1998). Nevertheless, depending on the parameter settings, such as r, any other unknown points are necessarily *estimated* based on the model. Also, since the resultant topographic surface is modeled from a single measured sample of n points, it needs to be understood as just one realization of the true topography.

Spatio-Temporal Dimensions and the Occam–Einstein Dimension

Natural processes, such as those which environmental modelers set out to understand and simulate, have a tendency toward characteristic scales in space and time. This was discussed in Chapter 2. We, therefore, can consider two dimensions, one being the time for completion of the process being

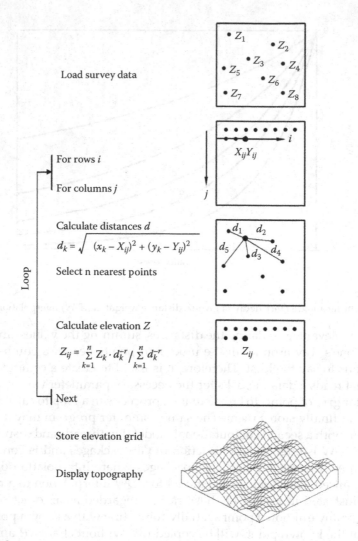

FIGURE 4.8
An illustration of a computational model for inverse distance weighted (IDW) interpolation.

modeled, the other being the size of area affected by the process (Figure 4.9). In the current context, these two dimensions are not independent as, in general, processes that are of short duration tend to affect small areas (e.g., soil detachment by a raindrop), whereas processes of long duration tend to affect large areas (e.g., tectonic processes shaping whole continents). There are of course possibilities for high magnitude, short duration processes to affect large areas (e.g., the eruption of Mount St. Helens on May 18, 1980 devastated some 500 km^2 in a matter of minutes, although one could argue that the real length of the processes leading to this event was much longer). Figure 4.9 shows an envelope of spatio-temporal domains where the shape shows both

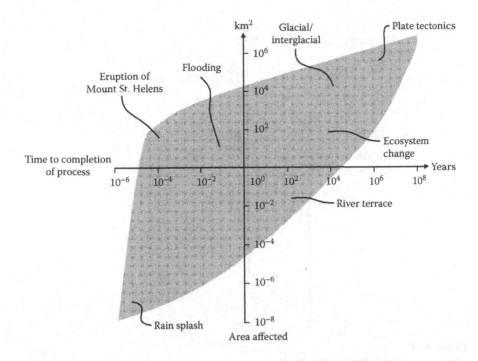

FIGURE 4.9
Envelope of spatio-temporal domains for process modeling. (Based on Delcourt, H.R., and Delcourt, P.A. (1988) *Landscape Ecology* 2: 23–44; Brunsden, D. (2002) *Quarterly Journal of Engineering Geology and Hydrogeology* 35: 101–104.)

the general trend and allows for short, high magnitude events. The model builder needs to determine at the outset, even implicitly, which portion of the envelope is relevant to the spatio-temporal domain of the process being modeled, as this has implications for data, parameters, discretization (see Chapters 5, 8, 9, and 10), model complexity, and so on. Even if done implicitly, based on the type of process under consideration, it is nevertheless worth recognizing as it helps maintain a certain consistency between model objectives and the various elements of the modeling.

However, there is another dimension we need to consider in model building: simplicity. On this particular axis is a continuum of models from the fullest possible complexity that approaches as near as possible to reality, to highly simplified models with gross levels of abstraction. This is entirely a modeler's choice within the available science and resources to determine parameters. One general goal of science, though, is not to make models more complex than they need be, that is, to apply the so-called Occam's razor. William of Occam was a fourteenth-century Franciscan monk who encouraged a scientific approach to research and is attributed with the maxim, "Entities are not to be multiplied without necessity," but, in fact, said, "It is vain to do with more what can be done with fewer" (in Russell, 1961), but

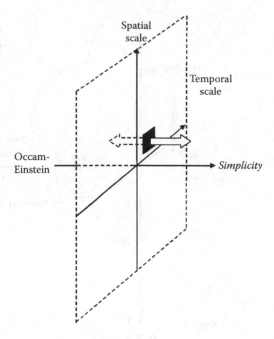

FIGURE 4.10
A conceptual model development space.

which has the same effect. (Of course, Occam's political and scientific views did not find favor with Pope John XXII and when forced to seek the protection of Emperor Louis at Avignon (France), he equally famously said, "Do you defend me with the sword, and I will defend you with the pen.") While unnecessary bulkiness of models should be avoided, Occam's razor needs to be applied carefully so as to remove only those features of reality that can be safely ignored. Taber and Timpone (1996), therefore, have named the simplicity dimension as "Occam's dimension," but since Einstein further encapsulated this concept by variously saying, "The best explanation is as simple as possible, but no simpler," I would prefer to call it the Occam–Einstein dimension. When coupled with the spatio-temporal dimensions (Figure 4.10), we have a conceptual 3D modeling development space. A model first needs to be appropriately positioned in the spatio-temporal domain. Then, it is moved along the Occam–Einstein dimension either because it is unnecessarily complex such that the number of parameters can be reduced, or because it is too simplified and more parameters need to be incorporated. Moving along the Occam–Einstein dimension always has implications for the number of parameters with relatively complex models requiring larger numbers of parameters (all of which will need to be quantified). As Kirkby (2000) points out, increasing the number of parameters will, in general, increase the accuracy of model outputs but, at a diminishing rate with each new parameter that is added and with a corresponding increase in complexity of calibration and finding optimum parameter values. In computational models, the Occam–Einstein dimension can also be applied to the efficiency of algorithms in terms of length of the code, speed of execution, amount of disk read–write, and so on.

Evaluating Models

Models must satisfy some desired level of utility. Such utility is not just concerned with providing acceptably "correct" levels of understanding or answers to problems, but includes notions of reliability, validity, beauty, and justice. We will consider these issues here in the general way that they affect approaches to modeling. Specifics will be dealt with at greater length in Chapter 9.

There are a number of terms dealing with the evaluation of models deriving particularly from the ecological sciences (Oreskes et al., 1994; Rykiel, 1996; Mazzotti and Vinci, 2007). The term *verification* refers to the correctness of the internal structure and working of a model while the term *validation* refers to the model performance and its applicability to a subject domain. A further activity, *calibration*, is the adjustment of parameters and constants to improve model agreement with an observable reality. Verification, validation, and calibration are integral to producing reliable models. Firstly, however, it must be recognized that there is no automatic means of telling when a model is finished. In the sense that theories are not verifiable but might indeed be falsified by new evidence and that new hypotheses continue to be established, a model should never be considered finished in an absolute sense—it lasts while its underlying theory stands and there is some utility in its use. Nevertheless, a sort of paralysis by analysis is equally undesirable as a model is made increasingly complex by the addition of hypotheses in the search for perfection even as its application is kept on hold. No model will provide perfect predictions. This is inherently so because models are necessarily simplifications of reality. What is desired is an acceptable fit with the observed reality that is being modeled. Issues surrounding this notion of fitness-for-use will be considered at depth in Chapters 8 and 9. In general, this involves a comparison between some observed data and the simulated outputs, often somewhat restricted to the statistical distribution of *residual errors* (observed values minus the expected values), to see if an acceptable goodness-of-fit has been achieved (Beck et al., 1993). While we all don't have the stature and confidence to brush off a poor fit by saying, as did the physicist Paul Dirac, "It is more important to have a *beautiful* theory, and if the observations don't support it, don't be too distressed, but wait a bit and see if some error in the observations doesn't show up" (Ferris, 1988), he does have a point—our observed data with which we validate models are rarely perfect either. The testing of output validity usually starts with *postdiction* (or *hindcast*), that is, being able to model some past event (for which data are available) to an acceptable level of accuracy. Having achieved this, models are often used straightaway for *prediction* of future events, but performance of the model should again be checked against reality when those events do happen.

There are two further issues that need to be discussed. The first of these is the reality of the process versus getting an answer that fits. This is *process validity*. For instance, the model of an Earth-centered universe favored by Aristotle and Ptolemy gave accepted answers (for several centuries) to the observed solar, lunar, and planetary motions, yet, as we now know, it was wrong. I'm not suggesting that the models we use today are so fundamentally flawed, but that in the process of abstraction, use of analogies, identification of appropriate mathematical and computational approaches, choice of parameters, and so on, there is always the danger that the resultant model no longer properly reflects the intended theory. Of course, a black box model in no way describes the actual processes at work, it merely transforms inputs to outputs, such as by using a regression equation. But even regression models can go astray in as much as the corelation between dependent and independent variables may be indirect (through an intervening variable) or may even co-vary according to some prior variable. The other issue is that of *reliability*, that is as interpreted here, the ability of models to give consistently similar results in similar situations or for similar events. For a simple deterministic model, one would naturally expect the same output values each time the same parameter input values are used. But for complex stochastic models using, say, a Monte Carlo simulation in one or more elements, then slightly different results within a known range can be expected. How any model reacts to slight changes in any particular parameter can be a complex matter to establish and is usually evaluated through methods of sensitivity analysis (see Chapter 9).

Finally, in evaluating models, one can consider the issues of beauty and justice (Lave and March, 1993). Paul Dirac has already been quoted as saying that it is important to have a beautiful theory. In modeling terms, *beauty* concerns not only aesthetics, but also the Occam–Einstein dimension of achieving fitness-for-use through simplicity. This is difficult to describe in words, but if we examine the quintessential example of Einstein's $E = mc^2$ which essentially says that matter is "frozen" energy, it is a small, neatly stated formula that expressed what was then a new and surprising view of the universe (a very large, complex phenomenon). Yet, it is also the broader implications of the formula, its generalization, that adds to its notion of beauty; examples being the thermonuclear processes of the sun and nuclear power (as well the atomic bomb). It is a combination of surprise and fertility. As for *justice*, models often play a practical role in informing policy and planning decisions and while we must hope that such application of our models are for the good of society (by providing valid outcomes to the debates) and are not used maliciously, we cannot ignore any unpalatable truths they may serve up. However, when models become used in the public domain, their creators unfortunately have little or no control on how they are applied. Thus, Einstein was appalled by the atomic bomb and what he expressed as the "drift toward unparalleled catastrophe" (Augarde, 1991). Indeed the 52,000 tons of spent nuclear fuel and the 91 million gallons of radioactive waste from plutonium-processing currently stockpiled in the United States alone (Long, 2002) could in some sense be seen to be a terrible consequence of the beautiful $E = mc^2$.

Applying Models

This is the ultimate goal of building models. But even here some caution is required. Harvey (1973) considers that the remark of Beach (1957) concerning the "history of economic thought [being] to a very large extent … a history of misapplied models" can be repeated to a greater or lesser degree for most academic disciplines. In our modern litigious society, applying models can bring liabilities with them and, where reasonable care has not been exercised, then a case of negligence can arise (Epstein, 1991; Miles and Ho, 1999). So, it is not just the art and science of building good models, but also the art and science of applying them well. An obvious first step is in identifying that a model can be appropriately applied to the specific problem at hand. This requires the modeler to understand the implications in any choice of theoretical approaches to the problem and how these have been incorporated into a model. Such can be the profusion of models these days that it is not surprising that practitioners can have difficulty in choosing from amongst them. The financial outlay in purchasing models for evaluation is likely to be prohibitive. This has led to the need for reviews, such as that of the ASCE Task Committee on GIS Modules and Distributed Models of the Watershed (1999). Model choice may not rest solely on the theoretical underpinnings, versatility, or whether it runs under Linux or Windows, but is likely to include considerations on the specifics of parameterization and the likely availability of the necessary data. In practice, one will tend to begin by applying simpler models with fewer parameters which, however, tend to be somewhat blunter instruments. As experience and confidence grows, the level of modeling complexity can be increased. Confidence in using a particular model (or it could be a suite of complementary models) comes from achieving consistently useful results first in replicating a process or system for known events in a particular situation, then for a variety of situations at the same and/or at different sites and finally across different scales of space and time. At any point, one must be ready to recognize the limits of the model and when these limits may have been reached or transcended.

Having chosen to apply a particular model, there are two critically important activities: *parameterization*, that is determining accurate values for the parameters of the model, and *calibration*, which is the adjustment of parameters to improve goodness-of-fit to known events (postdiction) prior to use for prediction. Since "the specific relationship between the general model form and the physical system being studied is gained via the model parameters" (Kirkby et al., 1987), the success of any application lies in the correct determination of parameters. Although these issues will be dealt with further in Chapter 9, they can be usefully introduced here by returning to the IDW modeling of topography described above. In applying IDW the topography must first be sampled. This in itself is by no means straightforward as there is considerable choice of random and nonrandom sampling approaches (Harvey, 1973; Walford, 1995) and then there is the question of sample size—how many data points to collect. A decision about these usually needs to be mediated between the nature of the topography (e.g., grain, degree of dissection, presence of break lines), the chosen method of data collection (e.g., ground survey, photogrammetry) and the way in which IDW models topography. Given a data set, running the model needs two further decisions: (1) the minimum number of known points to be included in the calculation of an unknown point and (2) the value to be assigned to the power weighting function r. The effect of these choices on the result can be illustrated in the following example.

Figure 4.11 shows a small area of our example topography. This topography has been mathematically generated so that we can know the true value z of the topography for any point on the surface. The formula that generated this topography and, of course, Figure 2.2 in Chapter 2, is

$$z = 100 \times (1 + \sin(0.5y + \sin(0.5x))) + 0.02y + 0.04x \qquad (4.8)$$

Use of the sine function produces the wavy periodicity in the topography of curving ridges and valleys, while the last two terms produce a general lowering of the topography from northeast to southwest. In Figure 4.12(a), a *random* sample of 25 points has been generated for which true z has been

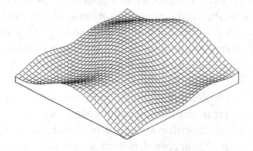

FIGURE 4.11
A portion of the example topography.

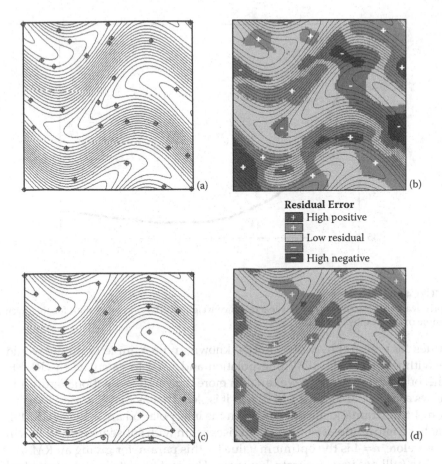

FIGURE 4.12
Samples for inverse distance weighted (IDW) interpolation and resulting residual errors: (a) random sample, (b) residual errors from IDW interpolation of random sample when $r = 4$ giving $RMSE = \pm 37.53$ m, (c) purposive sample, (d) residual errors from IDW interpolation of purposive sample when $r = 4$ giving $RMSE = \pm 19.49$ m.

calculated using Formula (4.8). To this has been added the four corner points so that the IDW interpolation can realistically cover the whole area of the map (more on this issue below). The sample for such a surface is quite small and it is unlikely that the IDW interpolation will produce a good fit, but this serves to illustrate a few issues in this chapter and in Chapter 8 and Chapter 9. In IDW interpolation, the parameter r needs to be set as well as the minimum number of sample points to be used in calculating an unknown point. Assuming a target grid of 25 m and the minimum number of points is fixed at 6, we can test the influence of parameter r by progressively increasing its value. For each interpolation, we can measure the deviation of the interpolated topography from the true topography using the *root mean square error* (RMSE) given in Chapter 8, Formula (8.1) by using map algebra. In practice, such true

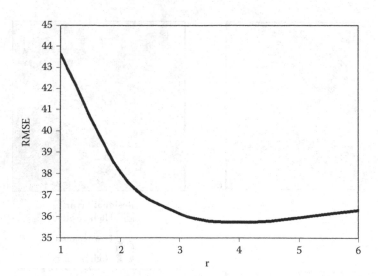

FIGURE 4.13
Optimization of parameter r in inverse distance weighted (IDW) interpolation for the given sample of topography.

values for topography would not be known and a sample of points would be withheld from the IDW interpolation and used to calculate the RMSE; with our true topography, we can do a more thorough evaluation. Figure 4.13 shows a graph of r against RMSE and it is clear that as r is increased in value from 1 to 3, dramatic improvements in the interpolation are achieved. RMSE reaches a minimum at $r = 4$ and then rises again with subsequent increases in r. Therefore, $r = 4$ is the optimum value for this parameter giving an $RMSE = 37.53$ m (still not very accurate, however). The spatial distribution of residual errors (*observed* minus *expected*) for $r = 4$ is given in Figure 4.12(b). To briefly look at the influence of sampling method, Figure 4.12(c) shows a *purposive* sample using the same number of sample points that have been placed on the topography to try and pick out better the ridges and valleys. For $r = 4$, the RMSE is now reduced to 19.49 m, nearly half of what it was before. The spatial distribution of residual errors is given in Figure 4.12(d) for comparison. It should be pointed out that while $r = 4$ has been found to be an optimum, this is only for this sample in relation to this topography. In modeling, the search of optimum parameters or near-optimum parameter sets needs to be carried out for each application.

This example also allows us to make the important distinction between interpolation and extrapolation when applying a model. *Interpolation* can be defined as the estimation of intermediate values from surrounding known values. The two key words here are "intermediate" and "surrounding." In other words, the estimated values must fall within the topological set of known values. In the current example, this set has spatial limits that can most easily be visualized as the *convex hull* of sampled points. A convex hull is similar to fitting an elastic band around all the known points in that it will form an outer polygon such that none of the internal angles will be greater than 180°. *Extrapolation*, on the other hand, is the estimation of values beyond the limit of known values. The estimation of these values is not bound within the topological set of the known values and may (or may not) fall outside the predictive limits of the model. Thus, in the IDW example, any grid values calculated outside the convex hull would have been extrapolated and may, depending on the nature of the topography, be wholly unreliable. This is why the four corner points were inserted into the samples in the above experiment to optimize *r* so as to avoid extrapolation. On the other hand, the extrapolation of say, the regression line in Figure 4.6 and Figure 5.5(b) in Chapter 5 beyond the given data would seem entirely reasonable and is often used for statistical prediction. Models of physical processes are usually designed to explain and predict behavior over a particular range of spatial and temporal scales. Each new application of a model is likely to be for a different size area and may need predictions for different time scales. Can a model work equally well across different scales; this is the issue of *scalability*. For example, a water quality model may have been designed to give reliable results for small- to medium-size drainage basins with daily calculations. The estimation of input parameters, while adequate for calculation of outputs at an aggregate daily level, may not be accurate enough to reliably calculate hourly outputs. Similarly, the same model might not work well for very large drainage basins. However, this is not just a matter of scale-dependence of parameter estimation or whether the physical processes at different scales might require more or less complex modeling, it is also about the validity of *aggregation* (lumping of either of inputs or of outputs) in that the usual outcome is to reduce heterogeneity and is likely to bias results. van Beurden and Douven (1999) have studied this dilemma in relation to national level results on pesticide leaching into groundwater for policy makers in The Netherlands (Chapter 9, Figure 9.6). They consider two approaches: aggregate parameters to a national level and use them in a model, or run models on local data and aggregate the outputs. Aggregating by averaging up resulted in the two approaches giving completely different answers. Given that policy makers prefer only one set of results to work with, the professional modeler has to have strategies for dealing with the inherent uncertainties of model application. This issue underscores Chapter 8 to Chapter 10.

A Summary of Model Development

Model development is more often than not a long and winding road, but one that nevertheless is amply rewarded in terms of our understanding of phenomena and how we might manage them. The process of how we arrive at a workable model and apply it dependably is, as is evident from this chapter, a multistage approach with its own positive and negative feedback loops. This model of model development is best expressed diagrammatically and is given in Figure 4.14. This chapter has been about generic modeling issues, the next chapter will focus specifically on the nature of environmental models.

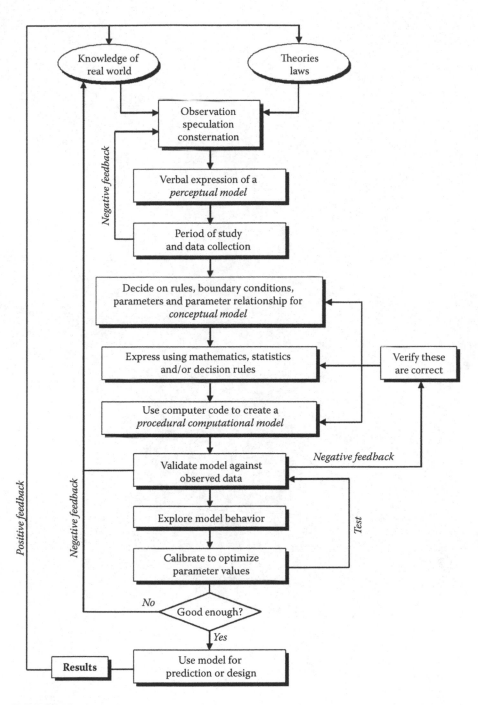

FIGURE 4.14
Schematic outline of the model development process.

5

The Role and Nature of Environmental Models

Models can only be fully understood within the context that they are built and used. In Chapter 4, we took a broad look at the purpose of modeling, the various forms they may take, and, in a generic way, how they are constructed and applied. Now we need to look more specifically at the nature of environmental models. Because models are abstractions of reality that assist our understanding and furthermore, because the environment includes the physical (natural and built), biotic, social, cultural, economic, and legal aspects of our world, there are potentially as many environmental models as there are things under the sun. What is more, all of these models together would encompass the whole of human knowledge. I leave that particular goal to the Wikipedia. In this chapter, we will start by looking at the context in which environmental models today are built and used. We will then look at some broad classes of environmental models as regards the way in which these models are structured and work. This chapter, thus, completes the framework for an understanding of the issues presented in Section III of this book.

Classifying environmental models into useful chunks for understanding how they are structured and work can be done in several ways, but whichever way you choose starts to get complicated as one recognizes exceptions, hybrids, and special cases. One common approach has been to recognize whether the underlying scientific logic of a model is *inductive* or *deductive* (Skidmore, 2002; Brooks, 2003). Inductive models are empirically based in which models represent generalized theories derived from observed data. The possible weakness of this approach was discussed in Chapter 4. Nevertheless, in cases where our knowledge of the exact processes at work or the mechanics of their behavior are not sufficiently well known in detail, inductive black- or grey-box models will often be adequate for our purposes. A statistical treatment of empirical observations and testing allows one to determine the probability or reasonableness of the results of a model (e.g., confidence limits of a regression line model). Deductive models on the other hand are based on physical laws, such as the conservation of matter and energy, which are either built up from axioms or on the basis of falsifiable hypotheses. Here, empirical data are used to test models rather than build them. Many environmental models that simulate physical or chemical processes are deductive. The problem with classifying models strictly in this way is that, in practice, many approaches to environmental modeling often require both inductive and deductive elements. At its simplest, GIS data handling in either preparing data for simulation models or creating models of the environment within GIS tends to be inductive in its approach. Remote sensing (RS) is commonly used to classify objects inductively. The numerical simulation model that uses such data may appear to be deductive in the way it handles the transport and fate of materials within a system, but may well employ empirically derived curves (transfer functions) embedded within them to make the models more tractable. Instead then, we will use a fourfold classification (wholly inductive) loosely based around how the models are derived and used: conceptual, empirical, knowledge-based, and process models. But before launching into these, let me set them in the general context of where, when, and for what purposes such models are used.

Context of Environmental Modeling

Some of the contextual issues were touched upon in Chapter 1. Here they need to be expanded upon and added to. Critical concerns over the management and custodianship of our planet and its resources began in the late 1950s and became established as a popular movement in the 1960s. This began to influence legislators, regulators, and policy makers from the 1969 enactment of the U.S. National Environmental Policy Act (NEPA) through the 1992 Rio Declaration on Environment and Development (Agenda 21) and up to the present. Even the industrial giants, the so-called "polluters," have had to promote green images to maintain their customer bases. However, in a modern, pluralist society, environmental issues have had to take their place alongside all the other competing objectives (Welford, 1995). Thus, from a European perspective, there is certainly a desire amongst the population to maintain high standards of living with inexpensive, good quality food and consumer products, affordable energy and transport on demand, and modern schools and hospitals with the latest technologies. And lest we forget, cheap holiday flights to our choice destinations. Twenty-first-century lifestyle for the individual seems to have become an inner balancing act between a genuine desire for greenness and sustainability on the one hand and having the necessary accoutrements of gratified hedonism on the other. For those nations still on the path toward these lifestyles, to be told by assisting nations and trading partners in the West of fears for the environment and damage caused by development, not surprisingly appears hypocritical. In the end, the environment of the twenty-first century will be the environment we make (Botkin, 1990). Consider Figure 5.1. It essentially says that any land use is the result of an interaction between the physical and biotic

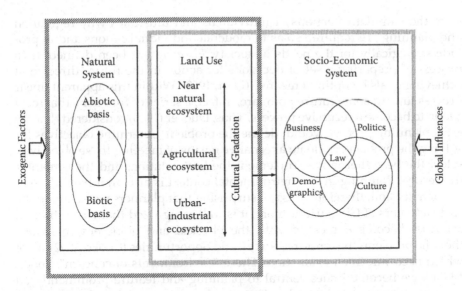

FIGURE 5.1
Shown is man–environment interaction. (Adapted from Messerli, B. and Messerli, P. (1978) *Geographica Helvetica* 33: 203–210.)

aspects of the environment and the intermeshed socio-cultural, economic sphere. Thus, the so-called "natural" landscapes remain more or less intact either because they present no opportunities for alternative use as yet, or precisely because their natural state affords some benefit (say, recreation). It is, of course, not just land use, but the quality of the air, water, fishing stocks, and so on. Either way, we humans dominate the agenda despite being probably outnumbered by the mass of organisms in just my back garden.

That the physical and biotic environments are critical to our well-being is well recognized. Since the 1960s, technological developments have facilitated inventories of the environment and measurement of impact and change with increasing ease and frequency and at higher resolutions. Key among these technologies is imaging using RS. From the earliest meteorological satellites, through the NOAA, Landsat, and Spot series to the latest IKONOS and EROS satellites, there are now in place a diversity of sensor types, bandwidths, swathe widths, and resolutions that can record a truly staggering range of environmental phenomena (for more details on sensors, see van der Meer et al., 2002). Such data have not only fuelled models, but also concerns. It was, for example, RS imagery (Nimbus 7) that first allowed scientists to identify the significance of and then study (model) the rate of ozone depletion in the upper atmosphere over Antarctica. The use of, say, an aerosol hairspray containing chlorofluorocarbons (CFCs) may have seemed an innocent action for an individual, but the aggregate cumulative result of many individuals over time has resulted in a "hole" in a critical layer of the atmosphere. This is but one example of how adverse impacts can arise from the cumulative actions of individuals and illustrates why regulation and planning are required so as to moderate individual actions and to provide strategically for the needs of society. *Planning* has been defined as "a process of preparing a set of decisions for action in the future directed at achieving goals by optimal means" (Dror, 1963). Within this apparent linear process toward some future point are, in fact, a series of functions that form a loop to become a recursive process. Thus, Dueker (1980) considered that the ideal planning process would: define the problem, determine objectives for solving the problem, evaluate alternative means of meeting those objectives, select the best alternative, implement the best solution, and then monitor the results. Planning in an environmental context is also an inherently spatial activity and, therefore, it is not surprising that planners were among the early adopters of GIS. Furthermore, it is well recognized that where human activities become enmeshed with the physical and biotic environments, there lie problems, hazards, resources, and opportunities (Cooke, 1992), all of which need informed evaluation. These "environments of concern" (Jones, 1983) have become issues central to planning and feature prominently on government and global agendas. The role of environmental modeling should already be apparent to the reader, but so that we can see the full spectrum, it is necessary to briefly run through three specific issues related to planning and management at the man–environment interface: environmental impact assessment, sustainable development, and issues of hazard, vulnerability, and risk.

Environmental Impact Assessment

In order to balance the increasing dilemma of maximizing our well-being while minimizing environmental abuses, a methodology of *environmental impact assessment* (EIA) was developed to assess the environmental implications (physical, biotic, socio-economic) that might result from the implementation of policies, plans, and development projects. The results of an EIA are embodied in an *environmental impact statement* (EIS). There is a considerable volume of literature on this subject, but in the context of this book, the reader is referred to Selman (1992), Welford (1995), and Therival et al. (2005) in the first instance for a fuller treatment. Environmental impact assessment was first mandated in the United States under the NEPA of 1969. This was gradually followed by other industrial nations and was adopted by the European Union (EU) in 1985 through Directive 85/337, which was then, for example, incorporated into British law through The Environmental Assessment Regulations 1988. EIAs, however, are not necessarily applied to all developments, but tend to be employed selectively depending on the type of development, its scale, and setting. Such selectivity is not left to whim, but tends to be stipulated in some detail by the relevant legislation (see http://ec.europa.eu/environment.eia for EU schedules of various classes of project in Directive EU 97/11). EIAs are generally structured as follows:

1. *Project brief:* In order to begin an EIA, there must be sufficient information about the nature, purpose, scale, and proposed location for the development project made available to the assessors. This description of the project, associated technical details, costing, and infrastructure implications (e.g., road upgrading, water supply) would also include any initial evaluations of environmental aspects carried out during the search for a suitable location for the project.

2. *Identification of impacts to be assessed:* Not all impacts need necessarily be assessed. Some may be trivial or of little significance to the outcome of the EIA. In practice, a limited set of issues is to form the focal point of an assessment. This process of identifying which issues are likely to be important is called *scoping.* An element of consultation with the authorities, interest groups, and local population may be used to draw up the list. Scoping may seem to be a straightforward judgment, but it begs the question as to what the priority values of society are or ought to be (Beanlands, 1988). Not only do values change over time, but in many projects there are tensions between the common good and local impacts (waste must be recycled or disposed of somewhere; meat at the dinner table necessarily needs abattoirs) and between local benefits and impacts (e.g., employment versus loss of woodland). And, of course, NIMBY (not in my backyard!) is often rife.

3. *Establishment of baseline conditions:* In the context of the scoping, it is necessary to establish the existing condition for two broad reasons. The first is to lay a foundation upon which the prediction of impacts can take place. If impacts are to be predicted, then those that derive from complex interactions must be modeled in some way and the models should take initial parameter values from and ought to be calibrated first to reflect existing conditions. The second reason is to have a marker against which later, postimplementation monitoring and audit can establish the degree of actual impact and the adherence to mitigation measures. There is likely to be a considerable amount of baseline data and much of it spatial. It is not surprising then that GIS have a major role in the organization, integration, and storage of the baseline data.

4. *Prediction and assessment of impacts:* There are two broad methodologies for completing this stage: matrix and checklist. The Leopold association matrix (Leopold et al., 1971) has 88 environmental characteristics (e.g., soils, surface water) on the vertical axis set against 100 actions that might cause an impact (e.g., irrigation, alteration of ground cover) on the horizontal axis. This gives 8,800 cells to be considered. For each proposed action in a project, a slash is placed diagonally across each cell opposite any environmental characteristic that might be impacted. In the upper left-hand corner, the

magnitude of the impact is rated –10 to +10 (no zero), while in the lower right-hand corner, the *importance* of the possible impact is rated 1 to 10. The completed matrix is then a summary of all impacts upon which decisions can be made. Needless to say, simplified as well as variant matrices have been devised. Checklists offer a less laborious approach in which a fairly generic list of possible effects of a project (e.g., on human beings, on flora and fauna, on water) is used to drive an assessment of impacts. In either methodology a range of value judgments, GIS-based analytical techniques, use of systems diagrams (Odum, 1971), and simulation models may be employed to predict impact magnitudes and assess their importance. Although EIAs contain considerable scientific content, gaps in available theory (Selman, 1992) and a host of other uncertainties (de Jongh, 1988) mean that assessments cannot be made purely on formulaic grounds. Value judgments, therefore, are a necessary element of this stage.

5. *Evaluation of mitigation measures and monitoring:* EIA decisions are rarely made before attempts have been made to mitigate against predicted impacts. It may be possible to avoid or reduce certain impacts by introducing certain measures. Such measures could include:

   - Changes to where the project is to be built (e.g., identifying a better location for a dam to be built).

   - Changes to the overall design (e.g., buried instead of surface pipeline) or specific details (e.g., nozzle size used on a coastal effluent outfall).

   - Stipulations on construction methods (e.g., use of mufflers and no night working to reduce noise impacts).

   - Stipulations on how the constructed facility should be operated over its lifetime (e.g., minimum downstream flow from a dam to maintain river ecology).

   It is also at this stage that the necessary postimplementation monitoring program, in order to measure actual impacts, would be designed. Monitoring may even be started at this stage in order to augment the baseline data for the key impacts.

6. *Consultation and review:* Prior to any final decision, consultations on the findings are usually carried with all parties (the public and the government) and since certain aspects of the assessment are necessarily judgmental, there are often differing interpretations on magnitude and importance of impacts and scope for review of the findings. Following this stage, a decision can usually be reached on the go-ahead (or otherwise) for a project, though not always unanimously.

7. *Implementation, monitoring and audit:* Because most environments are dynamic, it is often necessary to conduct monitoring of an

implementation against a reference site of similar situation, but unaffected by the project. Thus, the monitoring of impacts will tend to take the form of testing for any statistically significant differences from the original baseline data and reference sites. Audits are carried out after a sufficient period of monitoring to establish the general accuracy of the predictions made during earlier stages. This provides valuable feedback for use in other EIAs and establishes whether or not impacts for the specific project are within an acceptable range and whether or not further mitigation measures are required.

Again, the role of environmental modeling (physical, biotic, and socioeconomic) in EIA is obvious. Models as descriptive devices, embodying theories and hypotheses of the workings of processes, are used in the scoping exercise to identify likely key impacts and in the baseline stage informing the relevant layers of data that need to be collated. A wide range of simulation models are used to predict likely magnitude of impacts including air, water, and noise pollution, biotic models to predict ecosystem impacts, graphic models to assess visual impact, and, by no means least, economic models. But, despite the ample opportunity for employing models, EIA remains both science and art because much judgment is required. Nevertheless, insights provided by models are of key importance.

An Integrated Approach

There are considerable benefits to be gained by integrating planning, EIA, and preliminary design into a single, comprehensive process. This brings together multidisciplinary teams to evolve projects from conception to ensure compatible benefits for society, minimized impacts, and feasibility of construction. A case in point is the construction of new power stations and overhead power transmission lines in Hong Kong in the late 1980s and early 1990s (e.g., Power Technology, undated; Urbis Travers Morgan Ltd., 1992). With rapid economic growth and transformation of lifestyles since the early 1970s, Hong Kong's demand for electricity was growing significantly, particularly in the New Territories where eight designated new towns were under construction to house and provide employment for over a 3 million population by 1997 (Bristow, 1989) as well as growth in demand from industry and rural village renewal and expansion. The new transmission lines required to be constructed were 400 kV double-circuit with tower dimensions of up to 70 m high, 25 m diagonal leg span and able to withstand typhoons (hurricanes) in exposed mountainous terrain. The type of terrain is illustrated in Figure 5.2 and a transmission tower prior to stringing shown in Figure 5.3.

High land prices in low-lying, gentler terrain and a strong NIMBY effect due to the adverse impacts of transmission lines on the development potential of private land, meant that routing would inevitably have to be in the steeper,

FIGURE 5.2
Stereoscopic view of typical mountainous terrain in Hong Kong across which overhead transmission lines were to be constructed. (Photos courtesy of the author.)

FIGURE 5.3
Aerial oblique view of a double-circuit transmission tower, prior to stringing, constructed to withstand typhoons. (Photo courtesy of the author.)

mountainous terrain. However, most of the mountainous terrain in Hong Kong is of scenic beauty and is designated as Country Park. Such terrain is also prone to landslides and imposes engineering design challenges for the routing and construction of power lines. Clearly, a route of minimum environmental impact might not be capable of construction, at least not within a price that the community is willing to pay for its electricity. The solution was to carry out the environmental impact assessment and the preliminary design in one single study that worked toward minimizing environmental impact, minimizing the geotechnical constraints (slope instability, need for special foundations), respecting the design parameters of towers and cables, and without unnecessarily inflating the overall construction cost. During the process, where conflicts arose between the various facets, the effects were modeled and compromises were thrashed out through consultation. In the end, though, a solution was reached that all parties agreed to sign up to.

Sustainable Development

Are environmental impact assessments a panacea or a waste of time? They have certainly been hailed as both (Greenberg et al., 1978). A major flaw that has emerged is their project-by-project implementation. This was certainly my experience of the overhead transmission lines in Hong Kong even with the integrated process. As successive new power stations were built, even though they may be in different locations, the electricity still needed to be delivered to the same centers of population causing a sort of power line congestion. The first overhead transmission line to be built occupied the best route, with each new power line after that occupying a progressively suboptimal route with increasing scenic and technical constraints as the mounting number of power lines became increasingly obtrusive and must more or less run in parallel as crossovers are difficult to achieve. At no stage, in my opinion, did a study for any one overhead transmission line consider where any future capacity might go and whether the results of the current study might jeopardize such future developments. At the 1992 Rio Earth Summit, Brazil, the Rio Declaration on Environment and Development, Agenda 21 was adopted by 178 governments. Agenda 21 was all about *sustainable development* and the declaration set out 27 principles of how that might be achieved (United Nations, 1992). Key among these for a definition of sustainable development is:

Principle 1: Human beings are at the center of concerns for sustainable development. They are entitled to a healthy and productive life in harmony with nature.

Principle 3: The right to development must be fulfilled so as to equitably meet developmental and environmental needs of present and future generations.

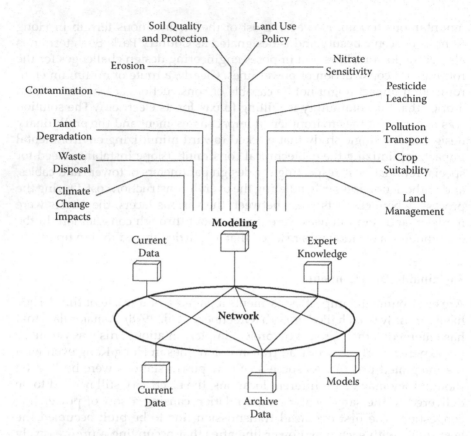

FIGURE 5.4
The concept of integrated information systems supporting planning and sustainable development. (Based on Hallett, S.H., Jones, R.J.A., and Keay, C.A., (1996) *International Journal of Geographical Information Systems* 10:47–64.)

The declaration incorporates both the temporal and the spatial and provides for a reconciliation of environment and economy (Welford, 1995) with ethics and politics emerging as central issues in applying the principles (Owens, 1994). Implicit is the wider use of models to simulate the medium- and long-term effects of development, and to be more integrative of these models in studying the interdependence of a range of phenomena. Such information systems are likely to be large and complex incorporating data, models, and expert knowledge (Figure 5.4). Indeed, this augurs well for GIS and environmental simulation modeling. But while the concept of sustainability has ushered in a paradigm change toward notions of planning and development, there has been little guidance on how to incorporate sustainability into the practicalities of the planning process at a local level particularly in how to define and measure the interests of future generations. The reality that has emerged is the recognition that sustainability is itself a dynamic concept (van den Brink, 1999). The knowledge, values, and norms that we apply today are likely to change over time. What we consider sustainable today, may not be considered so in the future. Given this shifting sand, sustainability is not a goal in itself, but a principle or litmus against which developments can be tested.

Hazard, Vulnerability, and Risk

Principle 1 of the Rio Declaration (above) seems to espouse a romantic view of man able to live in harmony with nature, a notion that sits nonconformably with the *International Decade of Natural Disaster Reduction* (IDNDR) that spanned the 1990s (Brimicombe, 1999a). The IDNDR was initiated by the United Nations in the face of the escalating human vulnerability to natural disasters in terms of the heavy loss of life in developing countries, huge economic losses in developed countries, and the growing possibility of mega-disasters due to population concentration (Hamilton, 2000). While insurance losses resulting from manmade catastrophes remained at a near constant level for the period 1970 to 1992, losses resulting from natural hazards increased nearly 10-fold (Swiss Reinsurance Company, 1993). The number of disasters per decade has increased fourfold since 1950 (Munich Reinsurance Company, 2001). Losses are predicted to continue increasing exponentially. Then there are the "creeping" disasters of climate change from which we can expect increasing intensity of rainstorms (including hurricanes/typhoons), droughts, and windstorms all of which threaten infrastructure and ecosystems with long-term cost implications (Saphores, 2004). Most natural hazards (earthquake, volcanic eruption, hurricane/typhoon, tsunami, landslides, avalanche, flooding, wildfire, pests) are characterized by nonrandom spatial distributions and can be expected to reoccur in the same locations along gradients of magnitude and frequency (high frequency of low magnitude events, low frequency of high magnitude events). Thus, inappropriate development in areas of known or suspected hazard consequent on the pressures of population growth may not immediately have catastrophic consequences, but can leave future generations vulnerable. Furthermore, the cumulative effects of incremental land use change can trigger adverse geomorphologic responses, such as increasing flood severity in urbanizing drainage basins (Morisawa and LaFlure, 1979), or, as has been increasingly experienced in England over the past few years, increasing flood severity due to changing weather patterns making themselves felt on urbanizing floodplains. With growing population numbers concentrated into ever larger cities worldwide, the propensity for disasters continues to grow. It is not just lives that are at risk from catastrophic events, but the economy and essential services.

Given our knowledge of physical processes, it should be within our power to model and manage. But this is over-simplistic. Let us take the case of Hong Kong: a small and wealthy region of just 1050 km^2, 60% mountainous terrain, and, hence, a severe shortage of easily developed land and a population of about 6 million. Average annual rainfall is 2,225 mm, but tropical depressions and typhoons can result in rainfall intensities that can reach 90 mm/hr. Four days of severe rainfall in May 1982, for example, resulted in more than 1,500 landslides (Brand et al., 1984). Since 1950, there have been more than 470 deaths from the failure of manmade slopes formed during construction of building platforms (http://hkss.cedd.gov.hk/hkss/). Yet, some two decades of cataloging, monitoring, and reconstructing these slopes, passing of new legislation and enforcing strict design criteria and construction codes, some deaths and injuries (though fewer in number) still occur, the most recent fatality (at press time) being in 2008 (http://hkss.cedd.gov.hk/hkss/eng/safemeasure/lpi/LPI_Chart_2008_Eng.pdf). Our knowledge of slope failure processes is good, but not perfect; slope failure models are good approximations, but cannot take account of all defects in a slope, particularly as we can only sample the physical characteristics of the soil and rock beneath the surface; we cannot predict when the next rainstorm that triggers a landslide will occur other than in rough probabilistic terms and, therefore, we cannot predict exactly where or when the next slope failure will occur nor who will get in the way. What we can do is identify the level of hazard, assess the degree of vulnerability, and determine the level of risk. But these terms—hazard, vulnerability, risk—which are in themselves models, need defining. In doing so, I have drawn upon Varnes (1984), United Nations Disaster Relief Organization (1991), Cutter (1996), Fernandez and Salas (1999), Kong (2002). and Raetzo et al. (2002) where the reader can find a fuller treatment:

Hazard: A hazard can be taken as any potentially damaging phenomenon for which there is a probability that it will occur within a given time period and in a given area. This can be applied to naturally occurring phenomena (natural hazards, such as those listed above) as well as to the economy (e.g., stock market crash), industrial processes (e.g., pollution), and even antisocial behavior (e.g., violence in a marriage). Explicit in the definition is the spatio-temporality of hazards and that within certain scales of space and time they can be modeled both as spatial and historical analogues (e.g., hazard maps) or modeled probabilistically by return period (see below). Recognition of a hazard, therefore, is dependent on data of prior occurrences and the ability to identify what and where they are and when they have occurred. In other words, a potentially damaging phenomenon about which we have no data and have not previously experienced cannot properly be called a hazard—it remains a conjecture.

Return period: Associated with any hazard is a probability of occurrence. This is usually expressed as the expected average time in years for an event of a particular magnitude to reoccur. For example, a severe flood may be said to be a "1 in 100 year" event or has a "100-year return period." This does not mean that a flood of this severity will have exactly 100 years between each event, rather the probability that a flood at least this severe will occur in any one year is 1%. From an engineering perspective, hydraulic structures, for example, will be constructed to cope with a "design flood," that is a specified return period of flood, which, if exceeded, may result in failure and, therefore, represents the likelihood of structural failure. The occurrence of natural hazards tends to follow highly correlated patterns of magnitude and frequency as illustrated in Figure 5.5(a). The pattern is one in which low magnitude events will have a higher frequency (shorter return period) while high magnitude (severe) events will have a lower frequency (longer return period). Calculating the expected return period for hazards is an important part of assessing the degree of vulnerability and risk a hazard may pose. Such probability calculations can be carried out from existing records of the magnitude and frequency of events over a long period or they may be simulated using Bernoulli trials (if events are considered to be independent) or Markov trials (if events are considered to be dependent) from shorter event records and plotted as log probability graphs to produce straight line relationships (Figure 5.5(b)).

Vulnerability: No less than 18 definitions of vulnerability are quoted by Cutter (1996). At its simplest, vulnerability is the degree of loss that may result from the exposure to a hazard. That loss may be rated from slight damage (low vulnerability tending to 0) to total loss (high vulnerability tending to 1). In human terms, should the loss of any life be the likely outcome of a specific hazard in a specific place, then it is usual to classify these individuals as being highly vulnerable. There is an obvious link between the estimation of the degree of vulnerability and the magnitude of event for which it is being estimated.

Risk: This can be considered in several parts. The first is *specific risk* expressed as the product of the probability of the hazard and the vulnerability, that is, the expected degree of loss arising from a hazard of particular magnitude. The *element risk* is the size of population, value of property and economic activity subject to the specific risk. The *total risk* is then the product of the specific risk and the element risk, that is, the expected number of lives lost and people injured, the damage to property, and the disruption to the economy arising from the event, either expressed monetarily or in descriptive terms. Eveleigh et al. (2006) provides an example of coupling systems engineering and spatial modeling in ArcGIS for natural hazard risk assessment in relation to the infrastructural elements for disaster response and recovery. Morita (2008) provides an example of flood risk analysis to determine optimum levels of flood protection in relation to design storms for different return periods.

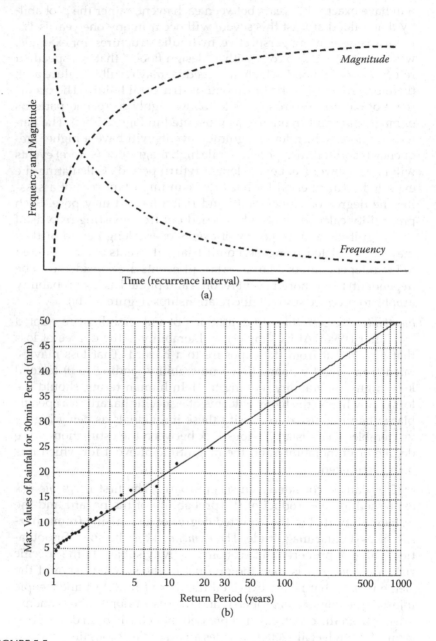

FIGURE 5.5
Frequency and magnitude: (a) the general relationship between magnitude and frequency of natural hazards, (b) an example of plotting return periods for event magnitudes. (Based on Table 4.2 in Thornes, J.B., and Brunsden, D. (1977) *Geomorphology and time*. Methuen, London.)

Decision Environment

Having identified pertinent hazards (or potential impacts arising from development) and assessed the degree of vulnerability and risks that these pose, in part through modeling, a decision on what to do about the situation needs to be made. Chapter 10 will look at some issues in depth concerning the decision-making process, but it is pertinent to introduce here some general aspects about the framework within which the GIS analyst and environmental modeler will participate in assisting the formulation of decisions. Decision making is often an iterative process. A problem may be recognized and objectives for its resolution set. Based on policy criteria, explicit or implicit in the objectives and any physical, economic, or social constraints that might exist, alternative strategies are developed and evaluated often with the use of models both to simulate the problem to fully understand it and to simulate the alternative strategies to study the effectiveness and consequences of any mitigation measures that might be adopted. As a result of this process, the understanding of the nature of the problem itself may be clarified or even change, the objectives may be modified; some residual problems may be outstanding from the main solution, any of which requires the cycle to be iterated. Even when a plan has been implemented, its performance will need to be monitored and evaluated and may raise issues about the initial problem that are not being effectively solved or new problems that have arisen, thus requiring yet further iterations. Developing alternative strategies, evaluating them and recommending one or more courses of action is far from straightforward and can itself be a politically hazardous journey. Returning to our theme of natural hazards, whether landslides or floods, there are a range of alternatives, even complementary strategies that could be put forward to reduce the level of risk depending on the degree of prevention or preparedness that society would be willing to accept (tolerate) and pay for. Figure 5.6 broadly summarizes these options.

The first thing to note is that mitigation in terms of prevention or preparedness is a spectrum from maximum risk reduction usually at a high financial cost (construction and maintenance) to a much smaller level of risk reduction usually at a much-reduced financial cost. Second, these are not mutually exclusive. Given the magnitude and frequency relationship of natural hazards, it may be prohibitively expensive or technically impossible, for example, to prevent loss from higher magnitude events, in which case some form of preparedness through zoning or early warning (such as for hurricanes) would be prudent. Lower magnitude events that happen more frequently on the other hand, such as a one-in-five-year flood, may indeed be preventable at an economic cost and allow land to be kept in productive use. Such decisions may well change from one area to another depending on the local level of risk. Decisions on where and when to employ structural or non-structural measures of mitigation and the mix that may represent a really good, affordable strategy are aided considerably by the use of simulation

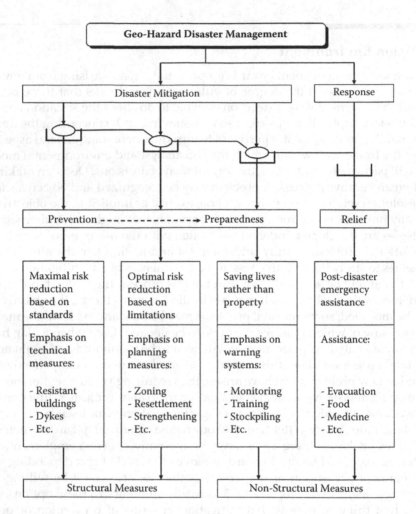

FIGURE 5.6
Broad approaches to risk reduction measures for natural hazards. (Adapted from the United Nations Disaster Relief Organization (1991) *Mitigating natural disasters—phenomena, effects and options.* UN Publication E.90.III.M.1, New York.)

models particularly where they are structured into decision support systems that facilitate "what if"-type queries and the analysis of multiple scenarios. This will be a theme we return to in Chapter 10; an example of decision support for basin management planning is given in Chapter 6. The point to note is that environmental simulation modeling has an important, almost defining role to play in such decisions alongside or actively in tandem with the use of GIS. Even evacuation strategies need to be evaluated and can be modeled using geographical micro-simulation (e.g., Chen, 2008).

Only a minority of problems can be solved by resorting only to simulation modeling. The environmental modeler and GIS analyst, while working

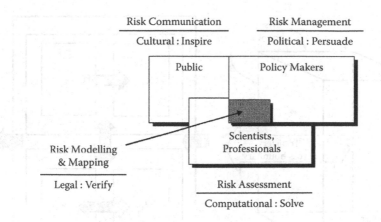

FIGURE 5.7
Reasoning about risks and solutions. (Adapted from Rejeski, D. (1993) *Environmental modeling with GIS*, ed. Goodchild et al. Oxford University Press, New York, pp. 318–331; and Berry, J.K. (1995) *Spatial reasoning for effective GIS*. GIS World Books, Fort Collins, CO.)

as scientists and professionals, need to interact with policy makers and the public and may need to resort to different approaches in getting their recommended solutions across to them (Rejeski, 1993; Berry, 1995). Perceptions of risk and the necessity for response can be quite different among these groups (Figure 5.7). The scientist and professional modeler first navigate a physical space to identify risks and then a decision space to identify solutions. The policy maker who has to manage risks is also navigating a decision space for what might be effective solutions, but also a social/perceptual space in judging politically acceptable solutions. The public navigates a perceptual space in terms of recognizing where they are at risk and whether or not they should worry about it. Communication between these different groups may require changing stances and certainly a recognition of the way in which facts and values interplay and may indeed come into conflict.

Environmental models, if employed perceptively, can be used to solve problems, verify outcomes, persuade policy makers, and even inspire changes in culture and public attitudes. There is everything to be gained. All this rests, of course, on the appropriate choice and deployment of models. We will now look at the details of four broad classes of environmental models: conceptual, empirical, incorporating artificial intelligence, and process.

Conceptual Models

Conceptual models contain a high level of abstraction. In Chapter 4, we looked at a couple of conceptual models as statements or simple formulae.

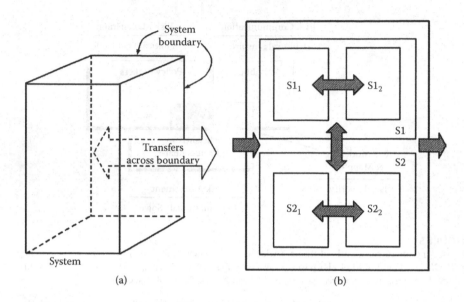

FIGURE 5.8
Systems thinking: (a) main elements of an open system, (b) the specification of subsystems within a system's hierarchy.

Frequently, conceptual models are also mappings of the main elements of a system together with how these elements are linked (Figures 5.8 and 5.9). These links may indicate the direction and strength of the interaction and the presence of any feedback loops. As we saw in Chapter 4, conceptual models are an early stage of progressing toward more fully specified working models and are used to describe the structure and dynamics of the system being studied. They are important in giving shape to our thinking and understanding of how particular aspects of the environment work. Therefore, it is worthwhile at this point to introduce some basic elements of *systems thinking* (Emery, 1969; Harvey, 1973; Deaton and Winebrake, 2000). A *system* can be taken simply as any part of the environment under primary consideration. A *closed system* is one that works with no interaction with its external environment. However, there are very few real world environments that could be usefully considered in this way. Dynamic systems usually have transfers of energy and matter into and out of the system and, thus, are modeled as *open systems*. An open system, as conceptualized, will have a boundary across which transfers can take place (Figure 5.8(a)). This boundary can be placed to restrict the number of processes under consideration and/or to limit spatio-temporal extent. Such boundaries are not necessarily real discontinuities (though it helps if they are), but can be artificially imposed to make the modeling tractable. Thus, in hydrology, it is common to define the system as a natural catchment area with the boundary defined by its drainage divide. For landslide models, the boundary may be an arbitrarily defined 3D section of slope. Boundaries can

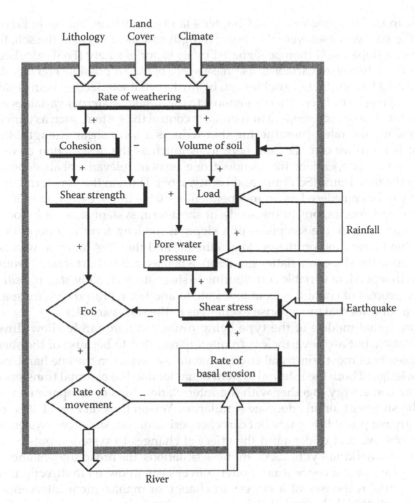

FIGURE 5.9
A conceptual systems model of a landslide. (Adapted from Brunsden, D. (1973) *Geologia Applicata e Idrogeologia* 7: 185–207.)

be placed specifically in order to study the transfers across them. Within the boundary can be described, if necessary, as a series of subsystems each with their internal boundaries and transfers between them (Figure 5.8(b)).

The *phase space* of a system is a multidimensional space of all *states* that a system could theoretically adopt depending on a range of states in the external environments, quantities, and rates of inputs, outputs, and so on. Clearly some of these states may be highly unlikely indeed, but if theoretically possible then they contribute toward the make up of the phase space. The actual state of a system at any particular moment (i.e., its position within the phase space with respect to an interval of time) is determined by its *state variables*, that is, the variables that characterize the properties of a particular system.

Thus, in the landslide model of Chapter 4 (and as a systems model in Figure 5.9), the state variables were the shear strength and shear stress of the soil; the state of a slope could then be signaled by its factor of safety (FoS) calculated from these two state variables, its phase space being all possible FoS that the slope could adopt. State variables can be further characterized as being *external* (exogenous) or *internal* (endogenous) to the system. External variables are considered to be independent and broadly control the system, such as climate controlling the rate of weathering that reduces a soil's shear strength. But, there is a further complication here in as much as the designation of variables as being dependent, independent, or even of no relevance at all, depends upon the time frame (Schumm and Lichty, 1965). Thus, in the long term, time itself can be considered as an independent variable in the evolution of landscapes and species, but in the study of short-term system states it becomes irrelevant. Again, the steepness of a slope in the long term is a dependent variable (depending on the geology, climate, and the time taken to weather and erode the slope to a flatter gradient), whereas in the short term, it would be an independent variable contributing to sheer stress. Finally, *state transition* is the process of changing from one state to another within the phase space over a time interval in response to changes in the state variables.

Conceptual models of the type being discussed here rarely allow direct simulation, but are nevertheless frequently resorted to because of the sheer complexity of most nontrivial environmental systems. On the one hand, our knowledge of both the internal system behavior and the external transfers of matter and energy together with the interrelationships of components may not be sufficient for an adequate simulation. Yet, on the other hand, it is frequently not possible to carry out direct experiments on, say, an ecosystem so as to observe and understand the effect of changes to system inputs on its behavior (Malkina-Pykh, 2000). In these situations, the broad understanding of a system that a conceptual model conveys can allow us to directly infer "ball park" responses of a system to change or management intervention and debate the likely level of risk.

Empirical Models

Empirical models are derived from observation and data, and from which conclusions are drawn on the effects or outcomes of processes. Thus empirical models rely heavily on induction for their construction. The models themselves are more often than not black- or grey-box models in as much as the details of the internal processes may not be known other than conceptually. What tends to be known are the inputs and the resultant outputs. Where these inputs and outputs have been adequately quantified, they can be subjected to a statistical analysis and used as either deterministic or stochastic

TABLE 5.1

Variables Commonly Used in a Map-Based
Inductive Analysis of Landslides

Slope angle	Geology
Slope aspect	Discontinuities
Slope height	Soil type
	Soil thickness
Rainfall regime	Distance to active faults
Undercutting	
Loading	Land cover

models. Otherwise, historical and/or spatial analogs can be developed. Let us deal with the analogs first. Empirical analogs that are inductively developed conform well with the way cartographic processing of map-based models is carried out using GIS. In Chapter 2, we looked at a landscape in which there were a number of landslides (see Figure 2.5). From the conceptual knowledge we have gained about the process of landslides in Chapter 4 and in the previous section on conceptual models, we could start to list the common variables that one might need to consider in a map-based inductive analysis. The list is quite long, as presented in Table 5.1, though not all of them might be relevant in every study (e.g., distance to possible earthquake epicenter in areas where there are only sporadic occurrences of minor tremors). An effective approach to evaluating landslide hazard and risk is terrain analysis using aerial photographic interpretation (API) supplemented by secondary data in the form of published topographic and geological maps and backed up by field inspection (Brimicombe, 1982; van Zuidam, 1985; Soeters and van Westen, 1996). Only recently has satellite RS imagery started to have sufficient resolution for this type of work. The terrain analysis would result in knowing the spatial distribution of:

- A range of slope instability events, usually classified by type (e.g., rotational slip, debris flow, rock fall) and apparent age (e.g., recent, ancient).
- The environmental setting of these features (e.g., geology, soils, hydro-geology, land cover).

An assessment of the condition and likely triggering mechanisms of these features (e.g., rainfall events, stream undercutting). On the basis of such data, key variables indicative of stable or unstable slopes can be identified. For example, this could take the form of any slope composed of colluvium (a relatively unstable soil type) of angle greater than 15° (relatively steep for colluvium) within 250 m of a stream (footslope position where saturation can occur during rainstorms, or is being actively undercut at the base). This then becomes the spatial analog that can be applied in two ways: (1) to extrapolate zones of likely instability beyond only those locations where historical landslides are visible, and (2) to rapidly apply the model in other areas that have a similar environment without the need to carry out a full terrain analysis. The approach is classically GIS-able and is a prime example of environmental modeling *within* GIS:

- A data layer for elevation is transformed into one for slope angle.
- A data layer of soils is invoked and colluvium selected.
- A data layer giving a 250 m buffer around streams is produced.
- If vector data sets, all three layers are topologically overlaid.
- Polygons (or cells) of unstable colluvial footslopes are identified and mapped through Boolean selection according to the relevant criteria.

The results of this type of empirical analog, for our example landscape, are given in Figure 5.10. By plotting the final result of Figure 5.10(a) in relation to villages and roads as in Figure 5.10(b), we have a means by which to start evaluating risks. The village in the northwest is clearly vulnerable as are parts of the larger village in the southeast. Some sections of road are also vulnerable, but these would carry a lower risk, say, to life. Further details of this approach for landslides can be found in Wang and Unwin (1992) and Berry (1995).

Statistical analysis of empirical data often results in the construction of regression line models, which can also be expressed as equations. These are numerous in environmental modeling and one such example has already been given in Figure 5.5(b) above. They are not difficult to construct and are standard functions of spreadsheets and statistical packages. For our example landscape, we can extract data, using GIS functionality, on gradient and landslide density, which can then be sampled and plotted as a scatter diagram for all densities above zero and a regression line fitted using a spreadsheet or statistical package (Figure 5.11). The regression line equation in Figure 5.11(c) of $y = 0.21x - 0.73$ could then be used deterministically to calculate likely landslide density (and, hence, the degree of likely hazard) for other similar areas using only a digital elevation model (DEM). The R^2 value of 0.67 represents the goodness-of-fit of the regression line to the data and, hence, its reliability as a tool for inferring y. In this case, the fit is reasonable for a single variable and shows the dominant effect in this instance of slope angle, though other variables, such as geology, would be expected to have an important influence too. Therefore, regression models can be created for single variables as in our example or for multiple variables. Moreover, regression models need not be linear, but can be curved in some way (in Figure 5.11(d), the second-order polynomial gives a better fit to the data). In environmental modeling, frequently used nonlinear models are exponential, logistic, oscillatory (e.g., sine curve), and higher order polynomials (first-order polynomials being linear).

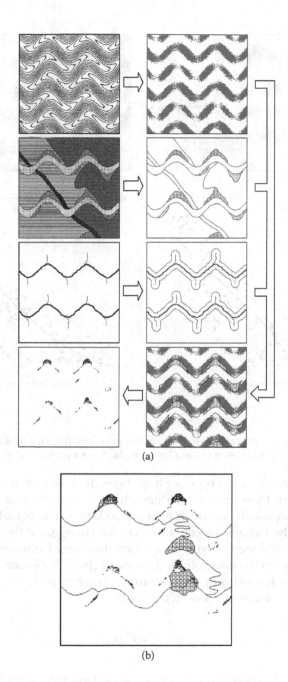

(a)

(b)

FIGURE 5.10
Application of an empirical analog using GIS to identify unstable colluvial footslopes: (a) GIS-based approach to evaluating steeper slopes, colluvial soils, and threshold distance from streams (see text for full explanation); (b) results plotted in relation to villages and roads.

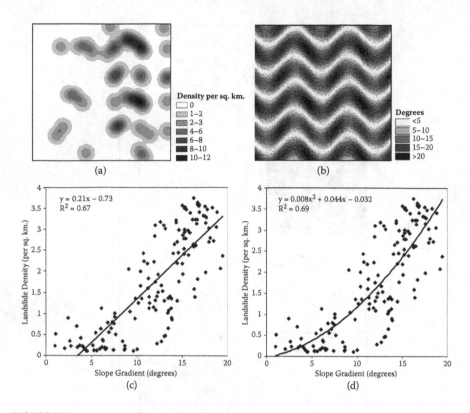

FIGURE 5.11
An illustration of an empirically derived regression model relating (a) landslide density to (b) gradient in the example landscape; (c) is a linear model, (d) is a polynomial model.

Empirical models of the type we have been discussing can be rather narrowly defined in their application depending on the nature of the phenomenon and the location(s) in which data are collected. The overall form of the equation and the variables it contains may not change, but the gradient and the intercept of the regression line equation may vary from one study to the next depending on the data collected. Gerrard (1981), for example, in citing a simple equation for soil transport by surface wash as being in proportion to slope length and slope gradient in the form:

$$S \propto x^a \tan^b \beta \qquad (5.1)$$

where S = soil transport ($cm^3/cm/yr$), x = slope length (meters), β = slope gradient (degrees), found that the exponents a and b as calculated by a number of authors, could range between 0.7 and 2 with most values, nevertheless, falling between 1.35 and 1.5. Furthermore, since Equation (5.1) is a proportionality ($\propto$), in order for it to be solved in any particular area, a constant needs to be determined that would reflect the ease of soil detachment according to soil type, rainfall characteristics, and so on, for that particular location or geographical unit. This simple example illustrates how modelers, in applying such an empirical model, would need to make decisions about appropriate values for exponents and establish an appropriate value for the constant. Such values may come from locally collected data, the scientific literature, or from published tables.

Empirically derived equations, if fully specified, can be solved cell by cell for a geographical area using the map algebra capability of GIS (Chapter 2). A data layer needs to be compiled or derived for each element of the equation in raster format. Suppose we take a much simplified wildfire hazard model in the following form and solve it for our example landscape:

$$B_i = \beta_0 + \beta_1 \, COVER_i + \beta_2 \, SLOPE_i + \beta_3 \, ELEV_i \tag{5.2}$$

where B_i = the propensity of the ith geographical unit to burn, $COVER_i$ = a scaling of land cover classes, $SLOPE_i$ = a scaling of gradient, $ELEV_i$ = a scaling of elevation as a surrogate for exposure to wind, β_0 through β_3 are coefficients. Assuming an intercept of $\beta_0 = 0$ and values of $\beta_1 = 0.6$, $\beta_2 = 0.3$, and $\beta_3 = 0.1$, and that COVER, SLOPE, and ELEV are scaled from 0 (no contribution) to 10 (maximum contribution) as in Table 5.2, the results of applying this model are given in Figure 5.12. Note in Figure 5.12(e) how the wildfire has kept to the higher ground in lower vegetation, thus reflecting Figure 5.12(d). Another way to express empirical data is in the form of an *index*. An index can in itself be a predictive model similar to multiple regression, it may be a means of reducing the number of data dimensions or a method of ranking, say, for prioritization. Malkina-Pykh (2000) considers that more environmental indices need to be developed to analyze environmental change and evaluate both interactions and cumulative impacts on multiple resources. Such indices can be calculated for systems without clear or exact geographical identity, or for specific geographical units. Probably the most widely used index of the latter type is the *normalized vegetation difference index* (NVDI)

TABLE 5.2

Example of Numerical Scaling of Variables for a Map Algebra-Based Wildfire Model

Land Cover		Slope Angle		Elevation	
Agriculture	2	0–5	2	50–150	2
Bare	0	5–10	4	150–250	4
Grassland	10	10–15	6	250–350	6
Shrub	8	15–20	8	350–450	8
Village	0	20+	10	450+	10
Woodland	4				

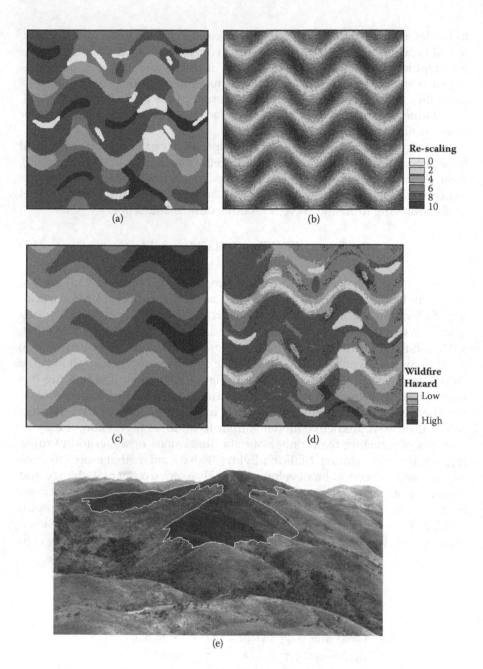

FIGURE 5.12
Application of a simplified wildfire model using map algebra techniques in GIS (see Chapter 7, Figure 7.8 for a flowchart of the model application): (a) rescaling of land cover, (b) rescaling of gradient, (c) rescaling of elevation, (d) result of solving equation (5.2) using map algebra, (e) burn scar from wildfire. (Photo courtesy of the author.)

used to assess relative amounts of photosynthetic activity for standing green biomass (or crop yield) from multispectral satellite imagery:

$$NVDI_i = \frac{NIR_i - RED_i}{NIR_i + RED_i} \qquad (5.3)$$

where NIR = reflectance in the near infrared channel and RED = reflectance in the red channel for the ith image pixel.

Models Incorporating Artificial Intelligence

The rise of *artificial intelligence* (AI) in the latter half of the twentieth century has been an "attempt to mimic the cognitive and symbolic skills of humans using digital computers in the context of a particular application" (Openshaw and Openshaw, 1997, p. 16). This covers an enormous area from robotics to fuzzy logic and computational linguistics, not all of which are relevant to environmental modeling. We will look at *knowledge-based systems* (KBS), *heuristics*, *artificial neural networks* (ANN), and *agent-based modeling* (ABM) here while fuzzy methods are best dealt with in Chapter 8. Overviews of AI are to be found in Wiig (1990) and Openshaw and Openshaw (1997), while the general benefits and risks of AI are reviewed by the Council for Science and Society (1989).

Knowledge-Based Systems

The terms *knowledge-based* and *expert* seem to be used interchangeably in much of the literature. There are, of course, subtle differences between using knowledge and the inference of expert judgments, but in the context of adding more "intelligence" to environmental models, we will, for the moment, refer generally to KBS. Within KBS there is the compilation of a knowledge base containing facts and rules about a specific problem or related set of problems. KBS can either give specific answers to situations or provide guidelines on what to try next either in a modeling context or in a management context. This may help professionals to grapple with complex issues or to guide less-experienced personnel as to appropriate forms of analysis and decisions. Let us briefly look at two examples. In Figure 5.10, we looked at the identification of unstable slopes. The empirical analog could be restated in the form:

GIVEN (antecedent evidence) THEN (consequent hypothesis)

which for the unstable slope problem would look like:

GIVEN (soil = colluvium AND slope ≥ 15° AND stream
proximity ≤ 250 m) THEN unstable

where the consequent hypothesis may be qualified by a probability statement depending on the strength of the evidence. In other words, suppose a slope was more than 250 m from a stream and, therefore, only two out of the three conditions for instability were met, then KBS might still conclude that a steep colluvial slope was unstable, but with less than 100% certainty. A complete series of such inductive rules may be developed to cover all combinations of situations. In Figure 5.12, we looked at a simplified wildfire model. Having identified areas of relative hazard, how to prioritize areas for protection and what management plan scenarios would be best for them? To answer these types of questions, Gronlund et al. (1994), for example, report on the development of a knowledge base and its use in the Crowders Mountain State Park, North Carolina, USA. Subsequent to the type of mapping carried in Figure 5.12, prioritization would reflect knowledge of the risk of fire to nearby infrastructure and amenities, endangered species and residential areas. The expert knowledge would also subsequently generate management plan scenarios to the level of detail necessary for allocating personnel and equipment. Another application of KBS to guide the choice and use of environmental models in modeling the changing effect of pH level in lakes on fish damage is given in Chapter 7.

Heuristics

Heuristics are not dissimilar to KBS rules, but have rather a different function in modeling. At their simplest, they can be regarded as rules of thumb on how to approach the solution to some problem. Sometimes in modeling there could be an extremely large number of possible solutions or scenarios that might need to be examined in order to find the best one. For example, in Chapter 4, we looked at the problem of topographic modeling and the use of inverse distance weighting (IDW) as one method of solving the problem. Whereas a KBS could be set up to advise a modeler as to which method of interpolation (IDW, triangular irregular network (TIN), kriging, etc.; see Chapter 8 and Chapter 9) would be the most appropriate in a particular set of circumstances (type of topography, sample size and type, honoring breaks), there is still the problem of what would be the optimal setting for parameters. In the IDW example, the power weighting (distance decay) function r in Equation (4.7) could take on an infinite number of values, all of which may conceivably need to be tested to arrive at Figure 4.13. In reality, heuristics can help us to find a solution faster. First, as a rule of thumb, it is conventional to start with $r = 2$ before trying others. Then, if a search for an optimum appears necessary, we would usually restrict the breadth of the search in the range 1 to 6 (Figure 4.7) and then we would usually be satisfied with an approximation of the optimum by an integer (rather than to one or two decimal places).

That cuts down the search for an *acceptable* value of r to a maximum of six trials instead of the theoretical infinite. The example given here is a relatively simple one when compared with models that use many parameters where the number of permutations and combinations can quickly escalate. Heuristics, when incorporated into computational algorithms, are thus effective as search procedures to find optimum solutions in large combinatorial problems. An introduction to Monte Carlo, simulated annealing and tabu search procedures can be found in Openshaw and Openshaw (1997). It is appropriate to mention here another heuristic approach, that of parallelism or parallel computing, which is designed not so much to find an algorithmic "rational shortcut" to a solution, but to speed up processing time through the use of multiple processors. Models that can be segmented into n independent activities that can then be executed concurrently (in parallel) on n processors can speed up the computation of the model by (theoretically) n times. Pirozzi and Zicarelli (2000) report using 512 processors to achieve a reduction in computation time from 30 days to less than one hour in modeling coastal thermal pollution from a hot water discharge. Such increases in speed for model execution will permit more comprehensive simulations to be carried out at finer temporal and spatial resolutions than previously possible.

Artificial Neural Networks

Artificial neural networks (ANN) are used to develop empirical models, but are dealt with here because the techniques used come under the definition of AI quoted above. ANN attempt to mimic the brain's capacity to learn by using a model of its low-level structure (Patterson, 1996). The brain is composed of about 10^9 neurons that are massively interconnected, averaging about 3×10^3 interconnections per neuron. These interconnections are not direct so that a signal from one neuron must cross a synapse (chemical-filled gap) before reaching another. Transmission of a signal across the synapse depends upon the strength of the signal and the efficacy of the synapse in relaying the signal. The receiving neuron only fires the signal onto another if it has exceeded a threshold level. Thus, in a learning process, the synaptic connection between certain neurons is considered to "strengthen" and more easily transmit the signal that is then passed on by other neurons. Put simply, a stronger connection is made in the brain. This learning process is mimicked by ANN. Consider the network shown in Figure 5.13. This shows an input neuron layer for m variables, an output neuron layer for n possible results (where $m > n$), and one or more intermediate "hidden" layers of j neurons (where $j > = n$). ANN are allowed to "train" on data sets of known results during which, through a process of constant stimulation, the weights w associated with the connections into each neuron are adjusted and, in turn, influence the transformation by a sigmoid function into the output data for each neuron. Training continues until inputs arrive at the correct results (or threshold accuracy of minimized overall error). ANN can

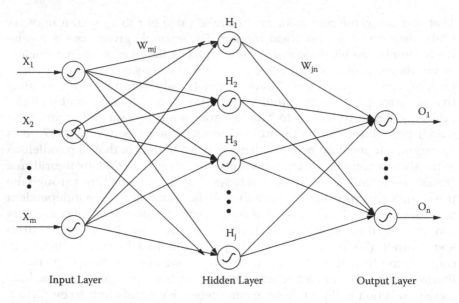

FIGURE 5.13
Structure of an artificial neural network (see text for details).

then be used to study the relationship between the input variables and the resulting phenomena and/or to process other data sets for which a result is required.

Taking again the theme of unstable slopes in our example landscape, a sample data set can be generated detailing geology, land cover, gradient, and distance from stream for slopes that can be categorized as either stable or unstable. A 1% sample is taken of the area. Part of this data set is held back for evaluating (verifying) the ANN model after each round of training, the rest are used to train the ANN. The results are given in Figure 5.14, which shows how the level of error in correctly predicting stable and unstable slopes falls with increased training. Eventually, there is no further gain in the training and we are left with about 10% error either

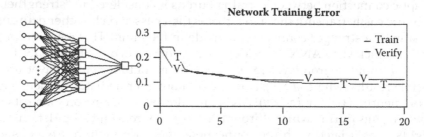

FIGURE 5.14
Results of ANN used to determine stable and unstable slopes in the example landscape (see text for details).

through misclassification or because there has been no classification as a result of detected ambiguity. Nefeslioglu et al. (2008) provide a comparative study using logistic regression and ANN for landslide susceptibility mapping in Turkey.

There are a number of different types of ANN to choose from. Perhaps the most popular is the multilayer perceptron, as illustrated in Figure 5.14, and can be simply interpreted as an input–output model. Others include radial basis function, probabilistic, generalized regression, linear, and Kohonen networks, the latter being particularly useful in exploratory analysis (StatSoft, 1998). ANN can be used not only on geographical attribute data, as in the above example, but also on time series data. This has led to a growing interest in the use of ANN in hydrology (Tokar and Johnson, 1999; ASCE Task Committee on Application of Artificial Neural Networks in Hydrology, 2000a; 2000b) for both rainfall runoff and water quality modeling. Drumm et al. (1999) report on the use of ANN in modeling the habitat preferences of the sea cucumber (*Holothuria leucospilota*) in the shallow-water ecosystem of the Cook Islands. Here ANN were considered to offer an advantage over traditional regression modeling because of the nonlinear nature of the ecological variables. ANN identified that higher species abundance was most associated with seafloor habitats characterized by rubble and consolidated rubble. Maeda et al. (2009) use ANN in conjunction with multitemporal RS imagery (NVDI values) to model areas of high risk of forest fire in the Mato Grosso, Brazil.

Despite the growing number of successes, there is a concern that ANN are being used as black boxes with "explanation" of physical and biological processes tied up in the hidden layers. Thus, ANN should not be viewed as a panacea substitute for a detailed knowledge of the relevant processes. Nevertheless, the number of models incorporating such elements of AI will undoubtedly increase as researchers and professionals seek to endow their models with greater levels of "intelligence."

Agent-Based Models

Agent-based modeling can be defined as "a computational method that enables a researcher to create, analyze, and experiment with models composed of agents that interact within an environment" (Gilbert, 2008). As for the term *agent* applied to software, there is as yet no universally accepted definition, and is taken here to be an autonomous, problem-solving, encapsulated entity in a dynamic and open environment (Woolridge, 1997; Jennings, 2000). Thus, a key characteristic of agents is their ability to direct their activity or change their state in response to changes in their surrounding environment or other agents with which they come into contact. Their behavior can be endowed with a certain level of intelligence. Where agent-based models have been used for spatial simulation, agents have tended to be deployed as spatial objects to computationally characterize the behavior of individual entities (ants, plants, people, vehicles) in order to study the patterns that emerge over time as a consequence of their microlevel behaviors and state changes (Batty and Torrens, 2005; Ferber, 2005). In other words, agent-based modeling is predominantly about studying emergent macro patterns of behavior. Such models can employ hundreds, thousands, or even millions of agents and are often referred to as multiagent systems. Recent reviews of agent-based models in GIScience are provided by Sengupta and Seiber (2007) and O'Sullivan (2008).

This concept of emergent macro behavior arising through micro simulation can be readily illustrated using a three-population Schelling model of segregation (Schelling, 1971) implemented using cellular automata (CA) software SpaCelle (http://www.spatial-modelling.info). CA are close in concept to ABM (Rodrigues and Raper, 1999) in that each cell can act autonomously making decisions on changing its state depending on the states of neighboring cells, thus resulting in patterns of emergent behavior. Under the rules of the simulation, the three populations are distributed randomly with a certain number of "empty" cells remaining. A cell at any one time can only have one state: one of the three population groups or empty. Based on a tolerance threshold, each populated cell will assess its neighbors and if they are sufficiently different to itself (other than empty), then it will change its state to empty (equivalent of moving out of the neighborhood). On the other hand, depending on a slightly different neighborhood tolerance level, an empty cell can change its state to one of the population groups (equivalent of moving into the neighborhood). Thus, with each time step, each cell makes its own micro decision based on simple rules on any change of state. Figure 5.15 shows the progress across 5,000 time steps of a run that starts with a random distribution. After 10 time steps, there is not much change, but by 100 time steps, a clustering of the three populations starts to emerge. At 500 time steps, very definite territories for each population emerges, but beyond that (given the parameter values in this realization) one population group starts to dominate so that by 5,000 time steps, it has occupied some 80% of the area. These then are the emergent patterns arising out of the micro behavior of each individual cell.

CA have been used to build a macroscopic collision model to simulate debris flow-type landslides calibrated against an actual debris flow in Campania, Southern Italy (D'Ambrosio et al., 2007). In another study (Almeida et al., 2008), ANN are employed in the parameterization of variables that control the state changes in a CA model that is then used to simulate land use change in a town near São Paulo, Brazil, so as to reveal emergent patterns of land use dynamics. Chen (2008) uses agent-based modeling to compare the effectiveness of simultaneous and staged evacuation strategies for hurricanes affecting Galveston Island, Texas. Agents in this case represented vehicles along evacuation routes. Through the agent-based simulation, it was found that a staged evacuation across the bridge to the mainland helped reduce total evacuation time.

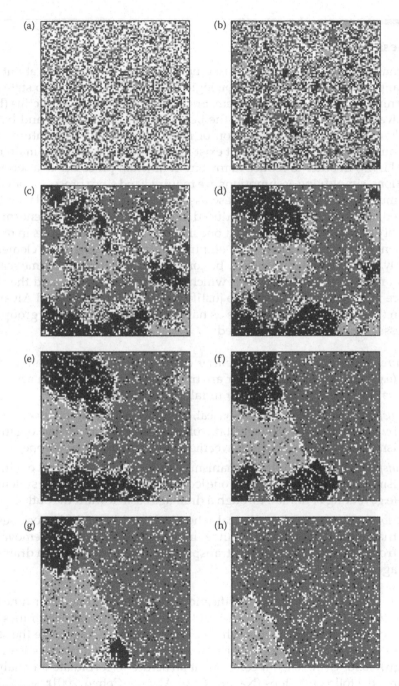

FIGURE 5.15
A three population Schelling model of segregation implemented using cellular automata: (a) random start point at time zero, (b) after 10 time steps, (c) 100, (d) 200, (e) 500, (f) 1,000, (g) 2,000, (h) 5,000. (*Note*: White cells are empty.)

Process Models

The construction of these models assumes that sufficient is known about the relevant physical, chemical, and biological processes to be able to state the governing equations and solve them. Such models tend to be deductive (but, see caveats on this distinction at the beginning of this chapter) and based largely on the laws of conservation of mass, energy, and momentum. The multiplicity of process models that exist have come about for three main reasons. Firstly, in order to reduce many environmental processes to a series of equations, simplifying assumptions are required and since process modeling is as much an art as a science, these assumptions are variously elaborated, evolved, dropped, new ones introduced, and so on. Secondly, the enormous diversity of real environments that one might wish to model results in many different formulations often with empirical and, increasingly, AI elements. Thirdly, that different models may be constructed depending on time scales, spatial resolution, the speed with which an answer is needed, and the confidence required in that answer to justify a decision. Nazaroff and Alvarez-Cohen (2001) view process models as having four key elements or groups of processes that should be considered:

Sources: These are inputs that either occur directly within the system (e.g., a source of pollution) or are transported into the system across its boundaries (e.g., input of rainfall to a drainage basin).

Transformations: These are the physical, chemical, and biological processes by which specified changes take place to the substances and organisms within the system (e.g., weathering of rock to soil on a slope).

Transport: These are all the mechanisms by which substances and organisms move or are moved from one location to another within the system (e.g., routing of rainfall through a drainage basin toward the outlet).

Removal: These are the physical, chemical, and biological processes that control the fate of substances in terms of their eventual removal from the system (e.g., evapo-transpiration of moisture out of a drainage basin).

To this we need to add the specification of parameters that determine the *initial state* of the system and the computational *time step* that determines the length of each time segment in which processes operate to change the state of the system. A diagrammatic visualization of these components is given in Figure 5.16. Using such a component structure to formulate an analysis requires the following steps (Nazaroff and Alvarez-Cohen, 2001):

- Identify the processes in each of the key elements above and translate each into a mathematical representation.

- Identify the known parameters and variables and quantify them.
- Identify the unknowns and assign symbols to represent them.
- Write the mathematical relationships that link the known and the unknown based on physical, chemical, and biological principles.
- Obtain a result by solving the mathematical equations.
- Interpret the significance of any results for the system being modeled.

It all sounds quite straightforward, but in reality, process models are invariably complex and, as a result of nonlinear relationships between inputs and outputs, which together with errors or imprecision in the data, may mean there is no unique solution, only approximations (Beven, 2001). Most process models, therefore, are heavily computational and need to be run by computers. The complexity and computational intensiveness of process models also tend to increase with resolution and dimensionality. It is common to classify process models by the number of spatial dimensions explicit within them—a temporal dimension (the time step) always being present in environmental models. Thus 1D models, also known *as lumped parameter models*, consider space to be uniform with all inputs and outputs to be uniformly distributed. These models are essentially nonspatial in as much as computation is carried out at discrete points. Two-dimensional and three-dimensional models, otherwise known as *distributed parameter models*, allow processes to be modeled across surface areas and with added depth. Thus, a 2D reservoir model will

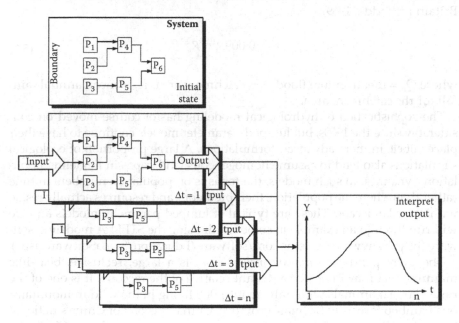

FIGURE 5.16
The broad components of a process model.

ignore variations in water quality with depth and concentrate on modeling horizontal movements of an averaged column of water through the reservoir while a 3D reservoir would consider both horizontal and vertical movements of, say, heavy metals through a reservoir. As can be expected, with each successive increase in dimension, there are more parameters and variables to be specified and thus model complexity increases.

Lumped Parameter Models

Perhaps the simplest equation that encapsulates a lumped input–output model is the "rational method" of calculating flood, so-called because it reflects the rational way peak discharge is expected to increase with catchment area and rainfall intensity:

$$Q = CA\bar{R} \tag{5.4}$$

where Q = peak discharge, C = a constant, A = catchment area, $\bar{R}$ = average rainfall intensity.

The average rainfall intensity is the input while the catchment area and constant transform that input into a peak discharge (the output). The calculation is for the point of discharge and any variability of rainfall within the catchment area is ignored. Clearly, parameters such as the constant need to be established and can be solved empirically. Thus, one solution for floods in Britain is (Rodda, 1969):

$$Q_m = 0.009 A^{0.85} \bar{R}_a^{2.2} \tag{5.5}$$

where Q_m = mean annual flood, A = catchment area, $\bar{R}_a$ = mean annual rainfall for the catchment area.

The sophistication of hydrological modeling has of course moved on considerably since the 1960s, but lumped parameter models continue to have their place, albeit, in more advanced formulations. A large proportion of ecological simulations also tend to assume homogeneous landscapes in modeling population dynamics. In such models, the focus is on population numbers in time rather than where the population, their predators, and resources actually reside within the landscape. These are typical of lumped parameter models and we will run through an example in some detail using the STELLA modeling software (http://www.iseesystems.com/software/Education/StellaSoftware.aspx).

The giant panda (*Ailuropoda melanoleuca*) is a large, reclusive, bear-like mammal that may in fact be a distant relative of the raccoon. It is one of the rarest large mammals with only about 1,000 living in the wild in mountainous bamboo forests in Sichuan Province, China. It is one of China's national treasures. The panda is a carnivore, but the wildlife in its habitat is scarce and difficult to catch for a slow 160 kg mammal. Instead it has settled into a sedentary, solitary life eating 10 to 15 kg a day of young and tender bamboo (predominantly *Sinarundinaria* sp.). The digestive system of the panda is not well adapted to digesting bamboo and, therefore, extracts little nourishment from it. This may well explain its relatively small range (about 8 to 10 sq km) for a large mammal and its general disinterest in mating. Pandas at birth weigh a mere 1/900th of the adult and remain with the mother for about 18 months before becoming independent. Because the female panda will only rear one cub at a time (regardless of two or more in a litter), the shortest birth cycle is about once every three years, but even then many cubs do not survive. The bamboo the panda feeds on may seem equally peculiar. Bamboo normally grows from a thick underground rhizome from which new shoots are produced. But once every 50 to 100 years or so, the bamboo will flower, produce seeds and then die *en masse*. Fortunately not all species will do this with the same periodicity and, hence, not simultaneously. This flowering occurred in two species of bamboo in Sichuan Province in 1970 and 1983. The latter event was particularly devastating to the panda, which, despite government organized rescue efforts, resulted in the death of an estimated 15 to 20% of the panda population. So, we will use this understanding of panda population dynamics (albeit, incomplete) and this type of bamboo flowering scenario to illustrate an approach to lumped parameter modeling.

The panda subsystem can be reduced to a few simple elements (Figure 5.17(a)). Central to the subsystem is the reservoir or stock of pandas (S_p), irrespective of where they are spatially. Inputs to this stock are panda births (B_p) and outputs from this stock are panda deaths (D_p). Thus, the size of stock of pandas after any particular time period t can be expressed as:

$$S_p(t + \Delta t) = S_p(t) + \{B_p - D_p\}\Delta t \qquad (5.6)$$

where $S_p(t)$ size of panda stock at time t, Δt the time increment, B_p, D_p = number of panda births and deaths, respectively. Thus, the stock after the passage of time t is the original stock at time t plus the net growth (or decrease) consequent on the number of births and deaths during that passage of time. Note that for simplicity we are ignoring in- and out-migration. Equation (5.6) is the *difference equation* for the stock $S_p(t)$. The actual number of births in Δt will depend on the size of panda stock and its birth rate. The dependency between these two elements of the subsystem to determine births is shown in Figure 5.17(a) by the arrow connectors. The birth rate b subsumes all the biological and psychological processes of the panda involved in mating and is expressed as the average number of births per individual in the stock. Thus, the number of births can be expressed as:

$$B_p(\Delta t) = P_b S_p(t)\Delta t \qquad (5.7)$$

where P_b = panda birth rate. And, similarly for the number of deaths. If perchance the birth rate P_b equals the death rate P_d, then the stock remains

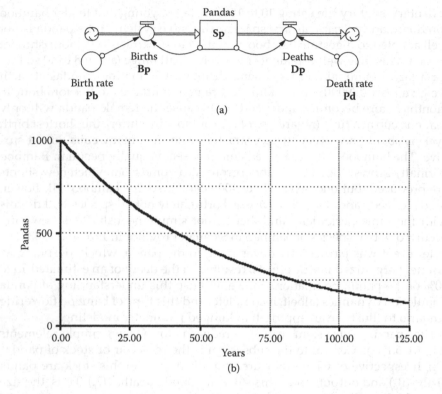

FIGURE 5.17
An illustration of (a) the panda subsystem and (b) a projection of panda stock over 125 annual time steps.

constant over time and the subsystem is said to be in a *steady state*. This is defined mathematically as when the instantaneous rate or *derivative* of $S_p(t)$ with respect to t is zero:

$$\frac{dS_p(t)}{dt} = 0$$

(5.8)

If the derivative is greater than zero, then the stock will increase and if less than zero, then the stock will decrease over time. Having defined the elements of the system and their dynamics, it is now necessary to determine the relevant parameters. We can set the initial stock at 1,000 since that is the estimated number of pandas. Pandas live for about 30 years in the wild, so in probability terms, we can estimate the annual death rate to be 1/30. The birth rate is a lot more problematic. First, the only known rates are those for pandas in zoos or in reserves and then many of these are by artificial insemination. Second, we know that even if there are multiple births, the mother will only look after one. So, from the available data (e.g., at http://www.wwfchina.org/

english/pandacentral/) and counting only the one birth that will be reared, a very broad guess at the annual birth rate would be 1/60. Now, this is not very good news for the panda, as illustrated in Figure 5.17(b), which shows an exponentially declining population with only 122 remaining at the end of the simulation, hence the amount of effort which goes into the protection and conservation of pandas.

The other subsystem within our lumped parameter model is the bamboo that the panda feeds off. This is constructed in a very similar way with a stock of bamboo S_b and its rate of new growth and depletion. Now these two subsystems could exist independently, but clearly they don't. The panda stock depends on the stock of bamboo as a source of food and the bamboo stock gets depleted by the panda eating it. These connections are shown in Figure 5.18(b) by the connector arrows joining bamboo stock and panda birth rate and joining the panda stock with the bamboo depletion. The first thing to note is that these connections are in the form of a feedback loop, that is, the subsystems interact with each other in a closed circle of cause and effect, one being:

$$S_b(t) \rightarrow P_b \rightarrow B_p(\Delta t) \rightarrow S_p(t) \rightarrow D_b(\Delta t) \rightarrow S_b(t)$$

The second thing that must be noted is that since the panda birth rate has a dependence on the stock of food, the birth rate should no longer be viewed as a constant, but as a logistics curve. The S-shape of this curve determines that when the bamboo stock is low, the panda birth rate will be low; as the stock of bamboo increases, so does the panda birth rate; however, as the stock of pandas reaches the carrying capacity of the bamboo (the maximum number of pandas that can be fed on the bamboo stock), the birth rate grows increasingly slowly toward the maximum birth rate. The same can be said for the rate of new growth in the bamboo in relation to the carrying capacity of the soil. Normally in such modeling it would be left to the bamboo stock to limit the panda stock through the panda birth rate. However, with the mass die-off of bamboo, there are direct panda deaths through starvation rather than just limiting the birth rate. Hence, the connector in Figure 5.18(b) from the bamboo stock S_b to the panda deaths D_b in order to regulate the panda stock more rapidly as a consequence of bamboo die-off. Figure 15.18(a) shows 125 years of logistic growth of the bamboo submodel (without grazing by the panda) with a flowering and 90% die-off every 50 years. This pattern shows initial unrestrained growth, which tails off as the carrying capacity of the soil is reached followed by the dramatic die-off; the cycle then gets repeated. In a joint simulation of panda–bamboo interaction, the initial stock of bamboo has been set higher than in Figure 15.18(a) and the logistics birth rate of the panda has been increased (out of interest) so that once the bamboo has reached 85% of its carrying capacity, it is assumed that there are sufficient tender, easily digestible bamboo shoots to afford the pandas a birth rate that matches its death rate (models allow us the possibility of experiments such

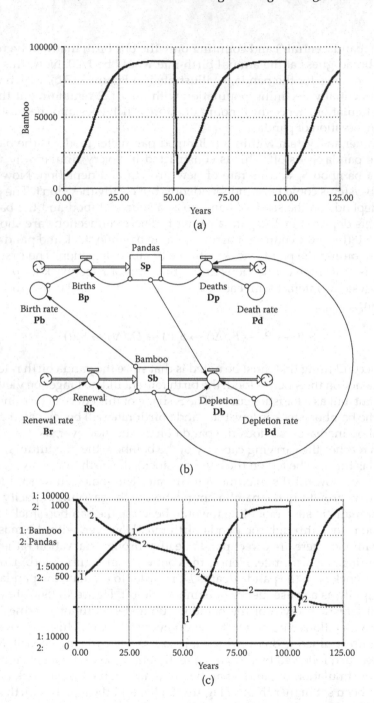

FIGURE 5.18
The simulation of panda–bamboo interaction: (a) die-off cycles in the bamboo subsystem, (b) linked panda and bamboo submodels, (c) simulation results.

as this). Figure 15.18(c) shows the resulting simulation. Note that the Y axis has two scales, one for bamboo and the other for count of pandas. The bamboo follows the predicted cycle of growth and die-off. The pandas, with a maximum birth rate of 1/30, initially decline as the amount of bamboo is below the threshold for the maximum birth rate. At the first bamboo die-off, the panda numbers decline rapidly in response due to starvation and fall in birth rate. However, as the bamboo pass the 80% soil carrying capacity threshold, the panda numbers achieve equilibrium at a stock of 365. Notice how the bamboo stock is greater in this cycle than the previous—due to less pandas consuming it. At the second bamboo die-off, the pandas are again affected, but because their numbers are smaller they are not as seriously affected and reach a new equilibrium sooner.

It needs to be stressed that Figure 5.18 should not be taken as a definitive model of panda–bamboo dynamics without further refinement of parameters and calibration. Nevertheless, it illustrates in some detail the nature of lumped parameter simulation models. Lumped parameter models need not always be used nonspatially. A series of models can be used together, each simulating the conditions found in a specific geographical area. Thus, models similar to the ones above could be established for the different subregions where the panda is found allowing parameters to be varied to reflect the specific conditions there as well as any migration between areas. Such an arrangement for hydraulic modeling for basin management planning is discussed in Chapter 6.

Distributed Parameter Models

These are the most numerically complex of environmental models that are framed computationally. With each distinct process resulting in its own model, it is only possible here to look at some broad principles. The first is discretization of space and time. Then we should consider routing over a topographic surface and finally transport through a medium. A more detailed look at parameter estimation and calibration is left for Chapter 9.

Discretization

In the lumped parameter modeling above, we have already had to consider the discretization of the temporal dimension. For the population dynamics of the panda, we fixed the modeling time increment *t* as one year. This would seem intuitive, allows for reasonable estimation of parameters (annual birth rates and death rates), and on modern PCs takes a matter of seconds to calculate a century's worth of data. We could have set the time increment at a decade, in which case we would probably be looking to simulate far longer time periods. While parameter estimation could have been more approximate, with only 10 results per century, the model may not have been sensitive enough to reflect say the details of a reduction in panda population in response to a bamboo die-off—its utility in this context would have been adversely affected. On the other hand, conceivably, we might have wanted to simulate with a time increment of one day. For this to be meaningful, our parameters would have to have had much greater precision (not just dividing annual rate by 365) and computationally may have become lengthy. Again, this may not be useful since panda dynamics are hardly observable on a daily basis. So, there is a compromise to be made and is nearly always a matter of judgment. In distributed parameter models, as well as time, there is space to be discretized. Again there are trade-offs to be made between the space–time domain of the specific process, parameter estimation, computation time, and utility of the results. Spatial discretization in a modeling context can either be sections of a network or a space-filling tessellation. Unless the modeling of a network is discretized at intersections or at regular distances or time intervals along the network, which is by no means always the case, then the discretization has to be effected manually. For continuous fields, it is normal to produce a tessellation, examples of which were given in Chapter 2, Figure 2.11. Because the distributed parameter model is not just concerned with inputs, outputs, and changes of state to the whole system, but with inputs, outputs, and changes of state of individual, discretized spatial units, then in calculation terms the geometrically simpler tessellations of triangles and raster cells have a computational advantage. In general, triangular and grid tessellations are used for different forms of calculation as discussed below. With heavy reliance on raster satellite imagery as a cost-effective means of data collection and with raster being one of the two principal data models in GIS, there is a tendency for distributed parameter models, which use GIS for data preparation, to use a grid cell tessellation. The size of the grid cell becomes critically important (Li, 2007). Although GIS software can quickly produce more or less any grid size, there are important trade-offs with the computation time of the simulation (even with today's fast PCs), which tends to grow exponentially as the size of grid cell is reduced. If the cell size falls below the spatial resolving power of the simulation model, then there is nothing to gain and the model may even become unstable. With the resolving power of the model as a lower bound, a general rule of thumb is for cell size (Δx, Δy) to be no larger than half the wavelength of the z to be accounted for. Modeling consequences of cell size are explored further in Chapter 9.

Routing across a Digital Elevation Model

A number of applications, which use distributed parameter modeling, employ a DEM for routing the transport of materials across a landscape. Obvious examples are surface hydrology and gravity flows of landslide and avalanche debris. If, for example, we look at the dominant variables in computing overland flow (Figure 5.19) where the continuity equation for each cell can be written as (Smith, 1993):

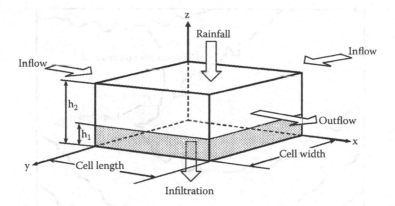

FIGURE 5.19
Variables used in computing overland flow in a distributed parameter model. (Adapted from Smith, M.B. (1993) *Hydrological Processes* 7: 45–61.) $h_1 = h(t)$, $h_2 = h(t + Dt)$ in Equation (5.9).

$$h(t + \Delta t) = h(t) + net\_rain + \left(\frac{\sum inflow}{cell\_area} - \frac{outflow}{cell\_area} \right) \Delta t \qquad (5.9)$$

where $h(t + \Delta t)$ = water depth at the end of the time increment, $h(t)$ = water depth at the beginning of the time increment, *net_rain* = incremental rainfall minus incremental infiltration; it becomes evident that topographic form is important in routing inflow into any cell and routing outflow from any cell. Note that a cell may have more than one direction of inflow from higher cells, but has only one outflow to a lower cell. Physical routing is dependent on gradient and aspect, in other words, the direction of maximum gradient of any cell. It is not surprising then that most GIS have built-in functionality for calculating these from a DEM for which there are a number of algorithms. However, before this can be done, it is necessary to clean a DEM of depressions or *pits*. Pits arise in a DEM where there has either been an elevation underestimate or overestimate, often as a consequence of rasterizing source data, which result in spurious depressions across which flow cannot be routed (since it is all uphill to adjacent cells), as illustrated in Figure 5.20. These are eliminated either by breaching across damming cells to form an outlet or by filling the depression to produce a continuous gradient to an outlet (Martz and Garbrecht, 1998). Once achieved, flow direction and flow accumulation can be derived (Figure 5.21). A review of methods of determining flow direction from DEM is given in Tarboton (1997). The most popular method, designated D8, was introduced by O'Callaghan and Mark (1984) and is implemented in a number of GIS packages. Each raster cell will be surrounded by eight neighbors (except for edge cells of a coverage, of course), either adjacent or on the diagonal, and the steepest gradient to a neighbor

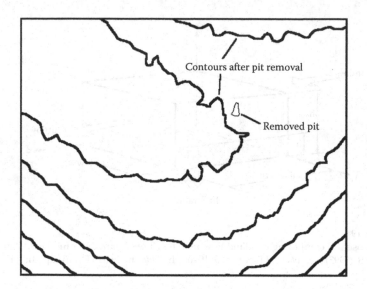

FIGURE 5.20
An example of a pit in a valley floor that has been removed from a DEM.

will be calculated and assigned a number in a base 2 series depending on that direction (Figure 5.22). This base 2 series translates into a binary series that becomes very quick to use computationally.

Decimal	1	2	4	8	16	32	64	128
Base 2	2^0	2^1	2^2	2^3	2^4	2^5	2^6	2^7
Binary	1	10	100	1000	10000	100000	1000000	10000000

From this data, the area of flow accumulation to each cell can be calculated and from which the drainage network and catchment boundaries can be derived (Martinez-Casasnovas and Stuiver, 1998). An implementation of much of this functionality in ArcView for inputs into TOPMODEL is described by Huang and Jiang (2002).

Many preparatory data processes relating to topography that were manually tedious or needed to be carried out by specially written modules prior to input to simulation models are now carried out efficiently by GIS. It may well be that the data models and algorithms implemented by the dominant GIS software vendors are beginning to dominate or even become a de facto standard in the way distributed parameter simulation models are structured to use DEM and prepare DEM derivative data.

Transport through a Medium

For many distributed parameter process models, the shape of the topography may be only one aspect of the routing problem. If, for example, we were

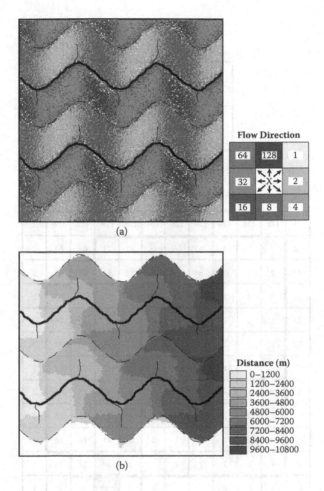

FIGURE 5.21
Routing over topography: (a) flow direction, (b) accumulated length of flow for the two stream outlets in our example landscape.

to consider the transport of heavy metals through a reservoir, the passage of smoke from an incinerator over an urban area, or the movement of an oil slick from a stricken tanker toward a shoreline holiday resort, then topography has some role in defining, respectively, the 3D shape of the reservoir, the flow of air over the city, and the morphology of the coastal zone. But central to the process model will be the need to mathematically predict the movement of substances or energy through the specific medium. Let us return for a moment to Figure 5.19 and suppose that the cell is instead a discretized element in a column of seawater. The inflows and outflows would then be determined by gradients within the seawater. These gradients might be in the temperature and pressure as a function of depth, due to local changes in salinity, as a consequence of the gravitational pull of the moon and sun

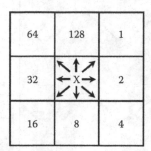

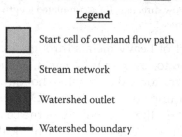

Legend

Start cell of overland flow path

Stream network

Watershed outlet

—— Watershed boundary

FIGURE 5.22
Numbering of flow direction in a cell and delimiting catchment areas. (Adapted from Smith, M.B., and Brilly, M. (1992) *Photogrammetric Engineering & Remote Sensing* 58: 579–585.)

resulting in tidal currents, directional friction of the wind producing surface currents, and so on. All of these gradients are subject to the physical laws of the conservation of matter and energy and are usually solved mathematically as partial differential equations because several variables are usually involved. Spatial discretization in 2D or 3D in the present context becomes part of the solution. Imagine looking at a smokestack belching out black fumes at a constant rate into a light westerly wind of constant velocity. At the exit of the smokestack, the fumes are dark and concentrated and clearly differentiated by eye from the surrounding clean air. However, at some distance from the smokestack, the fumes have diffused into a wider area, become less concentrated, and dissipated at the edges so it becomes difficult to differentiate between clean air and the fumes. If we were to cut two slices through the fumes perpendicular to the wind direction, one near the smokestack and one farther away, we would expect to find the same total amount of fumes in each slice, one concentrated and the other more dispersed and of lower concentration. Thus, there is conservation of matter, in this case the fumes. Incidentally, there is also conservation of the wind energy. If we wanted to know the concentration of fumes at any point, we would need to have a way of connecting the concentrations at the exit of the smokestack with the point we are interested in, so that the physical process of dispersion through air can be modeled. For this to be mathematically tractable, we need intermediate points to act as "stepping stones" in the solution so that, on the one hand, we don't have to solve the problem for an infinite number of possible points and, on the other hand, we can reduce what in reality are very complex shaped gradients to a series of linear equations between stepping stones that can then be solved by numerical methods. In other words, this can be viewed as a form of interpolation. Our starting point, then, is to express the general case of a physical process operating within a specific domain or study area Ω as:

$$\Omega: \ -\left(\frac{\partial}{\partial x}\beta\frac{\partial u}{\partial x} + \frac{\partial}{\partial t}\beta\frac{\partial u}{\partial t}\right) = f \tag{5.10}$$

$$\partial\Omega = \Gamma_0 + \Gamma_0' \tag{5.11}$$

$$\Gamma_0: \ u = \bar{u} \tag{5.12}$$

$$\Gamma_0': \ \beta\frac{\partial u}{\partial v} + \eta u = q \tag{5.13}$$

where x = location, t = time, u = unknown to be solved, $\bar{u}$ being a specific value of u, b = a coefficient > 0, f = the distribution of coefficients specific to the physical process being modeled, Γ_0 = the specified boundary condition,

Γ'_0 = the natural boundary condition, $\eta_r q$ = the specified distribution on the boundary, v = the perpendicular to the tangent at x, t.

Because in an environmental simulation, we are dealing with finite space and time, the specification of the domain Ω requires boundary and initial conditions to be set. For most physical processes, Equation (5.10) will represent a complex nonlinear relationship between inputs and outputs that may have no unique solution. A mathematically tractable approximation, therefore, is required. There are two main approaches: *finite element method* (FEM) and *finite difference method* (FDM), which are two forms of numerical integration of differential equations (Harris and Stocker, 1998). In fact, another method of numerical integration of differential equations—Euler method—was used by STELLA in the solution to the 1D panda–bamboo model. FEM and FDM both require a tessellation discretization in 2D or 3D and, in general, it can be taken that FEM uses a triangulation in 2D and hexahedron in 3D, while FDM uses a grid in 2D and a cube in 3D. In fact, FEM can accommodate almost any shape of discretized element, whereas FDM is restricted in its solution to grid and cube. In the description of these two methods that follows, we will restrict the modeling to 2D and we will be using hydrodynamic modeling (used, for example, in coastal oil-spill modeling, see Chapter 6) for simulating tidal currents in order to illustrate the workings of the two methods. Of course, it would be impracticable in a book such as this to present a complete numerical solution for each method. In the words of Beven (2001), "The approximate numerical solution to nonlinear differential equations is, in itself, a specialism in applied mathematics, and writing solution algorithms is something that is definitely best left to the specialist." In FEM, the domain Ω definition of the function is divided into a finite number of smaller subdomains (or subzones) Ωe in which the required function can be approximated by a balance equation based on the principle of conservation of energy; thus, for the general case Equation (5.10), the equivalent solution becomes:

$$J(u) = \iint_{\Omega} \left\{ \frac{1}{2} \left[\beta \left(\frac{\partial u}{\partial x} \right)^2 + \beta \left(\frac{\partial u}{\partial t} \right)^2 \right] - f u \right\} dx dt + \int_{\Gamma_0} \left[\frac{1}{2} \eta u^2 - q u \right] ds = \min \tag{5.14}$$

$$\Gamma_0: \ u = \bar{u}$$

Equation (5.14) should theoretically be equal to zero, but in reality because of the approximation of the domain function it is only possible to seek a solution that minimizes $J(u)$. The equation is in two parts where the first part integrates over the domain function (the study area) and the second part represents the boundary condition. Note also that, for FEM, only Γ_0 is required and not Γ_0' and is, therefore, simplified. Equation (5.14) can be rewritten as a series of linear equations that are computationally easier to solve using matrix algebra. Let us now have a look at FEM from a practical perspective of hydrodynamic modeling. Most dispersion models of coastal water pollution (e.g., oil and chemical spills, sewage outfalls, dumping of sludge) are developed on simulations of currents within the water. Reliable pollution predictions, thus, are based on accurate hydrodynamic modeling. Hydrodynamic models can be implemented using either FEM or FDM, which for coastal water and estuaries generally use depth integrated 2D shallow water equations. The governing differential equations for hydrodynamic tidal modeling are (Foreman and Walters, 1990):

$$\partial_\eta/\partial_t + \nabla[(H+\eta)\mathbf{u}] = \min \tag{5.15}$$

$$\partial_U/\partial_t + (\mathbf{u} \bullet \nabla)\mathbf{u} + \mathbf{f} \cdot \mathbf{u} + g\nabla\eta + k\mathbf{u}|\mathbf{u}|/(H+\eta) = \min \tag{5.16}$$

where t = time, η = sea level (x, y, t), $\mathbf{u}$ = horizontal velocity (x, y, t), H = water depth (x, y), $\mathbf{f}$ = Coriolis force, g = acceleration due to gravity, k = bottom friction coefficient, ∇ is a Nabla operator.

Equation (5.15) is the balance equation for the water mass and defines the change in sea level as a function of water flux. Equation (5.16) represents the momentum balance in terms of nonlinear velocity gradients, hydrostatic pressure gradient, Coriolis force, and bottom friction. Figure 5.23 shows a section of coastal waters that has been prepared for FEM hydrodynamic modeling. There are two important aspects to note. The first is that there are two types of boundary: the natural coastline and an open boundary that represent an arbitrary cut-off to the study area. Both are represented by a series of nodes in the triangular mesh. The coastline acts as a barrier across which there can be no movement, but the open boundary allows inputs to and outputs from the system. The second important aspect is that the position of nodes and, hence, the triangulation is not haphazard or random, but follows a specific model that takes into account the bathymetry. The Courant criterion couples the size of triangular elements and the critical time step of the model computation in the following way (Molkethin, 1996):

$$\Delta x > \Delta t\left(|\mathbf{u}| + \sqrt{gH}\right) \tag{5.17}$$

which can also be expressed as:

$$\Delta t < \Delta x/\left(|\mathbf{u}| + \sqrt{gH}\right) \tag{5.18}$$

where Δx = the size of spatially discretized element, Δt = the time increment, $\mathbf{u}, g, H$ are as in Equation (5.16).

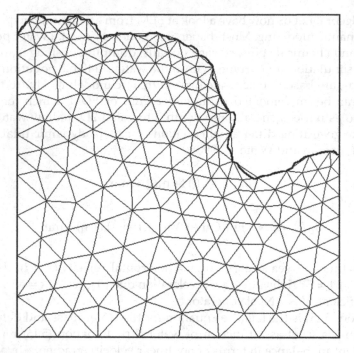

FIGURE 5.23
Illustrated is a section of shallow coastal waters and shoreline (400 km²) with FEM triangular mesh in preparation for hydrodynamic modeling.

On the one hand, simulation of tidal current in shallow water takes less time than in deeper water and in order to have a fixed global time increment the size of the triangle vertices should be in direct proportion to the square root of the corresponding water depth. On the other hand, a too large a time increment may cause instability and the simulation can generate erroneous results, while a too small a time increment would take too long to compute. Hence, the gradual enlargement of the triangular mesh in Figure 5.23 away from the coastline as the water gets deeper. At the start of simulation, the initial state is normally set as a current field of zero velocity. The simulation proceeds with input of a tidal model (Figure 5.24) across the open boundary. The tidal data shown in Figure 5.24 has two constituents:

1. M2: This is the tidal constituent caused by the gravitational pull of the moon with a periodicy of about 12 hours, hence, twice a day.

2. S2: This is the constituent caused by the sun and has approximately the same periodicy. Because in this example M2 and S2 are in phase, they will have a combined effect to produce a spring tide.

The effect of the tide data as input to the system at the open boundary is to force progressive changes in the sea level consequent on currents and their

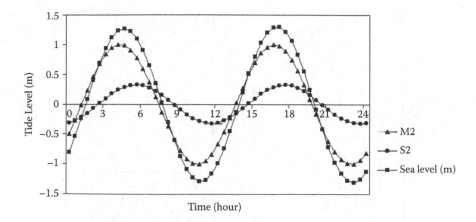

FIGURE 5.24
Example of tidal data with M2 and S2 constituents used at the open boundary.

velocities that are propagated through the elements of the triangular network by the FEM. To visualize the effect, imagine two people holding a stretched out sheet, one loosely holding still (the shoreline), the other moving the sheet up and down (tide data at the open boundary) causing the whole geometry of the sheet to rhythmically change. Over a 12-hour cycle, the tide will progressively rise from the open boundary causing flow toward the shore followed by a fall at the open boundary causing the water to flow away from the shore. This is illustrated in two hourly steps in Figure 5.25 where the arrows represent a vector giving the direction of current and its velocity.

The FDM takes a different approach to the solution of partial differential equations. The approximation is made using difference quotients in the form:

$$u_{i,j} = \frac{[u_{i+1,j} + u_{i-1,j} + u_{i,j+1} + u_{i,j-1}]}{4} \tag{5.19}$$

where $u_{i,j}$ is a grid node in which i, j represent inner nodes on a grid. Iteration through the grid is required in order to achieve convergence towards an acceptable accuracy. Figure 5.26 shows the same area of coastline with a grid prepared for hydrodynamic modeling using FDM. We have already discussed the issue of the Courant criterion on the modeling process and which becomes a drawback of FDM with the regular grid. The shoreline is less well-represented compared with the FEM and there are more nodes to be used in the calculation. Of course, the size of the grid could be enlarged to reduce computation time, but this will result in decrease in resolution and accuracy of the subsequent fate modeling of, say, an oil spill. This issue will be illustrated in Chapter 9. Figure 5.27 shows the results of the simulation for the initial time step alongside the FEM results.

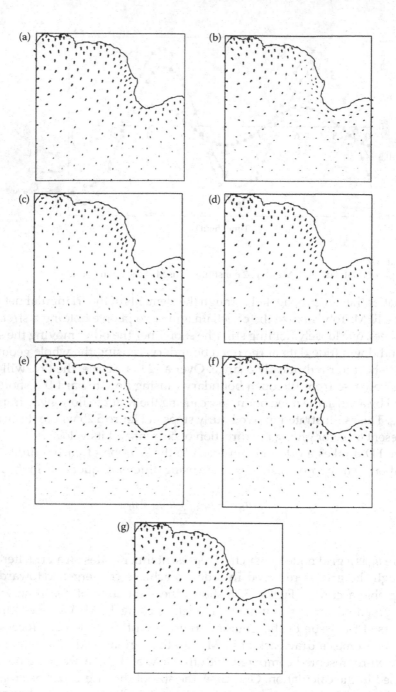

FIGURE 5.25
Results of hydrodynamic modeling using FEM with arrows showing current direction and velocity at each node: two hourly time increments with (a) at the initial time step $t = 0$ hours through to (g) at $t = 12$ hours.

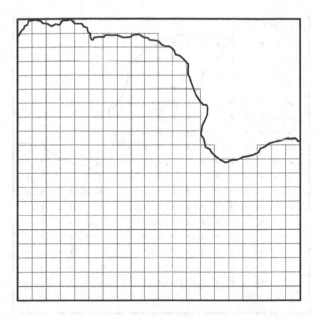

FIGURE 5.26
Illustrated is the same section of shallow coastal waters and shoreline (400 km^2) with FDM grid mesh in preparation for hydrodynamic modeling.

The obvious feature is that FDM neither gives results for open boundary nodes nor for inner coastline boundary nodes, which results from the method of calculation. This means that the shallow water environment where critical impacts often occur is not being adequately modeled. Some of the relative advantages of FEM versus FDM for simulation models have already become apparent in the above example: equalization of time step and results at the boundary. From a general perspective, the FDM solution to Equation (5.10) as expressed in Equation (5.19) may appear simpler and may indeed be quicker to calculate for small models of fairly simple physical processes. For larger, complex models (both geometrically and in the number of coefficients to express the physical process), FEM has the advantage in both calculating time and accuracy of the simulation. FEM uses symmetrical matrices in its solution of the linear equations, which help to minimize both computational time and memory requirements and, since convergence is assured, provides a robust solution. However, FEM treats simple and complex problems in exactly the same way, which explains its lower efficiency than FDM for simpler models. But this has a distinct advantage from a software implementation perspective. Each phase in the FEM solution is easily standardized in programming so that one single software implementation can cope with any geometric shape of the elements and any distribution of the coefficients relevant to the physical process. Data input can also be simplified. This makes FEM a powerful tool in both environmental simulation and solving engineering problems.

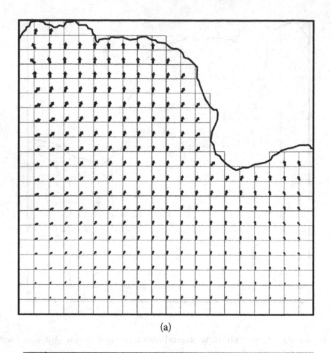

(a)

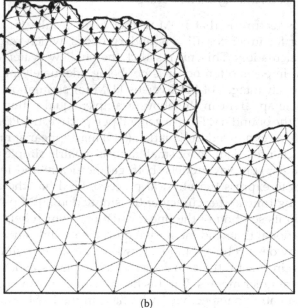

(b)

FIGURE 5.27
Comparison of results of hydrodynamic modeling using (a) FDM and (b) FEM at the initial
time step.

Section III

Section III

6

Case Studies in GIS, Environmental Modeling, and Engineering

This chapter will focus on a few case studies. Again, it is not my intention to produce here one example each of the full diversity of geographical information systems (GIS) and environmental modeling, but a selection that I have been involved in and for which identifying engineering solutions are very much the focus of the project. Other smaller case studies are used in the other chapters where specific issues need illustration, but the case studies presented here are intended to give some overall context to the previous chapters and to flag issues that are going to be dealt with in the chapters that follow. We will begin by looking at a taxonomy of modeling approaches to using GIS in environmental modeling and engineering. That model will provide a framework for looking at four case studies:

1. Landslides that could be potentially damaging to a reservoir dam.
2. Basin management planning using GIS and hydraulic modeling.
3. Coastal oil-spill modeling.
4. Forensic analysis of pipe failures.

An important characteristic of most real-world consultancy is confidentiality. The results of such studies may have important commercial or legal implications; the client may not want the world to know there is (or was) a problem, not from any sinister motive, but that the public may, for example, become unduly alarmed. So, for some of the case studies, I cannot be too specific about where they are (they are drawn from around the world) and, for the last case study, I have repeated the principle of the approach on sample data taken from elsewhere. Having read these case studies in relation to the taxonomy of modeling approaches, the reader will be in a position to conceptually "pigeonhole" other case studies on GIS and environmental modeling that they will find in the literature or may themselves be working on.

Modeling Approaches in GIS and Environmental Modeling

Figure 6.1 provides a taxonomy of modeling approaches. The three broad approaches represent how the existence or source of any hazard is modeled

147

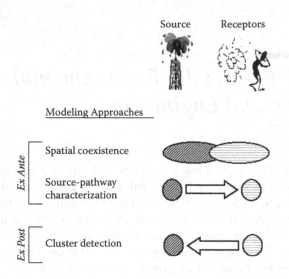

FIGURE 6.1
Modeling approaches in GIS and environmental modeling. (Adapted from Rejeski, D. (1993) In *Environmental Modeling with GIS*, ed. Goodchild et al. Oxford University Press, London, pp. 318–331.)

as risk in relation to vulnerable receptors. This implies that both the source and the receptors have to be identified and that there is some model of how that source impacts the receptors. A source may be identified as a specific object (a smoke-stack, effluent outfall), a zone (unstable slope from which specific landslides may occur, an area of seepage, a fault line), or diffuse over the whole area (strong wind event, heavy rainstorm). Generally, the first of these would be classified as *point* sources and the latter two as *nonpoint* sources. Difficulties arise over such an inductive classification because a nonpoint source in one area may originate from a point source in another. Thus, general air pollution in one area may originate from specific factories elsewhere. Depending on the scale of study, it may not be possible to model all the individual point sources, but treat the aggregate effect as a nonpoint source. Point and nonpoint sources need to be treated differently in GIS (point, line, polygon, or field) and are likely to influence the form of analysis. Receptors can be people, the flora and fauna, properties and land, again with different ways of representing these in GIS with consequences for the modeling. It is, however, the modeling of the means by which the source has the ability to impact the receptors that is distinctive within the taxonomy:

> *Spatial coexistence*: This approach assumes that there is a reasonably simple or obvious spatial link between sources and receptors. Thus, for example, by taking the floodplain and overlaying it on settlements one might infer that anywhere where polygon "floodplain"

and polygon "settlement" spatially coincided, people were vulnerable to a flood hazard. This approach relies heavily on conceptual and empirical models.

Source/pathway characterization: Not all spatial relationships between source and receptor can be conceived of as simply as is the case in the previous category. Where there is a significant and well-defined transport process linking source and receptor, then that transport process is explicitly modeled. Thus, this approach relies heavily on process models (deterministic or stochastic) that are either lumped parameter or distributed parameter.

Cluster detection: Whereas, the first two approaches in the taxonomy are *ex ante*, that is, are carried out to forward predict some event that can then be mitigated against in some way, this approach is *ex post*, that is, to detect that some event has or is happening to the receptors, to identify a source and then ensure that the hazard has ceased or needs to be mitigated. This is also empirical modeling using GIS, statistical techniques, and artificial neural networks (ANN) to search for space–time clusters in the event occurrences among the receptors that indicate that something has gone wrong.

Before coming to the case studies, we need to explore the difference between environmental modeling *within* GIS and environmental modeling *with* GIS. In the former, the entire environmental model is realized within GIS. In the latter case, GIS are linked with external models. Some would feel the distinction is unimportant; after all, it is possible to carry out the complete taxonomy of approaches within GIS. Well, yes and no. Both spatial coexistence modeling and cluster detection are heavily weighted in their approaches toward the spatial dimension of phenomena and are eminently suitable to be carried out in GIS or GIS-like modules. One problem here is that most commercial GIS software are poorly endowed, if at all, with cluster detection methods and that these need to be programmed either as externally linked programs or using the GIS internal macro language capability. Source–pathway characterization is somewhat different. While fairly simple transport processes, for the most part reliant on routing over topography, can be programmed as internal macros, the implementation of efficient finite element method (FEM), for example, is beyond these macro tools despite their growing sophistication. For many source–pathway characterization approaches, it still remains far more satisfactory to use specific environmental simulation software *with* GIS as an important complementary tool. This combined approach has grown in popularity due to the "dual recognition of environmental problems with compelling spatial properties, but also with a complexity that cannot be adequately explored through interrogation and recombination of geographic data alone" (Clarke et al., 2000). But, there is another important reason. There are many instances, for example, in engineering orientated applications where a client mandates the use of certain classes of model for which formal validation is available. In some places, such models are used as de facto standards and one's own macro implemented in GIS isn't going to cut much ice. They may even be quite specific about the nature of GIS software being used … and the database. This is a form of quality assurance where specification of tools of known quality means that the tools are not going to be the weakest link and are capable of supporting good quality work. Under these circumstances, tool coupling has to take place. This issue is considered in detail in Chapter 7.

Spatial Coexistence

This is the most common approach to GIS and environmental modeling and has been particularly popular throughout the history of GIS. After all, what is GIS good at—buffering, topological overlay, map algebra—all of which can be used to establish the spatial and temporal coexistence of sources and receptors. The transport mechanism remains implicit and largely unmodeled. If there is some spatial distance to be covered, this is usually modeled through buffering as a surrogate for the specific transport process at work. But spatial coexistence can also relate to the way in which variables come together spatially to form combinations that produce a hazard (factor mapping). Then, using spatial coexistence again with possible receptors, vulnerability and risk are established. A small example of this approach of establishing hazard and then risk was illustrated in Chapter 5, Figure 5.10. On the basis of a conceptual model arising from an empirically inductive understanding of the factors that promote landsliding in the specific study area, then by assuming these can be mapped in GIS, a combination of overlay and Boolean selection reveals those areas where a hazard exists. By further overlaying this result with the land cover layer, it would then be possible to identify which land uses were vulnerable, and if this included the class "village," then which people were vulnerable. If necessary, buffering of the unstable areas would indicate any village on or dangerously near unstable terrain. Mason and Rosenbaum (2002) provide an example from the Piemonte region of northwest Italy where, because of the nature of the geology, block slides develop along discontinuities and bedding planes (these weaknesses within the soil allow blocks of soil and rock to detach as landslides during heavy rainfall or earth tremors). Thus, slope failure is most likely to occur where the angle and aspect of a slope closely conforms to the dip and dip direction of the discontinuities. Using an equation in map algebra that resolves this relationship, Mason and Rosenbaum were able to map factor of safety (FoS) and, hence, hazard where FoS < 1 from slope and aspect raster layers derived from a digital elevation model (DEM). As stated before, such an example is not a universally applicable model, but one empirically developed from field data to reflect the specific conditions in an area. The spatial coexistence approach has also traditionally been used in flood risk studies (e.g., Cotter and Campbell, 1987) where geomorphological features associated with flooding, such as floodplains or flood extents calculated by engineers and hydrologists using simulation software, are overlaid with land use to identify settlements falling within these zones and, therefore, at risk.

The specific case study I would like to take you through in some detail concerns a small reservoir built using an earthfill dam at the head of a valley as the sole water supply to a small town farther downstream. Any damage to the reservoir could thus have potentially serious consequences for the residents of the town. The owners of the reservoir and their engineers were concerned about the slope stability of the reservoir rim during heavy rainstorm events and during rapid drawdown of the water level. There may be occasions when drawdown of a reservoir is either desirable or necessary. Drawdown is undertaken for periodic safety checks of a dam and other structures associated with a reservoir. It also occurs when the bottom of a reservoir is being scoured through a scour pipe in order to remove some of the accumulated bottom sediments. If a reservoir fills up with sediment, it decreases in water storage capacity and can eventually become useless. If the drawdown is too rapid, the effect on the side slopes is equivalent to an extreme rainfall event. Because the water table is suddenly lowered causing water to flow out of the saturated slopes, this increases pore water pressure, lowers the FoS below 1 and, thus, causes failure (Chapter 4). Figure 6.2 shows just such a slope failure at another reservoir in the region. Of particular concern to the engineers was the possibility of a large landslide entering the reservoir and causing a sufficiently large wave that would propagate through the reservoir and overtop the dam causing damage to or even failure of the structure. The engineers had carried out some preliminary slope stability calculations of theoretical cross sections and had arrived at an empirical model based on slope angle and soil thickness, summarized in Figure 6.3. The impasse for the engineers was in identifying which parts of the reservoir rim were "unstable" according to the model and which of these slopes might fail in such a way as to generate a large wave.

From a GIS perspective, this is a simple case of evaluating the spatial coexistence of slope angles with soil thickness. This resolves itself as four simple decision rules:

1. If slope angle < 20^0, then "stable"
2. If [slope angle ≥ 20^0 and < 35^0] and [soil thickness > $(0.58 \times$ slope angle $- 8.21)$], then "stable"
3. If [slope angle ≥ 35^0 and < 40^0] and [soil thickness > (slope angle $- 22.5)$], then "stable"
4. Otherwise "unstable"

FIGURE 6.2
An example of a reservoir rim failure due to rapid drawdown. (Photo courtesy of the author.)

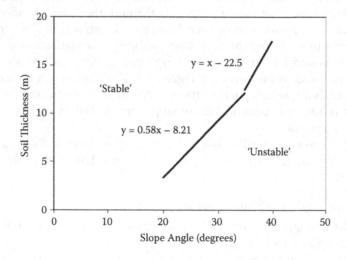

FIGURE 6.3
Empirical model giving relationship between slope angle and soil thickness for identifying "stable" and "unstable" slopes (specific to the geology and rainfall regime of the area).

Seemingly then, can two data layers—slope angle and soil thickness—be established? Slope angle should present no problem if a DEM can be constructed, but a soil thickness layer is a bit more challenging, to say the least. A DEM could be readily established using a triangular irregular network (TIN) data structure (Chapter 2, Figure 2.11(e)) from survey data collected during the construction of the reservoir from which a contour map could be constructed and checked for accuracy (Figure 6.4(a); this is not the whole of the upper catchment, but only those areas immediately around the reservoir rim). In this particular instance, the TIN data structure, as a tessellation, was also used to store attributes rather than further decomposing into raster. This was because the TIN was constructed using the theory of "high information" landscape features (Heil and Brych, 1978; Brimicombe, 1985) that produces a consistent representation of topography whereby the TIN elements form meaningful topographic and geomorphological units. Since each TIN element is treated as a planar surface subtended by its three vertices, it is a straightforward process to calculate the maximum slope angle for each TIN element (Figure 6.4(b)). To establish a map of soil thickness, a geomorphological mapping of pertinent features (Figure 6.4(c)) was carried out using aerial photographic interpretation (API), field inspection, and reference to the archive records of some exploratory drilling carried out prior to construction. From this study, it was possible to estimate soil thickness over the area (Figure 6.4(d)). At this point, it was possible to run through the decision rules and label slope elements "stable" or "unstable." Job done? Well, no. Though many GIS analysts may well stop at this point. The problem is: How accurate is the result? As we shall see in Chapter 9, there are a number of algorithms for calculating gradient, each giving a slightly different answer. But the main cause of uncertainty is likely to be the estimate of soil thickness. I can be superbly confident of my geomorphological prowess and it would be nearly impossible to check my estimates of soil thickness in the field, nevertheless, we need to recognize these attributes as best estimates and test the sensitivity of the results to reasonable changes in those estimates. In Chapter 8, we will look at the details of Monte Carlo simulation, but suffice to say here that the method involves repeated perturbations of the data and repeated calculation of the result in order to arrive at a best estimate of the true result. Thus, a series of both systematic and random changes were made to the soil thickness attributes and the model rerun numerous times. As a consequence of this sensitivity analysis (SA) (Figure 6.4(e)), it was possible to classify TIN elements as:

- "Very unstable": Those elements that were always classed as unstable regardless of the perturbation.
- "Unstable": Those elements remaining unstable at about the estimated soil thickness or less.

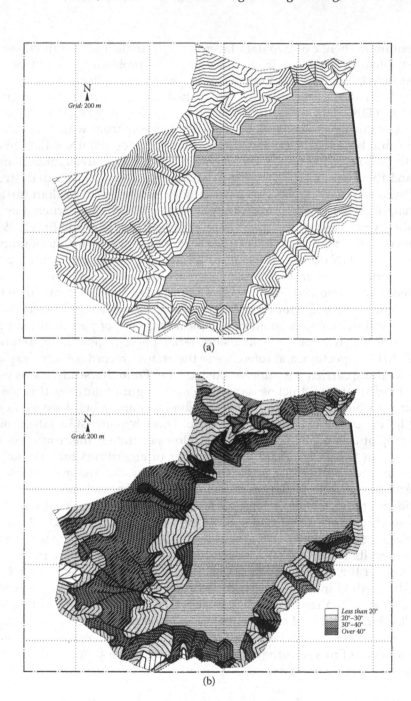

FIGURE 6.4
An investigation of reservoir rim stability involving a spatial coexistence approach on the basis
of an empirical mode (Figure 6.3). See text for explanation.   *Continued*

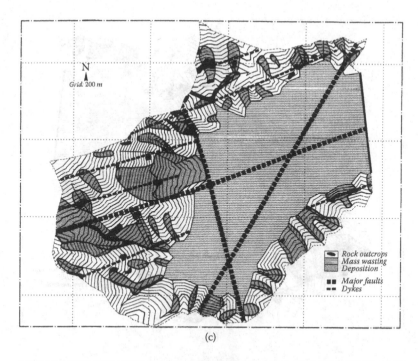

(c)

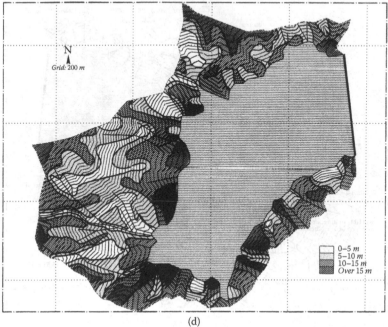

(d)

FIGURE 6.4
Continued.

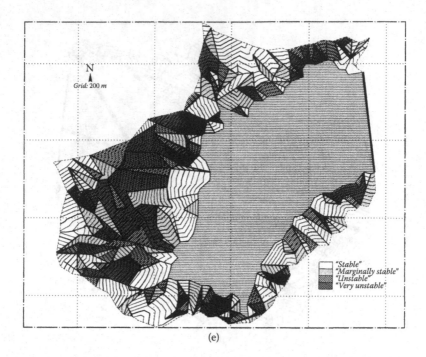

(e)

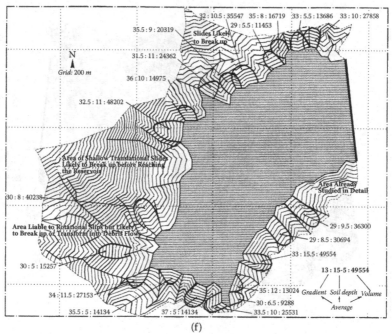

(f)

FIGURE 6.4
Continued.

- "Marginally stable": Those elements that were classified as unstable only after a substantial systematic reduction in soil thickness.
- "Stable": Those elements that were always classed as stable regardless of the perturbation.

Job done? Well, not quite yet. Figure 6.4(e) is not yet a product with which the engineers can work. It's the GIS spatial coexistence algorithmic answer, but it needs to be properly *interpreted* for meaning. For example, not all unstable slope elements may result in a single, large failure, some may be characterized by numerous small failures or failures that are likely to substantially break up before reaching the reservoir. This is the case on the higher slopes above the western end of the reservoir. An adequate interpretation is achieved by referring back to the aerial photographs in the knowledge of Figure 6.4(e) to identify potentially large landslide features for which the volume of their mass can be calculated from the GIS (Figure 6.4(f)). This then provides a product that the engineers can use to further their investigations. Following more detailed study by the engineers of specific features, they concluded that:

- Normal operation of the reservoir would not cause failures around the rim.
- Rapid drawdown through indiscriminate use of the scour pipe could cause failures and that appropriate operating procedures would need to be established.
- Severe rainfall events would probably result in failures, but that the waves generated were unlikely to cause damage to the dam.

Source–Pathway Characterization

The main difference between this approach and the previous one is that the transport mechanisms between sources and receptors or between inputs and outputs are explicitly modeled. This requires the relevant parameters to be quantified in order for the model to be calculated through for the entire process. There are many such process models in environmental science and engineering. In general, these either operate as lumped parameter or distributed parameter models, as discussed in Chapter 5. The lumped parameter models are either entirely nonspatial (e.g., the panda–bamboo interaction model in Chapter 5) or follow discrete spatial units, such as catchment areas. A drainage basin, for example, can be partitioned into a number of subcatchments where the parameters are being lumped for each subcatchment. Thus, if one is considering the effect of several large drainage basins, one effectively ends up with a distributed lumped parameter model of some complexity. This is the case in the first source–pathway study to be presented where we will be looking at the coupling of GIS and hydraulic modeling for basin management planning. The second study will be a fully distributed parameter modeling of coastal oil spills.

Basin Management Planning

This case study comes from Hong Kong and involves the coupling of GIS (Genasys II) and hydraulic simulation modeling (MIKE 11) into a spatial decision support system (SDSS). Although such couplings are now commonplace, when this project was started in 1991, the approach was a novelty. Presented here are the aspects of the project and its background that are already in the public domain (United Nations, 1990; Brimicombe, 1992; Townsend and Bartlett, 1992; Brimicombe and Bartlett, 1993; 1996; Drainage Services Department, 2008). We have already seen in the previous chapter that Hong Kong is a small region of just 1050 km² of which 60% is mountainous terrain and which, with a population of about 6 million, has resulted in intense development. Average annual rainfall is 2225 mm, but tropical depressions and typhoons can result in rainfall intensities that can reach 90 mm/hr. The steep terrain leads to rapid runoff concentration and flash flooding in lowland basins. In the period of 1980 to 1990, 84 major flood events occurred, many lasting two or more days. This compares with just 16 events in the 1960s, rising to 38 in the 1970s (www.dsd.gov.hk/ flood_prevention/flooding_problems/historic_data/index.htm). Flooding occurs mostly in lowland basins and natural floodplains in the northern part of the region (Figure 6.5). This increase in the number of flood events reflects the changing land use of the region over this period. There are three key elements:

Urbanization: Building of new towns in low-lying and floodplain areas (Figure 6.6).

Rural development: A combination of village expansion, the transformation of agricultural land to small industry and storage parks (containers, building materials, construction plant, and so on) and the abandonment of agriculture while speculating on development.

Floodplain reduction: The construction of ponds for fish and duck farming with bunds constructed to above flood level over substantial areas (Figure 6.7) has lead to a reduction in floodplain area.

These have all led to a scenario where not only was the number of floods increasing, but they were becoming more severe and taking longer to subside. Photographs of these flood events can be found at www.dsd.gov.hk/ flood_prevention/flooding_problems/flood_photos/index.htm.

FIGURE 6.5
The Hong Kong Special Administrative Region of the People's Republic of China showing flood prone areas in the lowland basins of the northern and western New Territories.

FIGURE 6.6
An example of new town development in low-lying basins; note backdrop of mountainous terrain. (Photo courtesy of the author.)

FIGURE 6.7
Pictured is an example of fish and duck pond development reducing the area of floodplain. (Photo courtesy of the author.)

In 1989, in recognition of the growing problem, the Hong Kong government established the Drainage Services Department (HKDSD) to formulate a flood prevention strategy through basin management planning. These plans might include a mix of river training, flood storage schemes, flood proofing, and land use development control. The Town Planning Ordinance was amended to curb unauthorized development and required all new developments to undertake a specific drainage impact assessment. In 1991, a "Territorial Land Drainage and Flood Control Strategy Study" was commissioned and would form the basis for drawing up 1:5000 scale *basin management plans* (BMP) for each drainage basin. It was in this context that a coupling of GIS and hydraulic modeling into a SDSS first took place and was to become established practice in Hong Kong.

In this project, the decision to use hydraulic modeling dominated the architecture of the system. Hydraulic modeling is used to simulate runoff along a drainage network in response to specified rainfall events. Models are constructed using true channel details (cross section, gradient), aggregated runoff parameters, floodplain storage mechanisms, and significant structures, such as bridges, culverts, and weirs. Thus, hydraulic modeling can be expected to give a truer simulation of channel capacities and their flows than hydrological modeling. Apart from studying the existing situation, hydraulic modeling is widely used as a design tool for remedial and mitigation measures and, therefore, can be used in "what if"-type analyses. The simulation is based on nodes (modeling points) linked by successive reaches of drainage channels into a topological network. Nodes are usually located at typical cross sections on a reach (rather than at a confluence) or used to represent locations of flood storage. Each node receives the cumulative flow from any upstream node and its own area of subcatchment. Parameters describing the flow accumulation within the subcatchment are lumped and attributed to the node. Thus, although technically this could be described as a lumped parameter model, there may be 100 or more subcatchments in a typically sized drainage basin for Hong Kong. The choice of GIS software rested on two key factors. The first was the need to handle large data sets in a timely manner, which, given the power of PCs and their storage capacity in 1991, meant that realistically the software needed to run on a UNIX workstation. Second, as will become evident below, the software needed to be able to handle fully topological vector data, TIN structures, and raster data in an integrated way. At the time, GIS and hydraulic modeling were only loosely coupled in as much as they were run independently with only the exchange of data between the two using reformatted ASCII (text) files. The overall pattern that developed was one of using GIS to integrate and preprocess data, which are passed on to the simulation and then to accept back the results of the simulation into GIS for postprocessing. By linking proprietary GIS and hydraulic modeling software in this way, the study team was able to assess flood hazard for current and projected land use scenarios over a range of rainfall events (1:2 through 1:200-year return periods) and for a variety of mitigation measures. On the basis of these multiple outcomes, well-founded decisions could be made regarding appropriate proposals and options for the BMP.

Figure 6.8 summarizes the range of data handled by GIS on this project and the preprocessing relationship with the hydraulic modeling. In 1991, digital maps of rural areas were not available. Typically, paper topographic maps are themselves prepared in layers with separate masters for contours, detail (roads, buildings, streams), symbols, and annotation (e.g., place names). Therefore, it was possible to go back to original masters, choose only the contours and the detail layers, which could then be scanned and vectorized. In this way, scanned 1:5000 scale topographic maps, which were then edge-matched into seamless base mapping, were used as the means to co-register and, therefore, integrate all the other data sets. There were five sets of other data:

Historic flood events: Two well-documented flood events (1988 and 1989) were mapped from oblique photographs of the floods taken from a helicopter, from press photos, and from questionnaire interviews of residents. Thus, the flood extents were mapped differentiating certain and uncertain boundaries. These could be used as part of the model calibration.

Drainage structures: This documented the location and attributes of all bridges, weirs, and culverts that would influence channel flow. These were collected by field survey during which channel cross sections were measured to be input directly to the simulation model.

Sub-basin parameters: These were a series of map layers derived mostly from API to produce the lumped parameters at modeling nodes. These are discussed in detail below.

Development scenarios: These were a series of future development scenarios that would replace current land use in the lumped parameterization of "what if"-type analyses and included various mitigation options, such as river training.

Floodplain elevations: Spot-heights were digitized from 1:1000 topographic maps in order to build up detailed floodplain DEM so as to extend the field surveyed channel cross sections to the edge of the floodplain, but also to act as a key data layer in the postprocessing of the simulation outputs.

The process of preparing the lumped parameters for the simulation modeling is as follows. A land use classification is created that reflects runoff characteristics and which can be consistently equated to the U.S. Soil Conservation Service's (SCS) runoff *curve numbers* (CN). These are numbers in the range 0 to 100 and can be loosely interpreted as the proportion of rainfall contributing to storm runoff depending on the nature of soils and vegetation. In the Hong Kong case, with intense rainfall on steep slopes that quickly exceeds the infiltration capacity, runoff is most influenced by land cover types rather than by soil types, which thus were not mapped separately. The highest CN is usually assigned to urban areas where there is little infiltration or storage, a low CN would be assigned to woodland, and the lowest to, say, a commercial fish pond where all the rainfall is retained within the pond. Eighteen classes of land cover were mapped by API and digitized into GIS. There is a tendency for tropical terrain to be characterized by a distinct break of slope between the steeper mountainous terrain and the gently sloping valley floors (noticeable in Figure 6.6 and Figure 6.7). This is an important feature in the simulation modeling that needs to be distinguished if realistic unit hydrographs are to be developed and so this too was mapped as a polygon boundary between "upland" and "lowland" zones. Land cover polygons together with upland and lowland areas for a small catchment are given in Figure 6.9(a). The engineers would identify the proposed locations of modeling nodes. These would be screen digitized into GIS and snapped to the drainage network. The coordinates of each node and along stream distance to the next node were output to the simulation model. From API, the subcatchments subtended by each node are identified and digitized (Figure 6.9(b)). The three layers—land cover, break of slope, and subcatchments—are then overlaid (Figure 6.9(c)) and tables produced giving

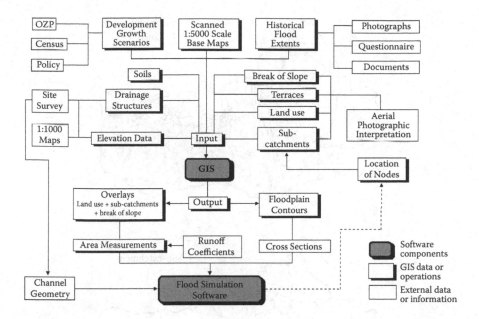

FIGURE 6.8
Illustrated is the role of GIS in data integration and preprocessing of inputs for the hydraulic modeling. (Note: OZP = outline zoning plans.)

total area for each class of land cover in the upland and lowland areas of each subcatchment. These are used to produce an area-weighted average CN for the upland and lowland portions of each subcatchment and are the lumped parameters used as input to the flood simulation modeling.

The hydraulic modeling used for the flood simulation was pseudo 2D, solving the full St. Venant equations to simulate variations of flow in space and time. The St. Venant equations assume that channel discharge can be calculated from the average cross-sectional velocity and depth and are based on a mass balance equation (6.1) and a momentum balance Equation (6.2), which assumes that water is incompressible (Beven, 2001):

$$\frac{\partial A}{\partial t} = -A\frac{\partial v}{\partial x} - v\frac{\partial A}{\partial x} + i \tag{6.1}$$

$$\frac{\partial Av}{\partial t} + \frac{\partial Av^2}{\partial x} + \frac{\partial Agh}{\partial x} = gAS_o - gP\frac{f}{2g}v^2 \tag{6.2}$$

where A = cross-sectional area, P = wetted perimeter, v = average velocity, h = average depth, S_o = bed slope, I = lateral inflow per unit length of channel, g = gravitational acceleration, f = the Darcy–Weisbach uniform roughness coefficient.

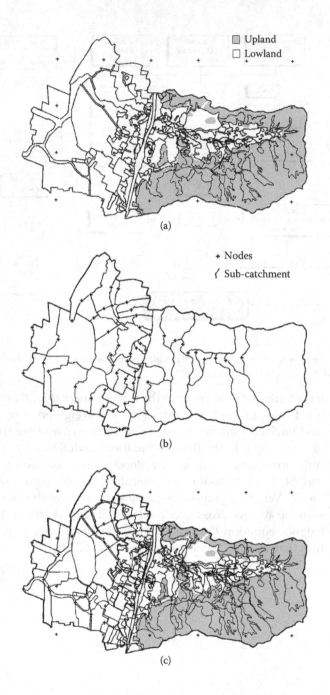

(a)

(b)

(c)

FIGURE 6.9
Preparation of lumped parameters from landscape characteristics: (a) land cover boundaries with upland and lowland areas, (b) modeling nodes and sub-catchments, (c) overlay of all three layers. Note: Tick mark spacing = 1000 m. (Based on Brimicombe, A.J., and Bartlett, J.M. (1993) *Proceedings of the 3rd International Workshop on GIS*, Beijing, China, 2: 173–182.)

In the pseudo two-dimensional approach, flow along reaches is modeled as 1D (at a node) using mean velocities, whereas overflow into floodplain storage and flow can also be simulated at representative nodes. Structures, such as bridges, weirs, and dams, are fully described. Calibration of the models was achieved using historical rainfall data and corresponding data from gauging stations giving velocity and stream level at a number of locations and by reference to the historic flood extents mapped in GIS. Output from the simulation is a time series for each node giving the height and the velocity of the flow. Since the simulation output refers to a series of points (the nodes), the data need further processing in order to visualize the flood extents and assess the impacts. Figure 6.10 summarizes this postprocessing carried out in GIS. At the time it was not possible to use the entire simulation output of height and velocity. Today, this could be resolved as a series of multiple maps brought together in an animation to show the progress of the flood from initial rise until it drains away. But, in 1991, that technology wasn't quite with us and with the need to evaluate multiple scenarios, time was tight anyway. With no satisfactory way of extrapolating velocities over the floodplain, postprocessing focused on flood height. Because the catchments are relatively small and all nodes reach their peak flow within a short time of each other, we could take the maximum flow height at each node and use this to extrapolate flood extent and depth across the entire floodplain. This is not as straightforward as it may sound. First, floodplains usually have a complex topography including levees and terraces. The extent of these features was mapped from aerial photography. Each terrace was modeled as a best fit polynomial from available spot heights with terrace scarps as sharp breaks in the terrain. This allowed a more accurate TIN of the floodplain topography to be created, which was then transformed into a raster DEM. Second, floods also have a subtle topography partly reflecting the gross characteristics of the floodplain, the ability of tributaries to drain into the main valley and damming effects of structures. It is entirely wrong to model a flood as a bath or pond with a hydrostatic water level. So, the flood topography has to be modeled. From Figure 6.9(b), it is evident that the number of nodes represents a sparse data set and merely triangulating them in order to contour maximum flood height is not very satisfactory. Besides, where the nodes follow a channel in more or less a straightline (eastern side of Figure 6.9(b)) triangulation is difficult (How do you create a triangle from three points in a straight line?) and the results would be unacceptable. The strategy that was adopted is illustrated in Figure 6.11. At each modeling node, a series of pseudo nodes having the same maximum flood height were created perpendicular to the direction of flow and extended to beyond the edge of the floodplain (Figure 6.11(a)). This assumes that at any modeling node, the cross section of its floodplain has the same flood height. This can then be triangulated into a TIN and a flood DEM thus extrapolated. The flood DEM and the floodplain DEM are then evaluated using map algebra (Figure 6.11(b)) using first a MAXIMUM function and then a SUBTRACT

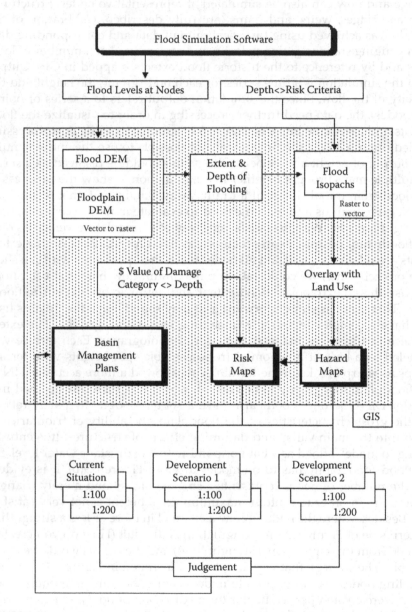

FIGURE 6.10
Shown is a GIS postprocessing of outputs from the hydraulic simulation modeling. Note: 1:100 is a 1 in 100 year storm event.

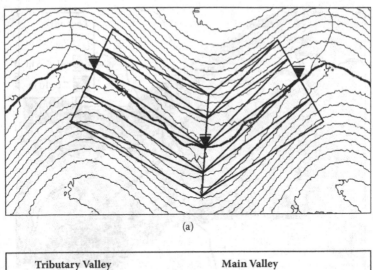

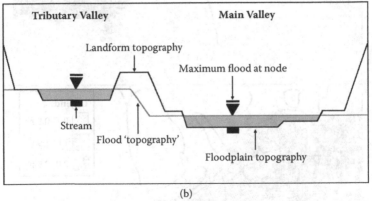

FIGURE 6.11
The principle of extrapolating of a flood extent from modeling nodes: (a) creating and triangulation of pseudo nodes, (b) cross-sectional illustration of the intersection of flood and floodplain DEMs.

function to arrive at a raster map that shows the extent of flooding (all non-zero cells) and where positive raster values represent flood depth over the floodplain. This latter value represents the degree of hazard for the relevant return period being simulated. Figure 6.12(a) shows a small portion of floodplain DEM from Figure 6.9 reflecting floodplain and valley-side terraces and Figure 6.12(b) shows flood depth for a simulated 1:10-year flood in relation to land cover polygons.

Not only is it possible to evaluate the hazard and risk associated with a range of return period rainstorm events (1:2-year to 1:200-year), but, in a cycle of GIS preprocessing, flood modeling, and GIS postprocessing, it is possible to evaluate a range of future development scenarios and mitigation measures as "what if"-type analyses for the same range of return periods. The multiple

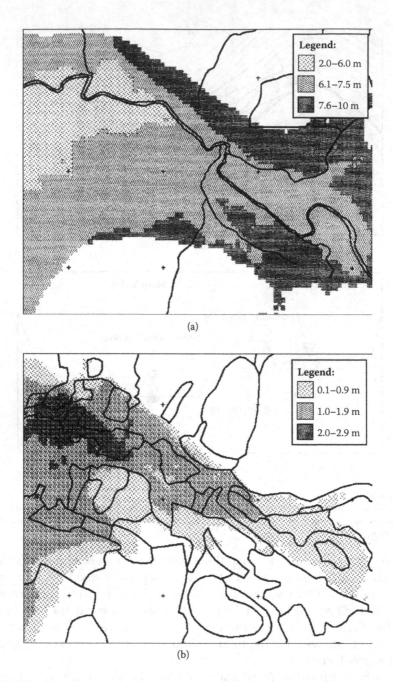

(a)

(b)

FIGURE 6.12

(a) Raster floodplain topography (10-m cells) to reflect terraces and (b) flood depth for a simulated 1:10-year flood in relation to land use polygons. Note: Tick mark spacing = 250 m. (b: Based on Brimicombe, A.J., and Bartlett, J.M. (1993) *Proceedings of the 3rd International Workshop on GIS*, Beijing, China, 2: 173–182.)

outcomes provide the basis for decision-making on the content of a BMP (Figure 6.10). In 1991, this approach was novel and showed a number of distinct advantages in coupling GIS and environmental simulation modeling:

Spatial data integration: For the compilation of BMP, data must necessarily be collected by a variety of techniques and from a large number of sources. The data inevitably comes in different formats and at different scales. This problem becomes acute where there is intensive land use that results in overlapping demands on land. GIS make the task of integrating the data into a common format possible and also promote more effective use of the data.

Quantification of parameters: Traditionally, paper maps and planimeters were used to measure areas. Measuring the area of all the polygons in Figure 6.9(c) by hand would be a daunting task and probably of questionable accuracy (and this is only a small catchment; larger catchments typically had tens of thousands of polygons after topological overlay). GIS can automate this process.

Flood extent and depth: Again, traditionally carried out by inspection of paper maps to find the flood extent. Such manual production of flood depths is difficult to achieve with any certainty. GIS can again automate this task for the production of hazard and risk maps.

Visual assessment of model calibration: The automated hazard maps provided the engineers with a new tool to visualize the outputs of their models. Not only could simulated floods be compared spatially with historical events during calibration, but they could more easily detect rogue parameter values (e.g., channel friction, bridge dimensions) due to counter-intuitive flood depths in the hazard maps.

"What if"-type analyses: The advantages cited above allow a wide range of options to be processed and evaluated in a spatial decision support framework.

Automated cartography: Because all the spatial data are integrated in GIS, the system can be efficiently used in the final reporting of a BMP by creating all the necessary cartographic products.

Coastal Oil Spill Modeling

Coastal oil spills are serious environmental disasters often leading to significant long-term impacts. From 1978 to 1995, there were in excess of 4,100 major oil spills of 10,000 gallons or more (Etkin and Welch, 1997). There has, however, been a downward trend in the number of major incidences from a peak in 1991, but with approximately 3 billion gallons of oil in daily use worldwide, a large proportion of which is transported at sea, the threat of coastal oil spills remains acute. On Friday, March 24, 1989, the Exxon Valdez, carrying 1.25

million barrels of crude oil, ran hard aground on Blight Reef, Prince William Sound, Alaska, spilling 258,000 barrels of oil mostly in the first few hours. Since contingency disaster planning had only focused on spills of up to 2,000 barrels, the spill could not be contained and instead spread to nearly 2,000 km of coastline devastating fisheries, wildlife, and scenic beauty. Restoration cost over $10 billion. On Thursday February 15, 1996, the Sea Empress ran aground in the entrance to Milford Haven, United Kingdom. Over the following two days, further damage occurred due to wind and strong tides. More than 7,200 tons of fuel oil were spilled, 5,000 tons of which reached the Pembrokeshire coast, Wales, polluting 200 km of shoreline including 100 beaches with devastation to wildlife and fisheries. The U.K. government's estimate for cleanup, salvage, and losses of fisheries and tourism was up to £43 million (Harris, 1997). On Tuesday November 19, 2002, the Prestige broke up and sank 210 km off Spain's northern coast. A few days earlier, the Prestige had been holed off Galicia, Spain, spreading oil to the nearby coast. The Spanish authorities had ordered the tanker towed away from the coast with perhaps inevitable consequences. For at least two months, oil continued to seep from the wreck to be spread by tide and wind onto the sandy beaches around the Bay of Biscay from Galicia to La Rochelle in France with again devastation to wildlife, fisheries, and tourism with severe social and economic consequences in many coastal towns and villages.

In a coastal oil spill incident, oil floats and spreads out rapidly across the water surface to form a thin layer—a *slick*. As the spreading process continues, the layer becomes progressively thinner, finally becoming a *sheen*. Complex interrelated physical, chemical, and biological processes depending on the type of oil, the hydrodynamics, and other environmental conditions govern the behavior of the slick and the sheen. Drifting of the oil is mainly by advection and diffusion due to currents (tide and wind). Weathering of the oil in transport leads to spreading, evaporation, photochemical reactions, dissolution, and sedimentation, which, in turn, lead to changes in the volume, mass, and physiochemical properties of the spilled oil (Sebastiao and Soares, 1995; ASCE Task Committee on Modeling of Oil Spill, 1996). Significant efforts are made worldwide in oil spill prevention, preparedness and impact assessment in which GIS adopt their usual role in the integration and handling of relevant spatial data. Analysis and assessment of the risks, however, rely heavily on numerical simulation of coastal oil spill behavior so that the environmental impact assessment (EIA) and contingency planning can be pertinent in the protection of sensitive areas and installations. Simulation of oil spill behavior requires more than one model, each having a specific task: hydrodynamic modeling, trajectory modeling, and fate modeling. The structure of typical coastal oil spill modeling is given in Figure 6.13, which represents a dynamic 2D distributed parameter model.

We have already seen the hydrodynamic modeling in Chapter 5. Figure 5.25 showed an example of currents simulated using FEM for hydrodynamic modeling. The tidal currents calculated from the forced inputs at the open

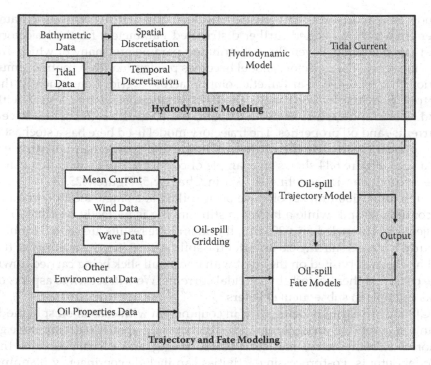

FIGURE 6.13
Typical structure of coastal oil spill modeling. (Based on Li, Y., Brimicombe, A.J., and Ralphs, M.P. (1998) In *Oil and hydrocarbon spills: Modelling, analysis and control*. Computational Mechanics Publication, Southampton, U.K.; Li, Y., Brimicombe, A.J., and Ralphs, M.P. (2000) *Computers, Environment and Urban Systems* 24: 95–108.)

boundary were shown graphically with the arrow size proportional to the speed of the current. The arrow direction, as shown in Figure 5.25, is actually a resultant vector calculated from two components: a northerly component U and an easterly component V. Each of these components has itself two components, which describe it: a velocity in m/s U_a V_a and a deflection in radians U_g V_g. For each tidal constituent in the hydrodynamic modeling, the current simulation can be represented by the following equations (Li, 2001):

$$U(\mathbf{x}, t) = U_a \cdot \cos(\omega t - U_g) \tag{6.3}$$

$$V(\mathbf{x}, t) = V_a \cdot \cos(\omega t - V_g) \tag{6.4}$$

where U_a = amplitude of northerly component, U_g = deflection of northerly component, V_a = amplitude of easterly component, V_g = deflection of easterly component, ω = angular frequency of the relevant tidal constituent, $\mathbf{x}$ = location, t = time step.

The purpose of the hydrodynamic modeling is to calculate a time series of currents from tidal constituents and bathymetry over the whole coastal study area and is in itself a means of calculating and distributing parameters to which are added further distributed parameters for the trajectory and fate modeling. Given these distributed parameters, many of which are in time series, the trajectory model becomes predominantly a routing simulation that requires an arithmetic solution across a grid. Consequently, the output from the hydrodynamic model is reinterpolated into a grid with additional inputs, such as mean current, wind direction (for wind-driven currents), and oil properties. The trajectory model used here has a stochastic component to simulate the random thinning and spreading of oil droplets in a slick. Figure 6.14 shows an example of coastal oil spill simulation using the hydrodynamic modeling shown in Chapter 5, Figure 5.25. It has been assumed that 100 units of oil have been spilled at a single location (though a continuous spill while a tanker is still moving is perfectly possible). The trajectory is modeled on a 200 m cell grid and the results are shown for half hourly intervals starting 0.5 hrs after the spill. After 3 hrs, the majority of the oil has been deposited on the coast with one small slick being carried down the coast and then easterly by the tidal currents. We will return to aspects of this example in subsequent chapters.

GIS are increasingly being used in conjunction with coastal oil spill modeling as a tool for integrating and preprocessing spatial data inputs (e.g., shoreline, bathymetry) and for postprocessing and visualization of the model outputs. Postprocessing activities can include contingency planning (Jensen et al., 1992), oil spill sensitivity mapping (Jensen et al., 1998), making operational decisions (Trudel et al., 1987), and damage assessments (Reed and French, 1991).

Cluster Detection

This *ex post* approach is structured quite differently from the previous ones. The initial phase is predominantly exploratory of spatial patterns either by statistical or geocomputational techniques and, once an abnormal concentration of some effect has been detected, some form of environmental simulation modeling may be used in order to confirm the transport and fate mechanisms from the revealed or suspected source to the receptors as a precursor to the implementation of mitigation measures. Many of these applications center around the investigation of diseases caused by unsanitary conditions and/ or pollution and have strong roots in spatial epidemiology (Lawson, 2001) and environmental engineering (Nazaroff and Alvarez-Cohen, 2001). But, as we shall see next, these techniques can also be applied to purely engineering phenomena, such as, for example, pipe bursts, landslides, subsidence, and so on. But first, we begin with some principles of the approach.

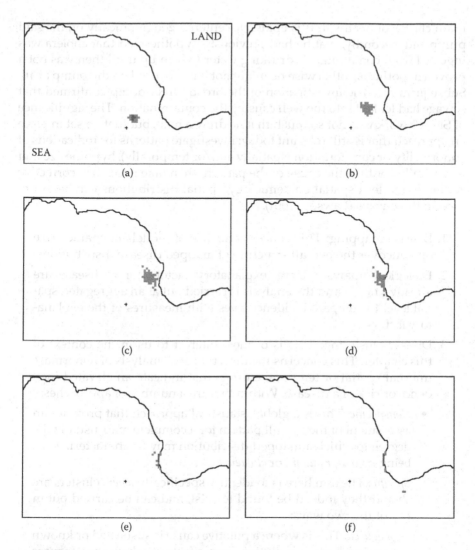

FIGURE 6.14
Simulation of an oil slick trajectory from a stricken tanker (location marked +) at half hourly intervals from 0.5 hrs to 3.0 hrs.

Recognition of spatial patterns to events have been a cornerstone of spatial epidemiology since John Snow (a physician to Queen Victoria) in 1854 determined the source of a cholera epidemic in the Soho district of London to be a pump on Broad Street. Once the handle of the pump had been removed, the epidemic subsided. Snow's revelation is often attributed to a mapping of the cases, but actually this map was only created after the event for a monograph recording his observations and analysis (Snow, 1855). Nevertheless, his correct deduction of the cause arose first from an observation that the main cluster of deaths in this epidemic centered geographically on the said pump and, secondly, that he had previously hypothesized that cholera was ingested from contaminated drinking water (which up until then was not a proven hypothesis), otherwise he might not have focused on the pump at all. Subsequent forensic investigation of the Broad Street pump confirmed that sewage had leaked into the well causing the contamination. The significance of Snow's work was not so much that he drew a map, but that he set in place an approach that is still relevant today: investigate patterns for indications of abnormality or concentration (spatially and/or temporally), hypothesize and forensically confirm the cause of the pattern, then take necessary corrective action. In modern spatial epidemiology, spatial distributions can be examined in three ways (Lawson, 2001):

1. Disease mapping: This concerns the use of models to characterize and uncover the overall structure of mapped disease distributions.

2. Ecological analysis: Here, explanatory factors for a disease are already known and the analysis is carried out at an aggregated spatial level to compare incidence rates with measures of the explanatory factors.

3. Disease clustering: This is of most interest to us in the context of this chapter. This concerns the detection and analysis of abnormal/unusual spatial or temporal clusters that indicate an elevated incidence or risk of a disease. Within this are a number of approaches:

   • *Nonspecific*: This is a global, statistical approach that provides an assessment of the overall pattern for a complete map, usually the degree to which a mapped distribution may be characterized as being *regular, random,* or *clustered*.

   • *Specific*: The aim here is to identify specifically *where* clusters are should they indeed be found to exist, and can be carried out in one of the two ways:

     – *Focused*: This is where a putative cause is suspected or known *a priori*, such as pollution from a factory, which then focuses the search for clusters.

     – *Nonfocused*: Where there are no *a priori* assumptions and an exploratory search is carried out to find clusters wherever they may occur.

An event, such as catching a disease, the occurrence of a landslide, or a pipe burst, can be treated as a binary event (0, 1) in as much as either it has happened or it hasn't. You don't get half ill and a pipe doesn't partially burst (unless you want to get pedantic and say it just leaks). Such binary events for the purpose of a specific cluster analysis are best treated as point data. Although there are a range of techniques for analyzing spatially aggregated data (e.g., Besag and Newell, 1991; Anselin, 1995; Ord and Getis, 1995), we will focus here on point binary events. Apart from spatial epidemiology, the analysis of such data has a long tradition in geography (Dacey, 1960; Knox, 1964; Cliff and Ord, 1981) and ecology (Clark and Evans, 1954; Greig-Smith, 1964) and has received renewed interest within GIS and geocomputational frameworks (Fotheringham and Zhan, 1996; Gatrell et al., 1996; Openshaw, 1998; Brimicombe and Tsui, 2000; Atkinson and Unwin, 2002), and more recently within spatial data mining (Miller and Han, 2001; Brimicombe, 2002; 2006; Jacquez, 2008). But what is a cluster? Unfortunately there is no standard definition, but, instead, two broadly defined classes of cluster:

- The first comes from the mainstream statistics of cluster analysis arising from the work of Sokal and Sneath (1963). Thus, clustering is an act of grouping by statistical means which, when applied to spatial data, seeks to form a segmentation into regions or clusters, which minimize within-cluster variation, but maximize between-cluster variation. There is a general expectation that the spatial clustering will mutually exclusively include all points and, therefore, is space-filling within the geographical extent of the data (e.g., Murray and Estivill-Castro, 1998; Halls et al., 2001; Estivill-Castro and Lee, 2002). With a spatial segmentation, further analysis of this form of clustering usually leads to aggregated data techniques (cited above).

- The other class of cluster is concerned with "hotspots." These can be loosely defined as a localized excess of some incidence rate and are typified by Openshaw's *Geographical Analysis Machine* (GAM, http://www.ccg.leeds.ac.uk/software/gam) and its later developments (Openshaw et al., 1987; Openshaw, 1998). This definition of a cluster is well suited to binary event occurrences. Unlike the statistical approach, there is no expectation that all points in the data set will be uniquely assigned to a cluster, only some of the points are necessarily identified as belonging to hotspots and these then remain the focus of the analysis. With this type of clustering, the null hypothesis of no clustering is a random event occurrence free from locational constraints and would thus form a Poisson distribution (Harvey, 1966; Bailey and Gatrell, 1995). Because the recognition of this type of cluster is in relation to some incidence rate, the significance of clustering is often evaluated against an underlying "at risk" or control population. This is a critical issue because misspecification is clearly going to lead to erroneous results. In some applications (e.g., data mining) the "at risk" population may be identifiable at the outset and for yet other applications (e.g., landslides, subsidence), the notion of an "at risk" population, such as all those parts of a slope that are vulnerable to failure, may have little meaning. Nevertheless, it is this type of cluster that is sought in cluster detection approaches to GIS and environmental modeling.

The eye is very quick to detect clusters of objects on the basis of proximity, concentration, and density change (Sadahiro, 1997), but in making objective decisions on comparative degree of clustering, such as differences between cluster patterns and separating out randomness, we remain fallible. We, therefore, need formulae and computational methods. One basic approach is the nearest-neighbor distance statistic R first developed by Clark and Evans (1954), defined as:

$$R = \frac{\bar{r}_a}{\bar{r}_e} = \frac{\sum\limits_{i=1}^{n} d_i \Big/ n}{\sqrt{(n/A)/2}} \qquad (6.5)$$

where $\bar{r}_a$ = average nearest neighbor distance, $\bar{r}_e$ = expected average nearest neighbor distance when random, A = area of bounded region of interest, d = distance of a point to its nearest neighbor, n = number of points.

R approaches 0 if the pattern is clustered, $R = 1$ if the pattern is random, and $R = 2.1491$ if the pattern is uniform. Greig-Smith (1964) developed quadrat analysis as a means of evaluating hypotheses about the nature of processes generating point patterns. Calculated from point counts in each quadrat, his *index of cluster size* (ICS) explicitly measures against the moments of a Poisson distribution (where variance equals the mean):

$$ICS = \frac{\sigma^2}{\bar{x}} - 1 \qquad (6.6)$$

where σ^2 = variance, $\bar{x}$ = mean.

If $ICS = 0$, the pattern is random, if $ICS < 0$, the pattern tends toward regularity, and if $ICS > 0$, the pattern tends toward clustering. R and ICS are both global statistics and both suffer from serious scale effects. For R, as A is increased in relation to a fixed distribution of points, so the statistic reduces toward 0 (clustered). For ICS, the scale effect is less straightforward and an example is given in Figure 6.15.

An approach that is increasingly popular in GIS is *density mapping*. An example of this was given in Chapter 5, Figure 5.11(a) in relation to the point occurrences of landslides. Density mapping is available as a function in the Spatial Analyst extension to ArcView, while Atkinson and Unwin (2002) have placed in the public domain a MapBasic code for density mapping in MapInfo. The simplest form of density estimation uses a "moving window" over each grid cell in turn to count the number of data points, divide by the area of the moving window and transform to a density measure per sq

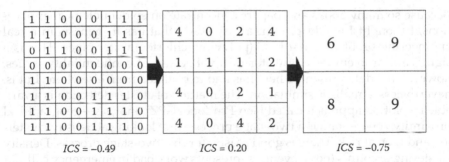

FIGURE 6.15
An example of scale effect on index of cluster size (ICS) with changing quadrat size. (Based on Tsui, P.H.Y., and Brimicombe, A.J. (1997) *Transactions in GIS* 3: 267–279.)

km or other spatial unit. A more sophisticated approach uses *kernel estimators* that uses either a fixed search area or are adaptive in their search area (Fotheringham et al., 2000). The kernel assumes the shape of an assigned probability density function, such as a bell curve, which provides an inverse distance weighting to the points that fall within it when calculating the local density. A parameter known as the *bandwidth* determines the lateral extent of the kernel. Specification of an appropriate bandwidth is by no means straightforward nor necessarily intuitive. If too small, the result becomes spiky; if too large, the result becomes overly smoothed. Some trial-and-error may be necessary though Fotheringham et al. suggest a starting point as:

$$h_{opt} = \left[\frac{2}{3n}\right]^{0.25} \sigma \tag{6.7}$$

where h_{opt} = optimum bandwidth, n = number of points, σ = standard deviation of point distances from the mean center of the point distribution (i.e., the standard distance).

The GAM of Openshaw et al. (1987) is an exhaustive heuristic search that has some resemblance to a "moving window" approach. Instead of fixed size, the moving window is a circle that when centered on a grid cell starts off small and is progressively increased in size to a maximum. At each increment, the points that fall within the circle are tested for significance against a background "at risk" population. If the result is significant, the circle is plotted. When all the circles have been tested, the moving window moves to the next grid cell and eventually around the entire map. In the current version, GAM/K (Openshaw, 1998), a kernel smoothing is applied to the incidence of significant circles to produce a density surface. When first conceived, this kind of exhaustive heuristic could only be run on supercomputers, but can now run on PCs.

Cluster detection methods are an area of ongoing research finding increasingly wider application in health, crime, education, business, and engineering.

Because so many books and papers concentrate on epidemiological applications, I thought I would present a case study that has its genesis in a real
engineering problem: burst PVC (polyvinyl chloride) water pipes. This is a
significant problem for any water utility. Because of confidentiality issues,
however, the data presented here has had to be taken from elsewhere, but is,
nevertheless, a realistic simulation. The method of cluster detection is a variable resolution approach called Geo-ProZone (GPZ) analysis—geographical
proximity zones—and is a two-stage process, first looking at density clustering and then at risk. There is good reason for this two-stage process. Density
clustering of counts (point events) represents workload in emergency call-out
to fix the problem. The results can be used for operational decisions about
crews and stockpiling. However, as might be expected, more call-outs can
occur where there are more pipes in the ground. The risk mapping, therefore, identifies areas of excessive incidence for a given unit length of pipe and
alerts the utility to the presence of some particular causal factors (manufac-
ture, method of laying, substrate materials, traffic vibration) that may signal
the need for pipe replacement or change in practice when laying new pipes.

The variable resolution approach has its genesis in resolving the scale
problems associated with handling many types of geographical data,
including point events, by using a computational heuristic technique (Tsui
and Brimicombe, 1997; Brimicombe and Tsui, 2000; Brimicombe, 2002; 2006).
Take, for example, the quadrat analysis given in Figure 6.15 where *ICS* value
changes in response to grid size, usually in unpredictable ways. Now compare it with Figure 6.16 where a variable resolution representation that is
quadtree-like gives an *ICS* that indicates some clustering. While a variable
resolution *ICS* = 1.13 differs from all the others in Figure 6.15, it is intuitive of
the initial pattern. But, we can test the idea. Figure 6.17 shows two point data
sets of 200 points each, one mathematically simulated to be random, the other

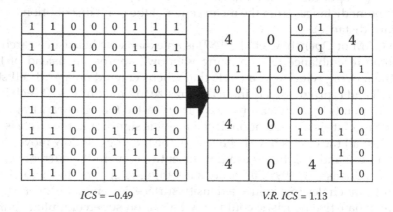

ICS = −0.49   V.R. ICS = 1.13

FIGURE 6.16
A variable resolution solution to the quadrat analysis scaling problems. (Based on Tsui, P.H.Y.,
and Brimicombe, A.J. (1997) *Transactions in GIS* 3: 267–279.)

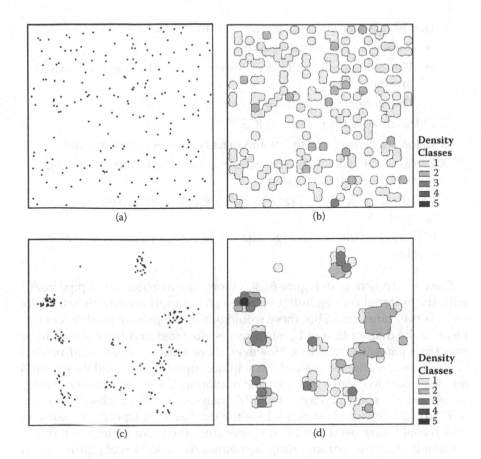

FIGURE 6.17
Geo-ProZone, geographical proximity zones on a random and a clustered point data set for verification of the technique. See text for explanation.

clustered. The random data set has R 1.047 (R 1 is random), the clustered data set has R 0.429 (R 0 is perfectly clustered with all points on the same spot). Variable resolution ICS 0.105 for the random data (ICS 0 is random) and is 4.818 for the clustered data. Further testing against other techniques can be found in Brimicombe and Tsui (2000). Visualization of the GPZ in Figure 6.17 allows any density clusters to be quickly identified.

The GPZ algorithm works as follows:

1. If there is no spatial boundary (e.g., administrative, catchment area), generate a 1% buffered convex hull around the data set.

2. Calculate the nearest neighbor distances.

3. Determine cell sizes in a hierarchical tessellation from the entire bounded data set down to a minimum (atomic) size mediated between median nearest neighbor distance and $\bar{r}_e$ in Formula (6.5).

4. Divide cells into successively smaller units unless:
   - It does not contain any points.
   - The smaller cells will have a variance that falls below a heuristic threshold.
   - The atomic cell size has been reached.
5. Calculate variable resolution ICS.
6. Calculate density per sq km and density classes per atomic cell.
7. Partition "at risk" data (if available) according to the cell arrangement from the previous step.
8. Calculate risk (see below) and robust normalized risk (see Chapter 10).
9. Choose class interval, merge adjacent cells of same class interval and display.

The case study data in Figure 6.18(a) shows the distribution of pipe bursts with the buffered convex hull. Usually such data sets contain thousands of records over large areas, but these would be hard to present in a book of this format. The distribution of bursts appears clustered and there seems to be two large patches of clusters. However, there may be some small patches where bursts, within the scale of the map, are superimposed and these would not be evident to the eye. The variable resolution $ICS = 4.01$, thus confirming an overall clustered pattern. The GPZ analysis of density classes is given in Figure 6.18(b), which shows where the localized "hotspots" are, some of which could have been identified by eye from the point pattern, but others register high because of superimposed bursts (i.e., a section of pipe has burst on several occasions). Some of the hotspots are spatially isolated, while others are part of a larger pattern of generally higher burst densities. In order to go beyond this and calculate risk, we need "at risk" data. This always needs to be carefully defined as it will have important consequences for the analysis. For a water reticulation network, what is it that constitutes the population: the number of links in the network, the number of discrete segments, or a unit length, such as per 1,000 m of pipe? In this analysis, a per unit length of in-ground PVC pipe was used as the "at risk" population. The risk calculation itself was the *relative risk* R_i, that is, the level of incidence in relation to what might be expected to arise given the level of "at risk" population:

$$R_i = \frac{n_i}{(N/P)p_i} \tag{6.8}$$

where n_i = local number of incidents, p_i = local "at risk" population, N = total number of incidents, P = total "at risk" population, and where a value of R_i > 1 indicates an incident rate in excess of the expected. The risk of a pipe

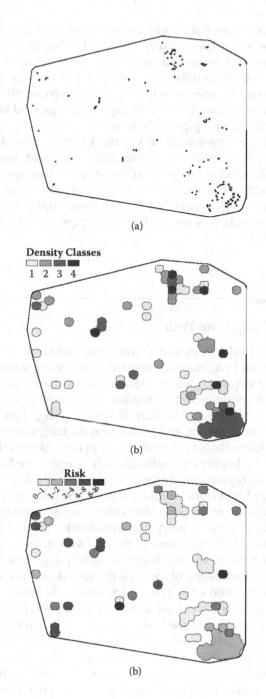

FIGURE 6.18
Geo-ProZone, geographical proximity zones analysis of PVC pipe bursts: (a) point event distribution with buffered convex hull, (b) point density classes, (c) risk classes.

burst is shown in Figure 6.18(c) and differs quite markedly from the burst densities. The areas that had initially appeared to be the dominant clusters are not the highest risk areas (although they do tie up most of the call-out resources) because of the number of bursts in relation to the amount of pipe in the ground. Instead, some areas that appear to have smaller overall numbers of bursts have comparatively little pipe in the ground and show some fundamental problem with pipe in these areas.

Further analysis, not presented here, would look at pipe attributes in the highest risk areas to identify any dominant causes and any lessons to be learned. This would include issues of size and age of the pipes, winter excess (freezing), contractor who installed the pipes, substrate that the pipe rests on, and type of fracture. Alongside this last attribute might be forensic laboratory testing to destruction of sample lengths of pipe to understand better the circumstances of failure.

... and Don't Forget the Web

In this chapter, an in-depth look at a number of case studies in GIS, environmental modeling, and engineering have been presented within a framework that classifies the approach when GIS are one of the coupled tools. When GIS and environmental simulation modeling are used in tandem, there are issues that arise from the architecture of the coupling, data quality, model quality, and algorithm choice, and in decision making where there is residual uncertainty. All of these will be explored in the chapters that follow. But, before leaving this chapter on applications, it is worth perhaps glancing at the Web—that vast repository of (almost) nearly everything.

I have just entered "environmental modeling GIS" into Google and received 366,000 hits. It sounds promising, but without going through every one of those hits, they appear to be predominantly books, journal papers, conferences, courses, and training. Some of this is likely to be a good source of further application examples, but there are hardly any sites that allow realtime environmental modeling of the type discussed in this chapter. GAM/K used to have a Web-based version to which files could be submitted, but it is available now only as a download to be run locally (http://www.ccg.leeds. ac.uk/software/gam). But, as the Web page says, it is "an experimental program and behaves as such." Also available for download is PCRaster (http:// pcraster.geo.uu.nl/pcrwin32/), which as discussed in Chapter 7 has strong environmental modeling capability. There are, of course, a growing number of sites that offer online full-functionality GIS through a browser, such as at http://www.onlinegis.net and at http://www.emapsite.com, but these are by subscription and are basically to allow corporate clients to access GIS and data anywhere at anytime without having to have GIS software or the data held locally. Then again, there are some Internet mapping sites that allow visualization of environmental data (as opposed to just street maps) for free. A U.K. example is MAGIC (http://www.magic.gov.uk). This was the first application in the United Kingdom to bring together all the key environmental data from different government departments and agencies into a single online interactive mapping tool. It allows selection of different layers of information to be viewed in relation to background topographical mapping right down to very local levels. It is aimed primarily at rural policy making and management and at best allows spatial coexistence to be visualized. A U.S. example can be found at http://nationalatlas.gov, which integrates all the relevant information in order to visualize continental-scale environmental issues. Figure 6.19 shows a landslide incidence and susceptibility map for the central United States from Utah in the west to Missouri in the middle of the country. The site aims in the future to provide global environmental information at 1:1,000,000 scale. Finally, a lumped parameter simulation model for use in planning reductions in transport CO_2 emissions is due to be made available at http://www.vibat.org/vibat_ldn/tcsim.shtml. This online real-time simulation will allow stakeholders to make decisions about different combinations and take up of, say, low emission vehicles, alternative fuels,

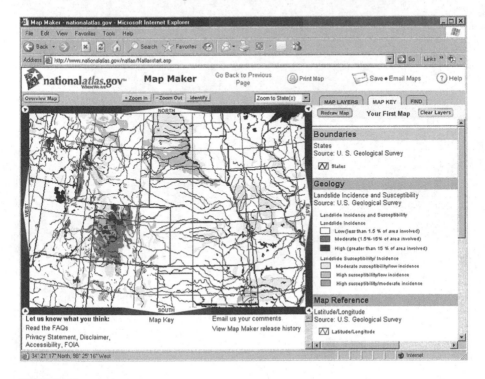

FIGURE 6.19
Landslide incidence and susceptibility in the central United States—an example of online environmental mapping from http://nationalatlas.gov (accessed June 6, 2009).

public transport, walking and cycling, urban planning (strategic and local urban design), information and communications technology (ICT) developments, driving and lower speeds, freight transport logistics and long-distance travel substitution (e.g., air to rail) in order to achieve target reduction in CO_2 by certain dates.

7

Issues of Coupling the Technologies

Geographical information systems (GIS) and environmental simulation models started to be used together around about the end of the 1980s. From my point of view, there was no particular application, no landmark paper that monuments the beginning of this fortuitous working together of the different technologies. It was more the case that the benefits of the idea independently sprung up on different types of projects in different parts of the world. It was, however, not inevitable that GIS and environmental modeling would come together (Parks, 1993). They are rather different technologies. GIS focuses on representations of location, the spatial distribution of phenomena and their relationships to one another in space. These are usually static representations. Environmental simulation models, on the other hand, are principally concerned with system states, mass balance, and conservation of energy, that is, focusing on quantities (populations, chemicals, water) in time. While, the distribution of "actors" (Fedra, 1993) within environmental simulation models are affected in their interactions and dynamics by their spatial distribution, many of the early models of the late 1960s, 1970s, and early 1980s did not treat the spatial dimension explicitly. If we treat environmental modeling in its broadest sense and include the logical models of land use suitability, then, yes, there has been a considerable tradition of treating space as explicit and it was these types of applications that propelled GIS forward in its earliest stages (see Chapter 2). Nevertheless, even with the advent of computer-based numerical simulation models in the atmospheric sciences, hydrology, biology, and ecology, the GIS link was not a foregone conclusion. To be sure, with the aid of GIS, hydrological models, for example, could more easily move from a 1D treatment of the drainage basin to a distributed parameter approach. GIS could be used to calculate gradient and aspect for a distributed model, discretize soil and land use coverages, interpolate sampled parameters, and thus handle spatial variability explicitly. This is not to say that GIS cannot have a role in 1D modeling (e.g., in producing spatially averaged parameters), it is just that in this mode, the two technologies can, if necessary, be used quite independently of each other and there are not nearly the same level of synergies to be achieved by getting GIS and environmental simulation models to work together. Nevertheless, in either case, it is absolutely critical to remember that GIS are not *sources* of spatial data, they are a technology for handling, manipulating, and displaying the spatial data provided to them (Fedra, 1993; Maidment, 1993b).

Some Preconditions

In changing over to environmental simulation models that make use of explicit descriptions of spatial variability, fundamental changes needed to take place that would assure GIS of their usefulness:

- That distributed parameter models would give more accurate or more useful results than 1D, lumped parameter models.
- That it was, therefore, worth the effort and expense of rewriting and testing models in distributed parameter mode.
- That the necessary reinterpretation of modeling parameters remained meaningful in terms of the reality being modeled.
- That the necessary data could be collected.

From a geographer's perspective, making spatial variation explicit would without hesitation lead to improved levels of explanation. But, that is not necessarily so from the perspective of environmental modelers. In hydrology, for example, after a decade of debate, the jury is still out (for a balanced summary of the discussions, see Beven, 2001). Physically based hydrological process models using distributed parameters, when subjected to blind validations tests (Ewen and Parkin, 1996) have produced limited success (Parkin et al., 1996; Refsgaard and Knudsen, 1996). While in theory, these models represent an appropriate catchment modeling strategy, difficulties arise in their application. In particular, there is still inadequate knowledge of the processes at the grid scale and, therefore, some of the process descriptions currently in use may not be appropriate. Moreover, the effective values of parameters may need to be varied with grid size and the methods of estimating the discretized values (e.g., by rasterization or interpolation) may be neither scale invariant nor provide adequate resolution to support a fully distributed model (Beven, 1996; Bronstert, 1999). Nevertheless, distributed parameter approaches are defensible on the grounds that they allow changes in runoff and water quality in response to subcatchment changes in land cover, groundwater use, irrigation, drainage, and so on, to be studied and predicted (Refsgaard et al., 1996). Such models allow management strategies to be developed, tested, and monitored in heterogeneous catchments. Efforts toward refining and applying such models are thus ongoing.

The change in scale, explicit in moving from 1D lumped models to distributed parameter models, has had far-reaching implications. In the early

1980s, when distributed parameter models began to move from being purely research tools toward regular professional use, computing and data collection technologies were at a much earlier stage and most of the current off-the-shelf GIS software were also at an early stage of commercialization (see Chapter 2, Table 2.1). Using process models that required finite element method (FEM) or finite difference method (FDM) approaches to the solution of distributed models increased the amount of computational complexity. This required a substantial development effort, as the model structures were completely different rather than an incremental change over previous models. Distributed parameter models took much longer to compute. Furthermore, reducing the size of the grid (to increase resolution) invariably led to an exponential rise in memory requirements and computation time. Though this has been more than offset these days by the operation of Moore's Law, it was an important consideration then. Data requirements fundamentally changed as well. For lumped models, parameters, input variables, and validation data could all be measured at points (or in small patches) that were representative of the larger landscape being modeled. For distributed parameter models, on the other hand, each cell or modeling unit (which could be grid, triangular irregular network (TIN) or irregularly shaped polygon) is assumed to be homogenous in its characteristics and, hence, needs its own parameter values determined for it. Rather straightforward GIS approaches to discretization based on vector to raster or interpolation may be adequate for land use or rainfall (subject to the uncertainty inherent in these techniques as will be discussed in Chapter 8), but are questionable, say, for many soil properties where the amount of variability is often at a much higher resolution than can be captured in sample data. GIS interpolation techniques cannot be applied directly to environmental data that are vectors (combined speed and direction), such as the tidal current data in the coastal oil spill example given in Chapter 6. Furthermore, finer resolution data usually requires process models to use more parameters in their solution. Taking a simple example again from hydrological modeling, when modeling large catchments, the characteristics of the channel network play a dominant role, while for small basins, the response is dominated by surface and subsurface flow on hill slopes (Beven, 2001), which requires the use of many more parameters to model and consequently more detailed field data. Problems of parameter estimation for an increased number of parameters for smaller units of discretization also serve to raise the level of uncertainty in model outputs and makes validation of the outputs almost as intractable. Other issues relate to the nature of GIS and state of spatial data technologies in the early 1980s. Vendors tended to offer either a raster or a vector solution. As was seen in the basin management application in Chapter 6, the ideal for environmental modeling is the flexibility offered by an integrated approach to both raster *and* vector. In the late 1980s, the options were pretty much limited to Genasys II or a combination of ArcInfo and ERDAS, all on UNIX workstations. Remote sensing (RS) data, which together with GPS have been largely responsible for overcoming the bottleneck in the availability in spatial data that existed not so long ago, tended to be still too coarse and presented problems in calibration for deriving quantitative data (McDonnell, 1996). Even so, the increased spatial resolution available today often increases scene noise as adjacent pixels respond to minor changes in ground surface properties and has not necessarily reduced uncertainty (Clark, 1998). Finally, the availability of digital elevation models (DEMs) used by GIS to prepare terrain data for environmental models were not nearly so readily available at that particular juncture.

Thus, in many ways, bringing together the two rather different technologies of GIS and environmental simulation modeling could have proved more troublesome and expensive than it was worth. Certainly in the beginning, consideration needed to be given as to how, in coupling the technologies, benefits could be derived through cross-fertilization and mutual support. For example:

- Making spatial representation explicit in solving environmental problems is indeed desirable and the way forward, but GIS lacked the explanatory or predictive tools (beyond logical manipulation) to analyze complex problems.

- Environmental modeling tools of the day generally lacked any spatial data handling and manipulation tools while GIS offered a reasonably standardized menu of spatial operators (Parks, 1993) mostly based on coordinate geometry and algebra. This would allow modelers to concentrate on modeling (Karimi and Houston, 1996).

- That the needs of both technologies was complementary and that by bringing them together they jointly become inherently more useful and robust in their solutions.

But, above all, the benefits needed to be forcibly pushed by researchers and practitioners in order to make them happen. In my own case, in Hong Kong, the arguments had to be repeatedly made and demonstrated to the client to be accepted (grudgingly at first, I felt) as an approach to environmental problem solving and which has now become common practice. Given that water moving through a landscape is highly sensitive to the spatial configuration and characteristics of that landscape, it is perhaps not surprising that hydrological and hydrogeological models were used most often to pioneer what became a paradigm shift for both GIS and hydrological modeling in the early 1990s (McDonnell, 1996). Indeed, as pointed out by Pozo-Vázquez et al. (1997), landform and land surface characteristics influence every hydrological process, land surface interactions, air temperature, wind, precipitation and so on. Over the past decade and a half, the issues in coupling GIS with environmental simulation models and the methods employed have changed in response to the evolving technological environment and our evolved thinking about what such a coupling means in the light of experience.

Initial Conceptualizations

In broad terms, GIS have provided environmental modelers with an ideal platform for spatial data integration, parameter estimation, and cartographic visualization, while environmental modeling has allowed GIS professionals to move beyond simple inventory and thematic mapping activities (Sui and Maggio, 1999). In practice, the degree of integration between the two technologies has tended to vary project by project. But, more to the point, what is meant by or how to specify any particular degree of "integration" is not so straightforward. Anselin et al. (1993) suggested a three-tiered classification based on the direction of the interaction between any two technologies: one-directional, two-directional, and dynamic two-directional (flexible data flow). Throughout the 1990s, however, it was generally accepted that for environmental simulation modeling, four levels of integration were possible using the following terminology: independent, loosely coupled, tightly coupled, and embedded (Fedra, 1993; Karimi and Houston, 1996; McDonnell, 1996; Sui and Maggio, 1999), although definitions vary slightly among these authors. These are represented diagrammatically in Figure 7.1.

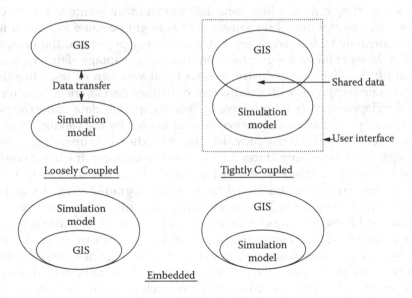

FIGURE 7.1
Initial conceptualizations of the levels of integration between GIS and environmental models.

Independent

This doesn't really represent a level of "integration" as such, but is included for completeness to cover those situations where GIS and environmental modeling are used together but independently on projects to achieve some common goal. In this context, GIS might be used to replace manual map measurements as traditionally carried out by modelers. Such measurements were invariably time consuming and prone to errors. Standard GIS functionality for measuring distances and areas could be used instead. GIS could also be used in parameter estimation for lumped models where, for example, dominant classes, spatially averaged, or interpolated values might be derived from relevant GIS coverage. The results from GIS usage would tend to be in the form of summary tables, which would then used as inputs to the environmental model.

Loosely Coupled

At this level of integration, GIS and an environmental model can share data files. GIS interaction with a dynamic simulation model is likely to be more than a once-only set of measurements or parameter determinations, particularly where the outputs of a number of scenarios may need to be visualized or further processed using GIS. Moreover, where parameter estimation is for distributed parameter models, a tabular approach to data exchange becomes extremely cumbersome. It is much better then to have some means by which both GIS and simulation models can share data files. More often than not, this entails exporting data files into some data format that is common to both GIS and environmental modeling software. This might be some formatted text file for attribute tables and raster matrices or, very popular at the time, the .dxf CAD format for vector graphics. One distinct advantage of this approach is that off-the-shelf and industry standard software can be used together, on the same computer, with a minimum of further development costs (even zero development costs if both have built-in compatible data import/export functionality). As each software becomes upgraded by the vendor, it can be brought into immediate use provided that a hardware or operating system incompatibility is not introduced in doing so (for example, the latest version of the GIS software might now only run under Windows XP or Vista, while perhaps the environmental model hasn't been upgraded since it was first compiled under Windows 3.11, or more likely, the chip and memory on your faithful workhorse PC cannot take the upgrade; even peripherals such as an older plotter might no longer be supported in the software upgrade). It is also possible to switch software completely as a result of new developments because of the particular characteristics of the problem to be solved or even for compatibility with some third party (research colleagues, client, or other consultants in a consortium). On the other hand, with each software running through its own interface, it becomes necessary to do GIS and simulation tasks one at a time in sequence, exchanging files and switching software at each stage.

Tightly Coupled

Under this level of integration, both software are run through a common interface that provides seamless access to GIS functionality and the environmental modeling. They may even share a common file format that avoids the need to translate files to an exchange format, but if not, a file management system provides seamless data sharing. There is a development cost in creating the common interface, but it brings about tangible advantages. First, off-the-shelf and industry standard software can still be used, as in the loosely coupled option, but avoids the need for the exchange files to be dealt with manually. This is important, as on large dynamic modeling projects, the number of these files can extend into hundreds, easily leading to mistakes in using the wrong file. This aspect plus the avoidance of alternately switching from one software to another can save considerable time and adds flexibility in running scenarios. Incremental development of the common interface and file management may be required with each software upgrade, which may not be at the same pace or timing for GIS and the environmental modeling. Also, if for reasons given above a different GIS package or environmental model needs to be substituted, the development effort has to be carried through again. Tight coupling in this way, therefore, tends to be implemented for stable situations where a large amount of work needs to be carried out over a period of time.

Embedded

A number of authors consider that an embedded level of integration is the same as tightly coupled and might indeed be so if defined as such. However, there is considerable difference in using GIS and environmental modeling through a common user interface and having either GIS functionality embedded in an environmental model or environmental modeling code embedded in a GIS package. For a start, one tends to dominate through the use of its interface as the only one used. Also, some of the embedded implementations can be partial, such as limited GIS functionality inside an environmental model. Embedded environmental models may also be in a simplified form. Often such embedding is carried out by vendors to make their products more attractive. Environmental simulation models are typically developed using mainstream programming languages (e.g., C++, FORTRAN, Visual Basic, Java) or advanced technical languages, such as MATLAB, which are not ideal, but offer a pragmatic solution for environmental modelers. However, a modeler would generally not want to attempt programming the spatial interactions and GIS functionality from scratch using these tools. Many GIS packages have built-in macro languages that allow modeling routines to be programmed or linked and are the means by which a number of environmental models have been implemented. These macro languages, however, are considerably slower than compiled code with any temporary or derived data layers having to be written to hard disk rather than stored in memory (Stocks and Wise, 2000). Consequently environmental modeling using these languages can be very slow in comparison with stand-alone models. Karssenberg and de Jong (2005a) conclude that, thus far, GIS have failed to become mainstream tools for the programming of environmental simulation models.

None of the above can be considered to be *fully integrated*, which would imply full GIS and environmental simulation modeling being developed as an integrated product. This would, however, be an expensive product to purchase and is yet to be achieved in any significant way for an entire range of issues discussed in the following section.

An Over-Simplification of the Issues

The types of coupling discussed in the previous section really only represent technical solutions as to how GIS and an environmental simulation model can share the same data rather than being integrated in terms of achieving compatible views of the world. In other words, this type of coupling has not necessarily led to an improvement in the scientific foundation of either GIS or environmental modeling (Grayson et al., 1993) even though it is generally recognized that there have been tangible benefits for both GIS users and modelers. The initial conceptualizations are also a rather simplistic view of the software/database environments that actually occur on projects with the need to work using spreadsheet, database, GIS software, statistical package, word processor, graphics package, RS image processing package, CAD, and environmental simulation model, not necessarily all simultaneously, but certainly at different stages of a project. There may also be more than one of each—specialized databases for specific types of data and perhaps several simulation models, one for each specific process; maybe even more than one GIS software package. The main difficulty in moving toward a more scientifically rigorous approach to the integrated use of GIS and environmental simulation models is their differing data models (Livingstone and Raper, 1994; Bennett, 1997a; Hellweger and Maidment, 1999; Aspinall and Pearson, 2000; van Niel and Lees, 2000; Bian, 2007). As discussed in Chapter 2, data models are abstractions of reality designed to capture the important and relevant features that will be required to solve a particular set of problems. In the case of GIS, the data model focuses on creating a digital representation of geographical space, the objects contained therein, and their spatial relations. The emphasis is on location, form, dimension, and topology. Environmental simulation models,

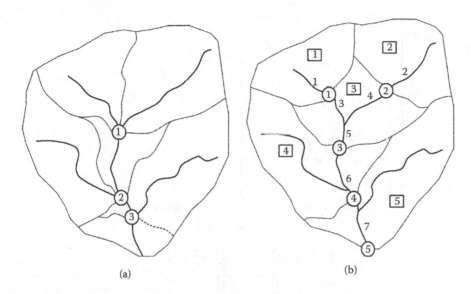

FIGURE 7.2
Views of a physical drainage basin: (a) geographical, (b) hydrological.

on the other hand, are predominantly concerned with spatial processes, their states, and throughput of quantities. One is a static representation, the other is concerned with dynamics. This means that their data models will be quite different and result in database structures (for the purpose of data manipulation) and databases (for the purposes of storage and retrieval) that are also quite different. Consider the examples given in Figure 7.2 and Figure 7.3 for a drainage basin. In both Figure 7.2(a) and Figure 7.2(b), the outer catchment boundary (watershed) and the streams within are the same, but the subdivision into subregions is quite different. Figure 7.2(a) gives the traditional geographical view of subcatchments as being the contributing area of overland flow to stream confluences (identified as 1 to 3). From a modeler's perspective, these subcatchments may not be homogeneous hydrological response units nor may the confluence itself represent a typical stream reach of relatively stable known properties (e.g., cross-sectional area, wetted perimeter) with which to model flow or from which to collect flow data. The modeler's view of the physical basin might thus resemble Figure 7.2(b). The geographical view in Figure 7.2(a) would perhaps result in a GIS representation of two data layers, one containing the streams as a network of lines and the other giving subcatchments as polygons (Figure 7.3(a)). The hydrologic simulation model, on the other hand, would require the elements of the drainage basin to be stored as subbasins, reaches, and junctions, as shown conceptually in Figure 7.3(b). In such an arrangement, the spatial dimension is only implicit with the data structure optimized for simulating how inputs (rainfall) are transformed into outputs (flow at the basin outlet) via their passage through the system.

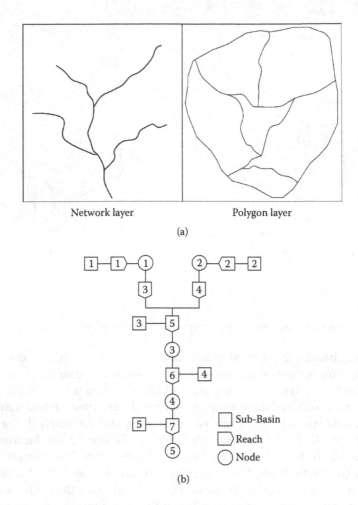

FIGURE 7.3
Contrasting data views: (a) as modeled using GIS, (b) in hydrological modeling.

Again, let us consider an ecological example. Figure 7.4(a) is a typical geographical view of vegetation mapped as succession communities. Each community is mapped with hard, nonoverlapping boundaries to make a polygon representation in GIS possible. While an ecologist might well be interested in the dynamics of the plant succession and its present state, there might not be recognition of distinct communities and they would rarely have abrupt boundaries. Instead, interest might well focus on species response to environmental gradients as in Figure 7.4(b). Furthermore, ecological simulation tends to assume homogeneous landscapes in modeling population dynamics (e.g., the panda–bamboo interaction model of Chapter 5). Within environmental simulation modeling, there is a wide range of data models because

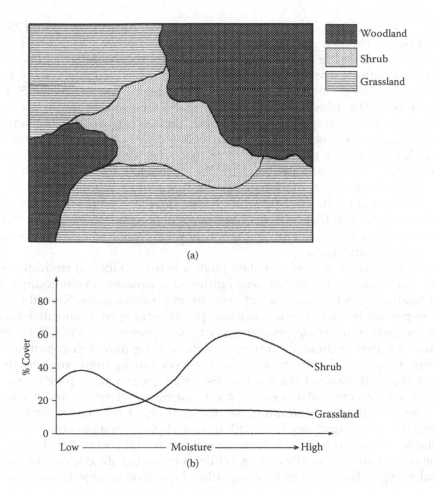

FIGURE 7.4
Views of vegetation: (a) geographical, (b) ecological.

each one will need to reflect the numerical methods used to solve the particular process model(s) being simulated.

There is another important way in which the data models of GIS and many environmental simulators differ, and that is the way in which they view flow or motion. For a modeler, there are two views to choose from, the Lagrangian view, which is dominant in GIS, and the Eulerian view, which is dominant in environmental simulation models (Maidment, 1993a; Sui and Maggio, 1999). Euler and Lagrange were both eighteenth-century mathematicians. A Lagrangian model of flow focuses on the object that is moving, such as tracking a car as it moves though the countryside. A Eulerian model of flow focuses on a fixed portion of space through or across which some motion takes place. This would be much like standing on the side of a road watching cars cross your field of view. A common analytical function in vector GIS is finding the shortest path (distance, time, or cost) between a starting point on a network and one or more destinations. Another, in raster GIS is tracing flow paths across a DEM. These are very much a Lagrangian view. But many simulation models of physical processes (as we saw in Chapter 5) are concerned with states and the changes in states for a specific length, area, or volume of bounded space. So, for example, a model might calculate changes in quantity over time for a specific area consequent on the rates of ingress and leakage across the boundaries of that area. This is a Eulerian view. While tessellations and networks in GIS are suitable spatial arrangements of bounded space for a Eulerian view of flows, there is a marked absence of corresponding functionality. Of course, there are some environmental simulation models that employ a Lagrangian approach, but, in general, there remains a marked dichotomy in the way flows are modeled in GIS and environmental simulations.

While, as we have seen, the data models between GIS and environmental simulation models and between different approaches to environmental simulation modeling stand apart, they are not irreconcilable. Nevertheless, as expressed by Livingstone and Raper (1994), making environmental models conform to the static, geometrically fixed representations of GIS for the sake of purely technical solutions to data sharing may well compromise them. Bian (2007) has discussed a similar issue arising from programming paradigms. Because of the need to discretize geographical space in order for it to be accommodated within a computing environment, and because of the prevailing programming paradigm of object orientation, there has been a growing tendency to regard all spatial phenomena as objects translatable into software objects. From an environmental perspective, however, the two do not necessarily equate. While some spatial objects, such as point and polygon features, can be encapsulated as software objects, such as in agent-based modeling (Chapter 5), caution should be exercised when applying a computing science paradigm to spatial phenomena best represented as continuous fields with attributes of spatial dependency. To treat such phenomena as discrete objects risks changing a fundamental conceptualization. In the example of flood simulation in Hong Kong given in Chapter 6, though the coupling was loose as a consequence of having to use commercially available state-of-the-art software both for GIS and the simulation modeling, the GIS data model for the project was largely dictated by and created specifically for compatibility with the data model of the hydraulic modeling software. In other words, a compromise was necessary, though GIS were largely there to serve the simulation modeling rather than the other way around.

Another criticism of the initial conceptualizations of technology coupling concerns the ability of GIS to include embedded environmental models. Notwithstanding the fundamental issues of divergent data models just discussed, the types of models that can be built within GIS are limited to what is possible within the software's internal analytical engine as accessed through the user interface (Wesseling et al., 1996; Karssenberg and de Jong, 2005a). The slowness of many in-built macro languages and constraints on dealing with intermediate states has already been mentioned. Few, if any, permit the type of feedback loops necessary to model dynamic systems. Also, as will be discussed in detail in Chapter 8, GIS are very limited in their ability to represent the fuzziness or imprecision inherent in most spatial data used in environmental modeling (Burrough and Frank, 1996). Thus representation of spatial phenomena in GIS are given an unnatural crispness and homogeneity not born out in reality and, hence, of limited use in studying trends and gradients (McDonnell, 1996). GIS analyses heavily rely on binary logic in which states become discreet (0, 1). GIS also have limited functionality in handling uncertainty inherent in the simulation and analysis of using spatial data (Burrough, 1986a; Goodchild and Gopal, 1989; Karssenberg and de Jong, 2005b). This will also be discussed in detail in Chapter 8. Its importance here relates to the ability of the user to assess the uncertainty in model outputs. Because these outputs can be the basis for making politically sensitive decisions, it is necessary to assess their reliability through uncertainty and sensitivity analyses (Heuvelink, 1998; Crosetto et al., 2002; Lilburne and Tarantola, 2009) for which vendor GIS packages have limited in-built functionality.

Other issues that need to be considered these days include the growing interoperability of software, such as within the Windows operating system, the fact that there is increasing use of network resources, and a growing number of geocomputational tools besides or in addition to GIS that can be used in environmental modeling (e.g., artificial neural network (ANN), agent-based models). All of the above have led to maturing conceptualizations of how GIS and environmental simulation models can be made to work more closely together within systems integration and geocomputational/geosimulation paradigms.

Maturing Conceptualizations

Despite the scientific separation between the way GIS and environmental simulation models are constructed and used, the continuing need to establish some level of integration beyond just being used together is driven by the growing recognition that integrated assessments of all aspects of the physical, biotic, social, and economic environment are required if sustainable solutions to problems are to be achieved (Clayton and Radcliffe, 1996; Aspinall and Pearson, 2000). One important change is a reconceptualization of where GIS and environmental simulation models stand in relation to each other (Clark, 1998). It should no longer be the case that one or the other represents the technological "heart" of the project. Rather, it should be the database or,

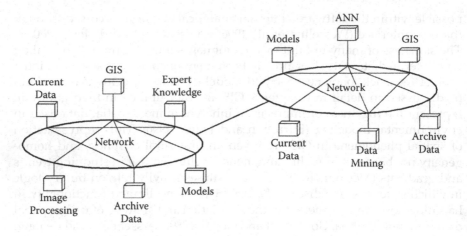

FIGURE 7.5
The network as the core technology.

more accurately, databases that form from which GIS hubs or portals, other geocomputational tools, and simulation models draw input data and submit processed output data. Such databases are increasingly accessible across networks and it is networks, particularly the Internet, that has become the key to the integrated use of diverse tools and data sets (Figure 7.5).

Integration versus Interoperability

First we need to make the subtle but important distinction between integration and interoperability. As we have seen above, the act of *integration* was to bring two rather different technologies or set of tools into closer proximity so they can work together, even perhaps as one. This could be done through sharing data and interfaces. Lilburne (1996) has suggested measuring the level of integration between GIS and another system with respect to three key components: user interface, data, and functionality. Of 104 cases studied, higher scores tended to be achieved for the integration of interface and data, but that, in general, these achieved poor scores when assessed on the integration of functionality. Given the arguments of the previous section, this should not come as a surprise. *Interoperability* on the other hand is the ability of client-side software applications to access a service (e.g., some specific functionality) from a server-side implementation such that it will respond as expected (Albrecht, 1996a). This is to do with software components that are interchangeable so that, within a specific hardware and operating system environment, groups of software can seamlessly operate together. Thus, for example, I can call MapInfo as an 'Insert, Object ...' right now as I am typing this chapter from within Microsoft Word (Figure 7.6). The toolbar becomes that of MapInfo and I have within Word some GIS functionality. Thus, interoperability is the bringing together of software at a more structural,

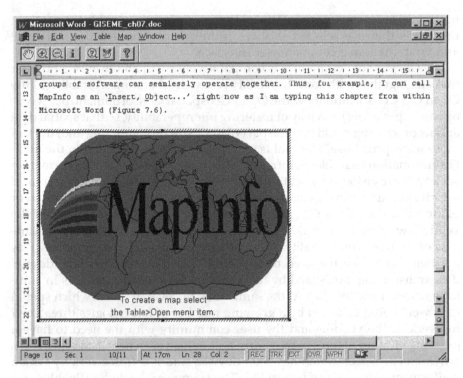

FIGURE 7.6
An example of interoperable functionality.

software developmental level than is usually achieved through the integration of independently developed software.

The dominance in the marketplace of a small number of operating systems, such as Microsoft Windows and Linux, together with the current prevalence of object-oriented (OO) programming and a high degree of standardization of Internet protocols for the access and communication of data, have all very much eased the convergence of once separate software types (database, spreadsheets, statistical packages, GIS, and so on) toward an environment of mutual interoperability. This brings with it three important advantages. The first is that much of the software tends to have the same look and feel, a shared principle of interface design using pull-down menus that have a common vocabulary and functionality as well as a familiar array of icons. For example, this icon: 🖼 almost universally means "open a file." Second, data access over networks is eased as the communication protocols, data extraction, and data transformation services have become largely transparent to the user. Data transformation services are important for environmental simulation modeling where the time steps and spatial scale of a data set in a data repository need to be transformed to that expected by the simulation model to be used. Thirdly, by using OO high-level languages, such as Visual Basic or Java, it is possible to construct programs that use selected services from one or more existing software and wrap them in a common user interface.

The GIS vendor community is comparatively small and in 1994 a number of them, together with some university research laboratories, established the Open GIS Consortium (now the Open Geospatial Consortium; http://www.opengeospatial.org) as a way of fostering interoperability, so that spatial data and its processing could become part of mainstream computing and, thus, of more widespread use. The goal is for GIS to become embedded in the main IT (information technology) infrastructure that supports today's information society. "Our vision is a world in which everyone benefits from geographical information and services made available across any network, application or platform" (OGC, 2002). GIS, as software, had been developed by each of the vendors with their own particular data models, data structures, and variants of the basic functionality. The further development of the industry was not being aided by the need to be continually transforming data (despite data transfer standards) and by having to understand the nuances in similarly named functionality. At the same time, the overall rate at which spatial data were being collected by a growing number of technologies threatened to overload the vendors and the user community with the need to have a growing number of tools to transform and upload each new data type. Thus, an Open GIS Specification has been devised and is a framework by which conformant software can be written. The framework includes (Buehler and McKee, 1998):

- A common means for digitally representing the Earth and Earth phenomena, mathematically and conceptually.
- A common model for implementing services for access, management, manipulation, representation, and sharing of geodata between information communities.
- A framework for using the Open Geodata Model and the Open GIS Services Model to solve not only the technical noninteroperability problem, but also the institutional noninteroperability problem.

The environmental simulation modeling community, on the other hand, is much larger and more diverse (multidisciplinary, in fact) and has not come together in the same way that the GIS community has. That is not to say that simulation modelers are not adopting open systems architecture or looking for greater interoperability with mainstream databases, spreadsheets, and statistical packages, it is more of an individual choice rather than an industry effort. Where does this leave GIS and environmental modeling? Both are working in increasingly interoperable environments with a growing commonality in their structural aspects. This makes deeper levels of integration more practicable particularly where multiple databases and models might be used in a computational framework. Wrapper interfaces have become much easier and quicker to design. With the GIS industry adopting an open systems architecture, software has been increasingly developed using a component-based approach. Such an approach has been facilitated by OO and means that instead of a monolithic software package it can be structured as a series of components that are independently developed, but ready-to-use units of software. These components can be called by a core program—usually driving the main interface—as and when needed. For that matter, components need not necessarily reside on the same machine, but can be called across a network. The component-based approach offers advantages when building complex systems (Bian and Hu, 2009), but it does mean that functionality and processes have to be capable of being decomposed down to primitive reusable units. GIS examples are ArcObjects from ESRI and MapObjects from MapInfo. Thus, using components it should be possible to directly incorporate selected GIS functionality that might be needed at particular stages of the modeling process, such as interpolation methods, into simulation models without having to "reinvent the wheel." However, many environmental simulation models remain closed monolithic systems limiting their flexibility and customizability. It also needs to be recognized that there are many so-called legacy systems (older software) still in use. Where such software was expensive and time-consuming to develop, the benefits of going through that effort again to produce an open, interoperable software environment may be hard to justify. In any case, the source code may be many thousands of lines and may either be unavailable (to protect proprietary ownership), written on media no longer readable with current equipment, or may appear an impenetrable, convoluted mass of unfamiliar code to a young researcher. These software programs carry on in use and still present, in such cases, very traditional problems of integration.

Environmental Modeling within GIS

Once again we need to make the distinction between linking GIS with established environmental simulation models and being able to carry out some forms of environmental modeling within GIS using cartographic processing and/or map algebra-type functionality. This distinction, though still relevant, has become blurred in recent years due to changes in the power of programming languages that are part of GIS and the advent of tools within GIS to assist programming efforts. The criticisms of speed and the difficulty of creating truly dynamic models discussed above still apply, but the effort to build embedded environmental models of increasing sophistication is ongoing. Stocks and Wise (2000) believe that the extended programming or scripting functionality (e.g., MapBasic in MapInfo) now available in one form or another for most off-the-shelf GIS packages, provide "an *a priori* case" for these tools to be used in implementing embedded environmental models. This is because these high-level programming languages make the input/

output, visualization, and computation of coverage quicker and easier to program than would be possible from scratch using FORTRAN, C++, or Java. One direction this has taken researchers and vendors is to provide GIS with flowchart-type model design front ends (similar to that of STELLA, as illustrated in Chapter 5, Figure 5.17(a/b) that allow users to graphically chain spatial operations into environmental models run *within* GIS. Prototype examples of these are the VGIS shell described by Albrecht (1996b) and SPMS described by Marr et al. (1998). A criticism leveled at these prototypes by Crow (2000) is that they only allow *static* cartographic processing, although it was noted that there was the intention to progress toward the inclusion of feedback loops so necessary for realistic environmental modeling and to implement links as OpenGIS compliant services. A vendor approach is the ModelBuilder first available in the Spatial Analyst 2.0 extension for ArcView and now in ArcGIS. This provides a wizard by which data and "processes" (raster data operations) can be chained together into complex models that automate all aspects of an analysis including input/output, data conversion (vector to raster), overlay, interpolation, and reclassification as well as map algebra-type operations. Figure 7.7 shows the ModelBuilder interface in which a very simple operation of obtaining a DEM and calculation of a slope map has been entered. Subsumed within this simple model of applying a function to data to derive a new data layer includes the specification of function parameters, applying class intervals, parameters for visualization of the

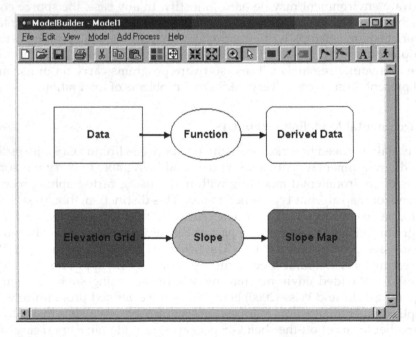

FIGURE 7.7
The ArcView ModelBuilder interface.

derived data, and its storage on disk. A more complex model for calculating wildfire hazard (first discussed in Chapter 5, Equation (5.2) and Figure 5.12), using two input data layers to derive three data layers that are numerically scaled (reclassified) and combined into a hazard model using weighted overlay, is illustrated in Figure 7.8.

The ModelBuilder, of course, is not restricted to environmental applications, but is certainly useful for static modeling of the environment within GIS. It is not yet a tool for building dynamic simulation models embedded in GIS. To date, the most successful GIS programming language for embedding environmental simulation models is PCRaster (van Deursen and Wesseling, 1995; Wesseling et al., 1996; Burrough, 1998; http://pcraster.geo. uu.nl/pcrwin32/). PCRaster is a scripting high-level language that allows either deterministic or stochastic modeling of dynamic physical processes over a landscape. It uses raster data and is an extension of map algebra and, in particular, the "cartographic modeling language" of Tomlin (1990). The extension functionality includes time series of changes to attributes at specific locations and the use of stacks of raster maps to represent the status of a model at different time steps with control on the duration of each time step. Illustrated in Figure 7.9 is three source layers—DEM, soils, and rainfall stations—that together with a time series of rainfall for each station, are used to calculate a series of six-hourly runoff maps (four of which are included here) that show the runoff increasing, reaching a maximum, and then abating in response to a rainstorm. Figure 7.9 has been compiled from PCRaster demo data. PCRaster has been experimentally extended by using Python in order to carry out dynamic modeling in 3D (Karssenberg and de Jong, 2005a). Thus, using a voxel structure (3D pixels), it is possible to model topographic (DEM) change in response to erosion and deposition over time.

Model Management

This development is one in which levels of integration between GIS and environmental simulation models is sought through the use of what Bennett (1997a; 1997b) terms "modelbase management," that is, "a knowledge-driven spatial modeling framework." This is built on the work of Armstrong (1991) who put forward a three-fold classification of types of stored knowledge in the spatial domain: *geometrical*, *structural*, and *procedural*, though this was originally in the context of cartographic generalization. Here these types of knowledge are given a broader GIS and environmental modeling context. Geometrical knowledge refers to the location, dimension, and spatial relationships (topology) of geographical features—the graphical element of a GIS database. Structural knowledge refers to the additional thematic (attribute) information that provides for a further understanding of the form and characteristics of the geographical features. Procedural knowledge refers to the understanding we have of how physical (and social) processes operate over time and space as expressed in simulation models. By including any

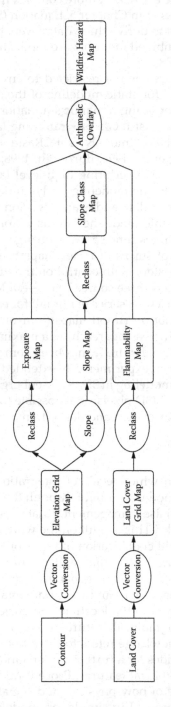

FIGURE 7.8
A wildfire hazard model using the ModelBuilder interface.

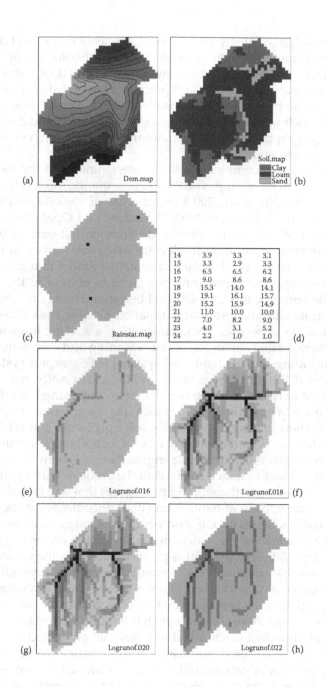

FIGURE 7.9
An illustration of progressive runoff mapping using PCRaster: (a) DEM, (b) soils, (c) rainfall stations, (d) time-series rainfall data for each rainfall station, (e) to (h) runoff at time steps 16, 18, 20, and 22, respectively. (PCRaster demo data from D. Karssenberg, E. Wesseling, P.A. Burrough, and W.P.A. Van Deursen, "A simplified hydrological runoff model." Available at http://pcraster.geo.uu.nl/pcraster/models/catsop/index.html.)

type of knowledge implicitly into a model, it makes that model difficult to modify or integrate with other models and/or technologies. By taking an OO approach to modeling features, say a subwatershed or a strip of coastline, the geometrical, structural, and procedural knowledge relevant to the specific object can be explicitly encapsulated into the object (see Chapter 2). A "menu" of such objects then can act as a database of model components from which geographically explicit environmental simulation models can be more easily created and modified.

Other research that complement these developments in the model management approach, have proposed the use of semantic models (Livingstone and Raper, 1994; Villa et al., 2009) and the use of various forms of artificial intelligence (AI) (e.g., Lam, 1993; Openshaw and Openshaw, 1997; Holt and Benwell, 1999). The fundamental differences that exist between GIS and environmental simulation models (as discussed earlier in this chapter) could be overcome by the use of a semantic model that reconciles the different meanings and representations given to objects in the two technologies. The semantic model is OO and becomes part of the interface that acts to integrate GIS and external simulation models, thus allowing entities in a database to be used independently of a specific implementation. AI is the automation of the processes of reasoning and comes in a number of guises including knowledge-based systems (KBS), expert systems, *case-based reasoning* (CBR), fuzzy reasoning, and the use of ANN. An expert system comprised of two databases, one database containing both declarative knowledge about a specific domain (much like a textbook) and procedural knowledge on how to develop strategies, solve problems, and achieve goals. The other database contains the known set of facts about a specific problem that needs to be solved. An inference engine then uses both databases interactively with the user to match the facts of the problem with the declarative knowledge and return both facts and procedures that will (should) result in a solution to the problem at hand. Lam (1993) documents the use of expert systems in conjunction with GIS and environmental simulation modeling. Fish damage in a lake can be calculated according to changes in pH level, but it is the calculation of pH that is problematic given the choice of using at least six possible combinations of simulation models depending on the specific characteristics of the lake. Lam used the expert system to determine the selection of the best combination of models to be used on any one lake. CBR is not too dissimilar except that the declarative and procedural knowledge are bundled or chunked as cases that are stored in a "library" of cases. These in one sense are historical and spatial analogs, using past cases to explain new situations and, hence, provide solutions or even going farther to critique some new solution on the basis of past cases to test its robustness before it is applied. Holt and Benwell (1999) use CBR as a means of classifying soils from basic soil survey and other landform parameters. Villa et al. (2009) propose moving beyond declarative modeling to semantic environmental modeling in which all concepts used in a model of a natural system are explicitly defined and embedded using ontologies (Chapter 3). They illustrate this concept using a predator–prey model not dissimilar to the STELLA model in Chapter 5. This knowledge-driven approach is likely to ease the path toward a more flexible component approach to building and integrating environmental simulation models.

Maturing Typology of Integration

Given the developments in technology and the evolution of approaches to coupling GIS and environmental simulation models, Brandmeyer and Karimi (2000) have proposed a revised hierarchical typology of integration (Figure 7.10). They frame their discussion around the use of "two distinct models" that are not framed necessarily as different data models, but as one or more programs operating within specialized computing environments, which could be GIS, database, or spreadsheet. For our present context, I would prefer to consider two or more distinct *technologies*, be these GIS, databases, simulation models, etc., where commonality of underlying data models and data handling cannot be assumed.

One-Way Data Transfer

This is the lowest level of integration in which technologies are used separately, usually because they have been compiled in incompatible languages, have different data storage formats, or need to be run under different operating systems. In these cases, all that might be possible is a somewhat laborious process of extracting and translating data from one technology to another in only one direction, such as from GIS to a simulation model. A small utility program may need to be written in order to do this. While cumbersome and sometimes error-prone, this is, however, a low-cost solution to using legacy software.

Loose Coupling

This level of technology integration is very similar to "loosely coupled" discussed above. Here a two-way exchange of data is possible, usually in an automated fashion, so that the technologies can be used dynamically. Thus, a GIS may be used to preprocess data, which are then passed to a simulation model; the results of the simulation are then passed back in turn to the GIS package for postprocessing as in the basin management example given in Chapter 6. Complementary simulation models can also be coupled in this way to provide a means of feedback between the models. Thus, Lofgren (1995) loosely coupled a general circulation model and a vegetation cover model with predicted climate used to determine equilibrium vegetation, the surface albedo of which was then used as feedback to the general circulation model.

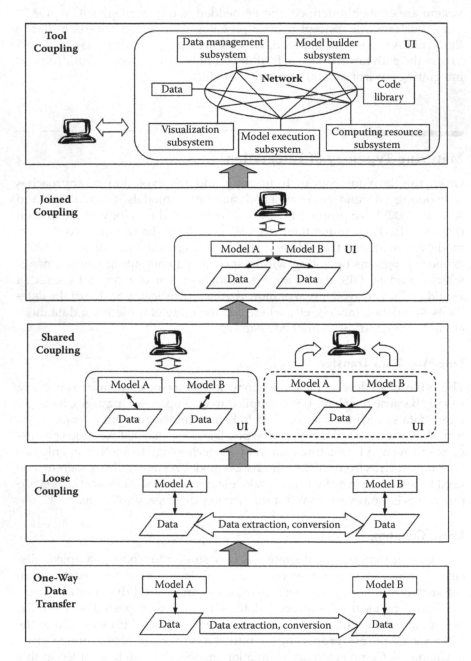

FIGURE 7.10
A maturing typology of GIS and environmental model integration. Note: UI = user interface. (Based on Brandmeyer, J.E., and Karimi, H.A. (2000) *Environmental Modeling & Software* 15:479–488.)

Shared Coupling

At this level of integration, the technologies share a major component in their architecture, usually either the data storage (*data coupling*) or the interface (*interface coupling*). Software languages, such as Visual Basic or Java, can be used to construct wrappers that provide a common interface to two or more separate technologies, particularly so where good interoperable potential exists between them. The technologies need not necessarily be resident on the same computer, but might be operated over a network (such as for a server-side database) or even over the Internet. The user becomes unaware that two or more technologies are in use. Under data coupling, the technologies continue to be used separately, but both share the same data that are used directly from the same database. This requires that the technologies will be using the same or very similar data models or that their separate schemas can be called from the same database. Windows provides a number of methods for interoperable data sharing, such as *dynamic data exchange* (DDE), *open database connectivity* (ODBC), and *object linking and embedding* (OLE). Of course, using these tools does restrict the data coupling to the types of data that these tools support (many geospatial data types are not well supported).

Joined Coupling

Here, as well as the features of shared coupling (at this level both common user interface and common database), the separate technologies themselves become further integrated either because one technology is embedded in another (*embedded coupling*) or because they are integrated as peers (*integrated coupling*). In the present context, embedded coupling usually occurs as simulation modeling *within* GIS using the built-in scripting languages. The integrated coupling is increasingly facilitated by interoperable services and initiatives, such as Open GIS, both discussed above, but to date few examples have emerged. Of note, however, though not incorporating GIS functionality, is CHASM (combined *hydrological model* and *slope stability model*) as integrated peers (Wilkinson et al., 2000), which allows the dynamic effects of pore-water pressure in response to specific rainfall to be incorporated into slope stability analyses.

Tool Coupling

This is the highest level of integration under the typology of Brandmeyer and Karimi. Tool coupling is really a modeling framework having integral subsystems accessed through a common user interface. The subsystems could be resident on the same computer, but could also conceivably be distributed over a network. Within the framework can be both joined and shared coupling within and between the subsystems. Subsystems can be specific to data management, spatial data processing, model building and management, model execution, quality management (see Chapters 8 and 9), and visualization. Such a framework could support a community of modelers working on a range of more complex and interrelated environmental issues rather than the narrow issues of a single discipline. A worthy goal indeed but accompanied by a high cost for framework design and development. Li et al. (2008) and Brimicombe et al. (2009) have proposed an agent-based distributed services approach to tool coupling. A software service is a piece of computer code that renders some form of assistance to a user in carrying out tasks. Obvious examples are procedural wizards that assist in setting up a new database, importing/exporting data, customizing views, etc. Agent technologies as multiagent systems can also be deployed to provide services in the spatial domain. Such agent-based services are an enhancement of procedural services in that they can have autonomous behavior, network mobility, goal-directed behavior, and work collaboratively to solve and carry out tasks. In the spatial domain, agent-based services have been primarily deployed in data integration and management over a network (e.g., Tsou and Buttenfield, 2002; Sengupta and Bennett, 2007). Brimicombe et al. (2009) demonstrate how agents can be used to find and assemble distributed components over a network in order to carry out specific modeling tasks, and if necessary to migrate tasks across a network should additional computing resources be required. Agent technologies are particularly suited to achieving interoperability in heterogeneous computational environments and in tool coupling GIS and numerical simulation models.

De facto Practices

In Chapter 3, we discussed an evolutionary model of operational GIS that moved from inventory activities in the early stages through analysis and on to management in the final stage (Crain and Macdonald, 1984). If the tendency is toward the use of GIS for (or with) modeling and simulation in a decision support system environment (Chapter 10), then the coupling of GIS with environmental simulation models becomes integral to that evolution. We are, however, not yet at the end of that particular road. There is a lot of research and experimentation taking place with shared, joined, and tool coupling that have some way to go before stable, tool coupled products appear in the market. In the meantime, current *de facto* practice is focused on using GIS and environmental simulation models as loose or shared couplings where the technologies remain distinct. In general, notwithstanding environmental modeling within GIS, the role of GIS to environmental simulation modeling has been (Goodchild, 1993; Clark, 1998):

- Preprocessing of spatial data to prepare them as inputs to simulation models. This can involve the integration of different data sets, transformations, such as vector to raster, buffering, or the creation of new layers through overlay, Boolean selection, or map algebra.
- Assisting in modeling tasks, such as calibration and scenario building.
- Postprocessing the outputs of the simulation for visualization and possible further analysis using cartographic processing or map algebra, e.g., to ascertain impacts on settlements, land uses, and so on.

While the role of GIS is not limited to these three functions, it forms the predominant mode of employment. The key issue of how the coupled technologies perform as spatial decision support system (SDSS) is left for Chapter 10.

8

Data and Information Quality Issues

Let us start off with a small thought experiment. To begin with, we should recognize that it is highly unlikely that any data set is 100% accurate. Then, suppose in the example landscape we have been using in previous chapters (e.g., Chapter 2, Figure 2.5), each layer is 90% (0.9) accurate. If we were to combine in an overlay of the geology and the land cover, we would end up with a map that is [0.9 and 0.9] accurate, which in probability terms would be $0.9 \times 0.9 = 0.81$, or 81% correct. Add another layer to the overlay and the result might theoretically be only 73% correct, and by the time we have used seven different layers in the analysis, our output product might be less than 50% correct. What then if this final map was used as input data for an environmental simulation? Of course, things are unlikely to be quite this bad in practice and, besides, plenty of errors (often unnoticed) were made in using traditional paper maps. Nevertheless, a good understanding of data quality issues is a key to informed use of geographical information systems (GIS) and environmental modeling. This chapter will tend to focus on issues of *spatial* data quality as these pose special problems in addition to those encountered in nonspatial data. As we saw in Chapter 3, spatial data quality is a fundamental concern of geo-information (GI) science. While considerable research is ongoing in this area, there is already a sizeable literature. For greater detail than provided here, the reader can refer to: Goodchild and Gopal (1989), Burrough and Frank (1996), Burrough and McDonell (1998), Shi et al. (2002), and Brown and Heuvelink (2008) for GIS perspectives; Heuvelink (1998) for an in-depth GIS and environmental modeling perspective; Li et al. (2000) for a process model perspective; Elith et al. (2002), McIntosh (2003), and Lowry et al. (2008) for an ecological perspective.

The Issue Is ... Uncertainty

A universal concern with all information systems must be the quality of the data contained within them, hence, the well-known adage of the computer age: garbage in, garbage out. Nevertheless, it should be recognized that "errors and uncertainty are facts of life in all information systems" (Openshaw, 1989). The process of describing aspects of reality as a file structure on storage media requires a high level of abstraction, as was illustrated

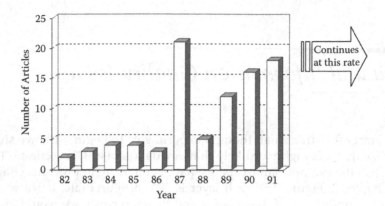

FIGURE 8.1
A graph showing the rise of interest in data quality issues within the GIS literature.

in Chapter 2. Thus, any attempt to completely represent reality in GIS, while no doubt resulting in robust and flexible data sets, would also result in large, complex, and costly data sets that would require a higher order of technology to handle them. Historically, a detailed consideration of data quality issues in GIS lagged considerably behind the mainstream of GIS development and application. This is evident from the growth of the relevant literature, which underscores a sudden vogue in spatial data quality research from 1987 onward, some 25 years after the introduction of GIS (Figure 8.1). This lag in concern for spatial data quality may be attributed to:

- The inherent trust most users have in computer output, particularly after some complex analysis.
- The possible lack of awareness among operators and managers from nonspatial disciplines of the sources of uncertainty in spatial data sets and the consequences of propagating them through analyses, other than the need to correct blunders.
- The growing desire in the late 1980s for remote sensing (RS) and GIS data integration, there having been already a body of research on accuracy assessment of RS data.
- The growth of GIS through stages of inventory, analysis, and management (Crain and Macdonald, 1984) such that a need to consider the consequences of uncertainty in outcomes on decision making may only become apparent after some years of system development.

Data are usually collected within a specific context and the design for any primary data collection is usually specified within that context. Surveyor and user may be the same individual, part of the same team or linked by contract. Thus, the chances for misinterpretation of outcomes or misconceptions concerning accuracy of the data should, in theory, be quite small. But, data are likely to have a life span (shelf life) well beyond the original context and may well be used as secondary data on other projects. Those who collected the data may be unaware of subsequent uses (or misuses) to which their data are put. Most of the early literature on GIS data quality was concerned with the accuracy of data sets, or more specifically, the recognition and avoidance of error. We will be taking a wider view of this issue by considering the level of uncertainty that exists in the use of spatial data and the fitness-for-use of GIS outputs.

Error is the deviation of observations and computations from the truth or what is perceived as the truth. This assumes that an objective truth can be known and measured. Statistically, errors may be identified as gross (outliers, blunders), systematic (uniform shift, bias), or random (normally distributed about the true value). *Accuracy*, the logical dual of error, is the degree of conformance of our observations and computations with that truth. *Precision* is the level of consistent detail with which observations can be made (often synonymous with the number of decimal places being used, though this can be misleading as large numbers of decimal places can be spurious). Imagine throwing darts at a dart board—getting a dart in the exact center is being accurate, getting all the darts on the board close together is high precision even though they may be nowhere near the center. As regards the quality of GIS products, however, it is necessary to make a clear distinction between the usefulness of data inputs and of analytical outputs. *Reliability* concerns the trust or confidence given to a set of input data on the basis of available *metadata* (data about the data: its lineage, consistency, completeness, and purported accuracy) and upon inspection of the data by the user. It refers to the assessed quality of the data on receipt. The user can then judge its appropriateness for use in a particular context. *Fitness-for-use*, however, refers to the assessed quality of the products of analyses used in decision making. Such evaluations and judgments must necessarily be the responsibility of the user (Chrisman, 1982). Where only a single theme is used, then fitness-for-use can be judged directly from the data's reliability. However, where data sets are integrated and themes are combined or transformed, then the analytical outputs are characterized by combination and propagation of the data reliabilities of the individual themes. In as much as research has focused on quality measures for data reliability, progress in the derivation of quality measures, meaningful in the evaluation of fitness-for-use, has been slower.

Uncertainty in its broadest sense can be used as a global term to encompass any facet of the data, its collection, its storage, its manipulation, or its presentation as information that may raise concern, doubt, or skepticism in the mind of the user as to the nature or validity of the results or intended message. Theoretically, this definition would also include mishandling of the data through improper analysis, inappropriate or erroneous use of GIS functions, poor cartographic technique, and so on. Thus, in the context of environmental modeling, Burrough et al. (1996) consider the quality of GIS

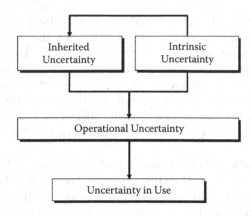

FIGURE 8.2
Main sources of uncertainty within spatial data.

informational output to be a function of both model quality and data quality. Modeling issues will be discussed separately in Chapter 9. The term *uncertainty*, therefore, will be used to refer to the inevitable inaccuracies, inexactness, or inadequacies that exist in most spatial data sets and their resultant propagation through analyses to adversely affect the usefulness of results and certainty in decision making. Four broad categories of uncertainty are given in Figure 8.2. *Intrinsic* and *inherited* uncertainties are those associated with primary and secondary methods of data collection (Thapa and Burtch, 1990), respectively. Secondary data (e.g., existing maps) will also have an element of intrinsic uncertainty. Once the data are used within GIS, intrinsic and inherited uncertainty will be propagated and additional uncertainty may be derived due to the nature of hardware and software. This is *operational* uncertainty. The resulting levels of uncertainty, if not quantified or in some way known, may lead to overconfident, uncertain, or erroneous decision making. Uncertainty or error in use may also derive from different perceptions or misinterpretation of the output information on the part of the user (Beard, 1989).

To say that GIS users were somewhat tardy in recognizing data quality as an issue is not to imply that data quality was altogether neglected. Geodetic, cadastral, and topographic surveying, as sources of GIS data, have a long tradition of assessing the accuracy of their numerical and graphic products. Cartographers, too, have long striven to reduce error to tolerable levels, though the task is somewhat less tangible because they must also address the effectiveness of their maps in the communication process (Muller, 1987). In RS, the appropriate methods for assessing the accuracy of classified imagery were subject to debate from the mid-1970s onwards (see, for example, Hord and Brooner, 1976; Turk, 1979; Congalton and Mead, 1983; Story and Congalton, 1986). Initially, this debate was separate from GIS, but came together in the early 1990s with the move toward integrating RS and GIS data (Lunetta et al., 1991). Nevertheless, despite the progress made, in practical terms it remains an issue today in that many users do not routinely assess the uncertainty of their data and of their analytical results.

Early Warnings

MacDougall (1975), writing in the context of paper maps, provided an early warning of what to expect from cartographic processing in GIS: "It is quite possible that map overlays by their very nature are so inaccurate as to be useless and perhaps misleading for planning." The concern was that maps, while having individually adequate planimetric accuracy and "purity," may when combined produce a seriously degraded product. Chrisman (1987), however, considered MacDougall's analysis to be "dangerously oversimplified." Another early concern lay with the nature of the data themselves. Most GIS are reductionist and support parametric forms of enquiry, which force data into well-defined categories even though in reality they may lie along a continuum (e.g., mutually exclusive land use classes or soil types). Boundaries (within and between layers) can be unintentionally misleading in that they imply both spatial homogeneity within a polygon and equal homogeneity for all areas of the same class (Robinove, 1981). Even though methods of classification, categorization, and boundary definition may be known for each layer, their effects on the validity of results due to combination may be very difficult to assess with scope for misinterpretation of the results.

Many of these problems already existed in paper thematic maps. Thus, "the essence of mapping is to delineate areas that are homogeneous or acceptably heterogeneous for the intended purpose of the map" (Varnes, 1974). This again leads to a reductionist process of defining a hierarchical structure of classes, assigning each individual to a class, and placement of the classified individual in its correct position (Robinove, 1981). However, "the objective is not to accurately represent the real world, but rather to show a simplified model of reality, an abstraction which helps separate the relevant message from the unwanted details" (Muller, 1987). Not only will such maps contain factual information, but may also include hypothesis and synthesis. Deviations from reality are intentional and controlled and can be achieved through appropriate use of symbolization and scale so as not to compromise the authority of the product. Though there are no accuracy standards for thematic maps, they are less open to user abuse (though not immune) in paper form due to the symbolization and scale being fixed. Fixed scale also applies to paper topographic maps and accuracy tests are scale dependent (see below). Transferred to a digital environment certain safeguards inherent in paper maps, such as the fixed scale, were removed, though traditional attitudes toward spatial data persisted: If it wasn't a problem in the past, so why is it now (Openshaw, 1989)? There is also evidence that many users poorly understood the accuracy of paper maps, often attributing them with higher levels of accuracy than warranted. "Data for which no record of its precision and reliability exists should be suspect" (Sinton, 1978). While there were accuracy standards developed for primary data collection, in secondary data collation (e.g., digitizing of existing maps) the assumption seems to have been that since the source documents were originally compiled to the prevailing standards (or were somehow authoritative), they could also be converted to GIS and used without problem. The ability to change scale in the digital environment and combine data sources at will was viewed only as a positive advantage. There appears to have been little regard for, or understanding of, the cumulative effects of these data combinations on the quality of the informational outputs.

Three widely reported studies seemed to clinch it for the GIS community. Blakemore (1984) tested an actual database of employment office areas and manufacturing establishments for which it was necessary to identify which manufacturer (a point) fell within which employment office area (a polygon); in other words, a standard point in a polygon test. By making basic assumptions regarding data input accuracy for both the points and the digitized polygon boundaries (ε or epsilon band, as illustrated in Figure 8.3), Blakemore found that only 55% of points could be unambiguously assigned. Newcomer and Szajgin (1984) tested error accumulation as probabilities when combining raster data layers. Their conclusion was similar to that of MacDougall (1975): a general rule of thumb that the result of an analysis would generally be less accurate than the least accurate layer used. Then Walsh et al. (1987) extended Newcomer and Szajgin's analysis by using typical data sets for that time. By using more than two layers and by varying cell size, they found that

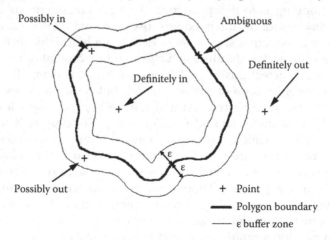

FIGURE 8.3
ε accuracy band for a digitized boundary and resulting uncertainty in a point in a polygon test. (Based on Blakemore, M. (1984) *Cartographica* 21: 131–139.)

errors inherent in the data and additional errors introduced through combining layers led to a sufficient total error to render composite maps highly
inaccurate. For two-layer combinations, the highest accuracy was 29% and
the lowest 11%, while for three layer combinations, the accuracy ranged from
11% to just 6%. Though the situation today with higher resolution and more
accurate spatial data is unlikely to result in such poor analytical products,
such studies were nevertheless a very clear indication of the potential seri-
ousness of the problem.

So, How Come ... ?

To begin with, no observation of geographical, environmental, or social phenomena is perfect. There are a number of quite valid reasons for this that can
be classified as follows:

- Imperfect measurement.
- Digital representation of phenomena.
- Natural variation.
- Subjective judgment and context.
- Semantic confusion.

Imperfect Measurement

Equipment used for measurement, whether it be the time-honored tape measure, weighing scales, or advanced uses of the laser beam, all have their design
accuracy and level of precision, some equipment requiring calibration before
use. Most equipment, if used for repeated observations of the same static
object or phenomenon, will result in a cluster of measurements normally
distributed around the true value and spread in relation to the equipment's
precision. There is a tendency for this to be compounded by errors or bias in
recording measurements. A gross error or blunder, such as putting the decimal point in the wrong place or writing too many zeros, should be noticeable
as an extreme value or outlier when all the measurements are graphically
plotted, such as by using box plots. Bias often occurs through unintentional
rounding of the observations, say upward to the nearest 0.5 or to the nearest integer. Bias can occur if equipment is poorly calibrated or is adversely
affected by temperature and/or humidity resulting in a shift in the readings.
Bias can result through the use of an inappropriate sampling scheme for the
phenomenon being measured. Harvey (1973) and Hirzel and Guisan (2002)
constructively summarize sampling issues. Bias in measurements can also
result from temporal and spatial autocorrelation (Goodchild, 1986; Griffith,

1987) in which there is a tendency for spatially or temporally neighboring observations to have either a markedly greater (positive autocorrelation) or markedly lesser (negative autocorrelation) similarity that might be expected from a purely random association of such observations. This arises out of the issue of spatial dependence discussed in Chapter 4 in relation to modeling topographic surfaces. These biases tend to result in systematic errors that are more difficult to detect and correct, but which, as we will see later, can compound during analyses and raise the level of uncertainty. It is also worth mentioning here the 'small number' problem. This often crops up, for example, in population dynamics when working with proportional data. A population of two that increases by two has increased by 100%, whereas an increase of two in a population of 100 increases by only 2%. Though not a measurement error, this is a measurement scale effect of working with small numbers where proportional increases can appear disproportionately large, but nevertheless result in biased analyses.

Digital Representation of Phenomena

In Chapter 2, we saw how the digital representation of reality was driven by the data model, which is then translated into a data structure and finally into a file structure. In GIS, we have basically two ways of representing spatial data: raster or vector. In the raster approach, accuracy is going to be limited by cell size, and while the concept of accuracy is essentially independent of the issue of resolution, cell size will limit the minimum error that can be measured (Chrisman, 1991). In vector, points that subtend lines will be recorded with high precision regardless of rounding at the time of measurement. Thus, a line ending at a point with x coordinate measured as 12.3 rounded to the nearest 0.1 nevertheless may be stored implicitly as 12.300000 and would be considered as not joining to another line ending at 12.300001, though this would not be warranted within the initial precision of measurement. When lines are snapped to form polygons, the software forces the point of snap to be exactly the same number for both lines to the nth decimal place. However, different hardware may handle floating-point arithmetic differently, so, for example, I have found in moving data sets from an IBM server to a Sun server and vice versa (both using the same GIS software in a UNIX operating system) that some polygons no longer snap—it's the nth decimal place. The same goes for attribute data. Whether vector or raster is used, space needs to be partitioned into discrete chunks in order to be handled digitally. A cell or a polygon needs defined boundaries whether implicit or explicit. Each cell or polygon is homogeneously defined as belonging to a particular class leading to abrupt change from one class to another at the boundaries. This reflects the reductionist nature of GIS already mentioned above where there is a tendency to reduce the complexity of the real world to discrete spaces characterized by a number of discrete classes. Another effect of digital representation is that you can zoom in, and zoom in yet farther and all the lines remain pixel-thin implying a high level of accuracy that is again not warranted. I call this the "infinite zoom" problem. This leads users to believe that they can overlay data surveyed at, say, 1:1000 scale with data surveyed at 1:100,000 scale, whereas they would never attempt to do this with paper maps.

Natural Variation

Natural variation exists in our landscapes because of the complexity of the systems at work and the multiplicity of causal factors that operate. These causal factors vary both singly and in combination along environmental gradients. They are also reinforced or dampened by internal feedback loops. Temporal variability exists due to fluctuations in the external environment. Our tendency is to handle such complexity through inductive generalization into discrete, mutually exclusive, homogeneous classes (the dominant GIS mode of data modeling discussed above), but we are merely creating an interpretation of reality. Just as models are simplifications of reality, so too are our mapped representations. Hence, Burrough's (1986a) statement that "many soil scientists and geographers know from field experience that carefully drawn boundaries and contour lines on maps are elegant misrepresentations of changes that are often gradual, vague, or fuzzy." It is our intuitive need to distinguish boundaries within continua that lead to many spatial data problems (see Figure 8.4).

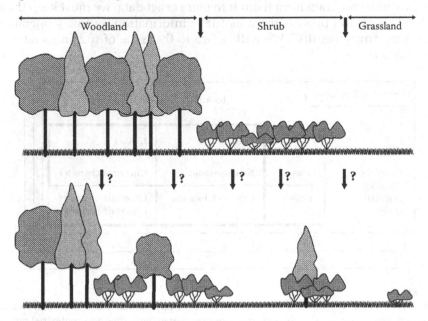

FIGURE 8.4
Problems of interpreting natural variation into discrete classes.

Suppose we had decided to map vegetation in three classes: woodland, shrub, and grassland. In the upper part of Figure 8.4, our job is easy; there are homogeneous groups of the three vegetation types to which we can affix boundaries with reasonable certainty. In my 30 years of mapping vegetation, such "convenient" natural landscapes are infrequent. Most of the time it looks like the lower part of Figure 8.4, but the mapping task still needs to be done. We could introduce more classes, such as "shrub with trees" and "shrub with grass," and all the other possible combinations that exist. However, this adds a level of complexity that doesn't necessarily ease our problem because how many trees do we need before we abruptly change from "shrub" to "shrub with trees"? Our maps may end up with a myriad small polygons that make analysis orders of magnitude more difficult. Some would say it's a matter of scale, you just need a finer resolution. I defy anybody to out and unambiguously peg on the ground the boundary to some woodland: Do you take the tree trunks, the extent of the crown, what about the roots? No, more often than not we just have to deal with it and interpret some boundaries. This then leads to series of polygons that users then interpret as homogeneous woodland, shrub, and grassland when in reality there is a degree of heterogeneity. Figure 8.5 gives a contingency table of uncertainties that can arise consequent on our treatment of natural variation. As Bouille (1982) states: "Most of the phenomena that we deal with … are imperfectly organized, incompletely structured, not exactly accurate, etc. In a word, the phenomena are 'fuzzy.' However, we must not reject fuzzy data, we must not transform them into more exact data; we must keep them as fuzzy data and process them as fuzzy information by fuzzy operators producing fuzzy results." We will return to the issue of fuzziness later in this chapter.

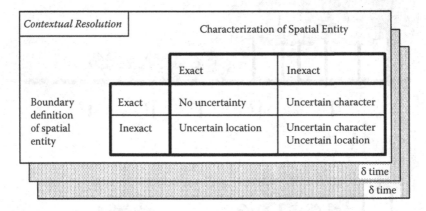

FIGURE 8.5
Shown is a contingency table of uncertainties arising from boundary definition and characterization of spatial objects. These uncertainties may change with time and contextual resolution. (Modified from Robinson, V.B., and Frank, A.U. (1985) *Proceedings from the AutoCarto 7*, Washington, D.C., pp. 440—449.)

Subjective Judgment and Context

We have already touched on this in looking at natural variation. While some methods of data collection may appear highly objective (e.g., laboratory measurement of soil pH values from field sampling), the interpretation of what these results mean is often a matter of judgment. Other forms of data collection are often highly subjective and dependent upon expert knowledge. Let us take aerial photographic interpretation (API) as an example. A lot of spatial data is collected in this way. A number of authors have studied the consistency and correctness of API. Congalton and Mead (1983) tested five interpreters on classes of tree cover. Although the results differed, they were found to be not significantly different at the 95% confidence level. Yet Drummond (1987), on a more varied test of nine land use classes carried out by five experienced, midcareer professionals, found that the superimposed results had considerable variability. Where contrasting land uses were juxtaposed (e.g., an area of agriculture in a woodland clearing), boundary conformity was high. Villages, on the other hand, which in this area tend to have diffuse boundaries as well as classes, such as "fallow bush" (which can easily be confused with other cover types), tended to have low boundary conformity. Fookes et al. (1991) were able to qualitatively compare eight interpretations carried for the Ok Ma dam site, Papua, New Guinea, where a 35-million m^3 landslide occurred as a result of construction work. The interpretations, which required a high level of skill, were very different both in style of presentation and in their conclusions. Fookes et al. went on to note that the more correct and informative conclusions were based on interpretations that deduced the active processes rather than relying solely on the recognition of individual features. Carrara et al. (1992) found some 50% discrepancy between interpretations of individual landslide features mostly due to uncertainty in mapping old, inactive landslide bodies. However, once the individual features were extrapolated over landform units, the results were felt to be acceptable (83%) despite loss of resolution. One could conclude from these studies that when you ask n experts, you'll get n somewhat different opinions, but that if you combine these opinions you may well have a model of uncertainty from which to work. Finding n experts is not always feasible, is likely to be time consuming, and, above all, an expensive way to collect data. This leads us to consider the contexts within which such experts work.

GIS and environmental modeling can be both a research and a professional activity. It is, however, predominantly an applied activity often with the analyst working in a consultant–client relationship, which has important ramifications. Whereas a researcher's primary concern is with understanding, the goal of a consultant is action on his or her recommendations. To achieve this goal, the consultant should exercise judgment, based on experience and intuition, and focus predominantly on the principal variables that are under the client's control (Block, 1981). Blockley and Henderson (1980) consider the major dissimilarity between science and, say, engineering to be in the consequences of incorrect prediction. Whereas the scientist is concerned with precision, objectivity, and truth, and attempts to falsify conjectures as best he or she can; for the engineer, falsification of conjectures and decisions means failure, which must be avoided. The engineer strives to produce an artifact (road, dam, bridge) of quality (safe, economic, aesthetic) and, therefore, is "primarily interested in dependable information ... is interested in accuracy only to the extent that it is necessary to solve the problem effectively" (Blockley and Robinson, 1983; see also Frank, 2008). Thus, context (science versus engineering) has important pragmatic quality implications with regard to the collection and use of data in GIS and environmental modeling.

Semantic Confusion

Salgé (1995) defines semantic accuracy as the "pertinence of the meaning of the geographical object" as separate from its geometrical representation. The exact meaning of common words used to characterize classes of objects is a frequent problem in database integration. Let me provide a real example. In Chapter 6, we looked at a basin management planning project in Hong Kong. As part of this project, I mapped the land cover over large areas of the territory and one of the classes was "village." The definition of "village" here was a cluster of predominantly residential buildings within a rural environment and was constructed as a separate class to differentiate the runoff characteristics from those of the surrounding fields. The Planning Department, on hearing that villages had been mapped digitally approached me to explore acquisition of the mapping. However, I was aware that their definition of "village" was quite different and referred to traditional settlements (as opposed to more recent informal settlements) with the boundary extending 100 m beyond the outer buildings of the settlement. This is not what had been mapped and there was considerable scope for confusion, inadvertent misuse of the data, and eventual dissatisfaction with the data provider. Fortunately, this situation was avoided, but it is very easy to occur where commonly used words can have ambiguity and where precise definitions of feature classes are not available.

Finding a Way Forward

Following on from the above discussion, we can now flesh out Figure 8.2 with much more detail (though by no means exhaustive) on the specific causes of the four main types of uncertainty (Figure 8.6). This covers all aspects of primary data collection, deployment of secondary data, processing by hardware and software, and the final use of the analytical products. It's a veritable minefield.

The concept of fitness-for-use has already been mentioned. The data quality debate has been too narrowly focused on error in data rather than the

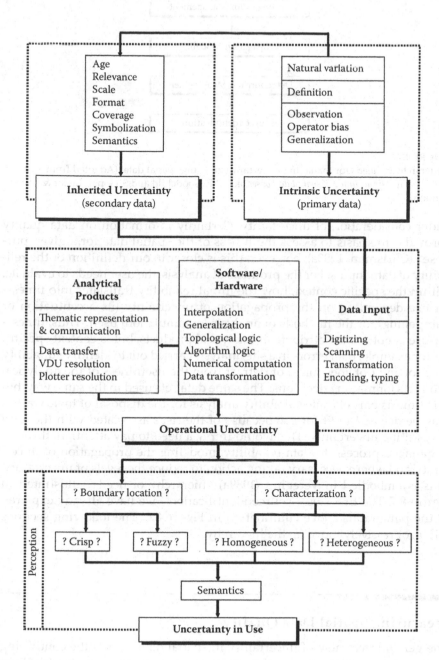

FIGURE 8.6
Causes of uncertainty in spatial data and their effect.

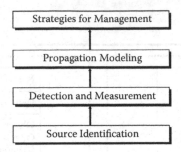

FIGURE 8.7
A hierarchy of needs for managing uncertainty in using spatial data. (Adapted from Veregin, H. (1989a) in *The accuracy of spatial databases*, ed. M.F. Goodchild and S. Gopal. Taylor & Francis, London.)

wider consideration of uncertainty. Certainly information on data quality "provides the basis to assess the fitness of the spatial data for a given purpose" (Chrisman, 1983a); however, this is closer to our definition of the reliability of data inputs. For the products of analysis, the user needs to evaluate, within the specific context, how the initial reliability translates into fitness-for-use depending on the propagation of uncertainty (its accentuation or dampening) for the methods of analysis (or simulation) used. Thus, fitness-for-use is not a fixed attribute of analytical products, but is context specific. So, for example, data errors in a site analysis carried out to identify suitability for growing potatoes may not detract from the usefulness of the analytical products for making decisions. The same data sets used in the same way, but this time as part of a site suitability analysis for the disposal of toxic wastes, may be deemed unfit for use because of the risks associated with the level of resulting uncertainty. This would imply a need to pay attention, through a managed process, to data reliability, modeling the propagation of uncertainty and where necessary taking action to reduce the levels of uncertainty. This is embodied in Veregin's (1989a) "hierarchy of needs" illustrated in Figure 8.7. The first step—source identification—we have already explored in the paragraphs above culmination in Figure 8.6. The following sections will progressively move up the hierarchy.

Measuring Spatial Data Quality

The general treatment of uncertainty in spatial data reflects the continuing conceptual closeness of digital maps to their paper roots in the minds of users with an overriding emphasis on accuracy and error. Thus, adopted wholesale from the mapping sciences has been the testing of geometrical accuracy based on *well-defined points having no attribute ambiguity* (Bureau of

Budget, 1947; American Society of Civil Engineers, 1983; American Society of Photogrammetry and Remote Sensing, 1985). These are usually stated in the general form: "90% of all points tested shall be correct within X mm at map scale." In some countries, such as Australia, the law—Survey Co-ordination Regulations, 1981—allows plans to be classified from AA through DD according to a range of permissible plotted positional errors (reported by Millsom, 1991). Testing is supposed to be carried out by reference to a survey of higher order and is usually reported as a root mean square error (RMSE) for x and y dimensions (which we have already used in Chapter 4):

$$RMSE = \sqrt{\frac{\sum_{i=1}^{n} e_i^2}{n}} \qquad (8.1)$$

where e = the residual errors (*observed – expected*), n = the number of observations. Vertical accuracy is generally treated in the same way with separate statements in the general form: "90% of all interpolated elevations must fall within one half of a specified number of contour intervals." Thus, it is also possible to classify vertical accuracy over a range by increasing the number of contour intervals (e.g., one, one and a half, two contour intervals). Whereas the accuracy testing described thus far is predicated on point samples, an alternative measure of map accuracy using line intersect sampling is given by Skidmore and Turner (1992). The sampling is used to estimate the length of boundaries on a map that coincide with the true boundaries on the ground. The results can be converted into a percentage accuracy statement for the map. An example of map accuracy for U.K. Ordnance Survey topographic maps is given in Table 8.1. Attributes tend to be treated separately from location geometry. Appropriate testing of attribute accuracy depends on the measurement class used. Continuous data, such as digital elevation model (DEM), can be tested for horizontal and vertical accuracy as with most point sampling described above either through interpolating contours or interpolating to known points. An alternative is statistical analysis of expected and observed values. A number of case studies are reviewed by Shearer (1990).

TABLE 8.1

Absolute Geometric Accuracy of Ordnance Survey Topographic Maps

Scale	RMSE	95% Confidence	99% Confidence
1:1,250	< 0.5m	< 0.8m	< 1.0m
1:2,500	< 1.1m	< 1.9m	< 2.4m
1:10,000	< 4.1m	< 7.1m	< 8.8m

Source: www.ordnancesurvey.co.uk

Reference Data

		W	S	G	Total
	W	28	14	15	57
Classified Data	S	1	15	5	21
	G	1	1	20	22
	Total	30	30	40	100

Producer's Accuracy	User's Accuracy	Overall Accuracy
W = 28/30 = 93%	W = 28/57 = 49%	$\frac{28+15+26}{100} = 68\%$
S = 15/30 = 50%	S = 15/21 = 71%	
G = 20/40 = 50%	G = 20/22 = 91%	

W: Woodland	S: Shrub	G: Grassland

FIGURE 8.8
A numerical example of a confusion matrix from which three values of proportion correctly classified (PCC) can be derived. (Adapted from Story, M., and Congalton, R.G. (1986) *Photogrammetric Engineering & Remote Sensing* 52: 397–399.)

Other quality issues for DEM are the nature and quality of source documents (if digitized), the sampling interval and orientation (if on a grid) in relation to the configuration of the terrain, and the intended use of the data. Where attributes are recorded on nominal scales or in discrete classes, the use of classification error matrices is widely used. Such techniques are of particular importance in testing the classification of RS imagery according to spectral response. A number of indices can be derived to summarize the matrix, such as *proportion correctly classified* (PCC), the Kappa statistic, and GT index (Turk, 1979; Congalton and Mead, 1983; Rosenfield, 1986; Hudson and Ramm, 1987).

Debate has not only centered around the appropriate derivation of indices, but also on whether these should reflect accuracy from the producer's or user's point of view (Story and Congalton, 1986), as illustrated in Figure 8.8. Testing is usually carried out on point samples using a suitable sampling scheme. Difficulties arise because classification schemes rarely have the necessary mutual exclusivity to avoid ambiguity, boundaries are often avoided (as in soil sampling), and the position of the sampling point must be correctly located on the ground. Middelkoop (1990) puts forward an alternative approach whereby a confusion matrix, generated by having several experts carry out interpretation of the same test area, is used to study boundary uncertainty. Gopal and Woodcock (1994) also use expert opinion to generate a matrix to assess thematic map accuracy. An increasing number of studies have utilized fuzzy measures and fuzzy logic in the interpretation of RS imagery as a means of better reflecting natural variation (techniques summarized in van Gaans and Burrough, 1993). Fuzzy measures are also used to reflect the quality of the product, a topic discussed further below.

Data conversion from analog to digital format is likely to introduce an element of error due to inaccurate placement of the digitizer cursor (Keefer et al., 1988) and due to simplification arising from point sampling of linear features (Amrhein and Griffith, 1991). Tests of human performance using digitizer pucks would indicate accuracies of 0.25 mm at map scale or half of this if a magnifier is used (Rosenberg and Martin, 1988). Trials reported by Maffini et al. (1989) showed that 90% of discrete entities digitized from 1:50,000 scale maps fell within 0.4 mm. An elaborate experiment by Bolstad et al. (1990) determined that 90% of well-defined points digitized from 1:24,000 scale maps fell within 0.5 mm and, thereby, met map accuracy standards. Error due to registration of the map to the digitizer was a large component. Errors were found to be significantly different from normal with differences between operators also statistically significant. Dunn et al. (1990) found that the interaction of scale, quality of source documents and digitizing operator could result in unexpectedly large amount of error. Vector to raster conversion or rasterization necessarily involves a degree of generalization (loss of precision) as a function of cell size (Veregin, 1989b). Rasterization is often an integral part of database creation or used as an adjunct to layer combination techniques, such as map algebra. An example of the effect of vector to raster and raster to vector conversion can be seen in Figure 8.9 where source contours have been converted to raster and back again. The discrepancies are

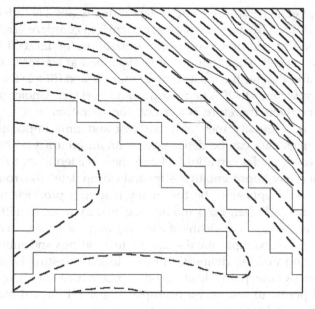

FIGURE 8.9
Example of distortions in data that arise from vector to raster and raster to vector conversion.

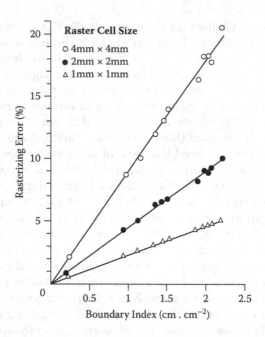

FIGURE 8.10
Relationship between central point rasterizing error and boundary index BI. (From Bregt, A.K. et al. (1991) *International Journal of Geographical Information Systems* 5: 361–367. With permission.)

obvious. Bregt et al. (1991) used a boundary index *BI* (total length of polygon boundaries divided by map sheet area; cm/cm²) to compare error resulting from central point and dominant unit rasterization for 1:50,000 scale maps for three sizes of cell. They found that *BI* explained at least 99% of variance (Figure 8.10) and that for the most complex maps tested (*BI* > 2) the incremental error ranged from 5 to 20% depending upon grid size. Choice of grid size, which is user driven, therefore, is a critical consideration.

Improving data quality inevitably has cost and time implications. "Few natural resources data can be determined with an accuracy ±10% at a price resource survey agencies can afford. Hence, there is a tendency to study them intensively at a few 'representative' sites and extrapolate" (Burrough, 1986b). The reality would appear to be that many mapping products are just not tested for accuracy (Fisher, 1991) and become instead an act of faith. Recourse to higher-order surveys as a means of checking may increase uncertainty. The object "village," for example, breaks down into buildings and subland uses at higher resolution making delineation more difficult. Testing based on well-defined, unambiguous points leads to other problems. First, there may be no well-defined points to test as, for example, in a flood hazard map. Second, well-defined, unambiguous points are likely to be more accurately mapped in the first place leading to a biased evaluation of accuracy. The use of error matrices has also been criticized as inadequate (e.g., Lunetta et al., 1991) since they do not address the spatial distribution of errors and the relationship between error and class boundaries. Grundblatt (1987), for example, identified that errors increase along boundaries. Methods of sampling may also introduce bias because, for example, in soils mapping, testing is frequently carried out purposefully in the interior of polygons and not near the edge. Methods of deriving measures of data quality usually result in global measures of accuracy either for an entire coverage or for individual classes within them. These measures can provide limited information on spatial variation in quality since they assume that error is uniformly distributed. In reality, this is unlikely to be the case. Concepts of statistical accuracy and error are not so easily transferred to the notion of uncertainty in the analytical products derived from the base mapping. This requires us either to be able to model error propagation through to the products or have ways of assessing the fitness-for-use *per se*.

Modeling Error and Uncertainty in GIS

It should be noted at the outset that there is no single best approach to modeling uncertainty for the various data handling and transformation functions that source data might be subjected to in GIS. This is partly because we do not yet have a single, generally accepted theory of uncertainty in GIS (Heuvelink, 1998) and partly because different GIS functions operate on the uncertainty in different ways. Hence, we shall approach the problem of modeling error and uncertainty from the perspective of specific GIS functionality to begin with (topological overlay and interpolation) and then go on to some wider generic issues (fuzzy concepts and uncertainty analysis). Given the wide range of possible GIS functions, I have been selective here. For other overviews, see Goodchild and Gopal (1989), Heuvelink (1998), and Hunter (1999). This is an area of ongoing research and the reader is urged to consult relevant journals on a regular basis.

Topological Overlay

Topological overlay is one of the functions that characterize GIS from other types of software. The overlay operation requires that two or more data layers are superimposed or combined to produce a new, composite map. Identification of co-location of objects or feature classes through overlay is fundamental to many forms of spatial analysis. An example is the spatial co-existence approach to environmental modeling discussed in Chapter 6. The overlay operation can be carried out on both raster and vector data and may take a number of forms. For example, layers having numerical attributes can be combined using arithmetic operators (map algebra) while categorical

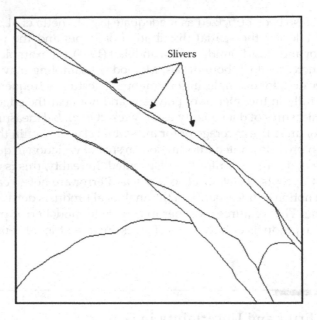

FIGURE 8.11
Example of sliver polygons in the overlay of geology and land cover in our example landscape.

attributes can have Boolean operators applied to them. For vector polygon overlay there are three fundamental components: (1) the determination of geometrical intersection, (2) reconstruction of topology, and (3) the assignment of attributes. Implementations may differ between vendors—ArcGIS and ArcView, for example, combine overlay and Boolean selection in one command (IDENTITY, INTERSECT, UNION). Where individual data layers contain geometric discrepancies, polygon overlay results in the creation of spurious polygons commonly known as slivers (Figure 8.11). These small, often numerous polygons tend not to reflect reality and are derived mostly from data processing. The original discrepancies may arise from digitizing, numerical rounding, generalization, changing map projection, and from poor conflation (matching of common boundaries in different layers). Numerical errors introduced by geometrical operations on objects represented by floating-point numbers (Hoffman, 1989) will result in perturbations and creep in vertices and edges during the overlay process (Pullar, 1991). Veregin (1989a) provides the following formula for the maximum number of slivers to be expected from overlaying two polygon layers:

$$s_{max} = 2 \cdot \min(v_1, v_2) - 4 \qquad (8.2)$$

where s_{max} = maximum number of sliver polygons, v_1, v_2 = the number of vertices in each data layer.

TABLE 8.2

Estimated ε (mm at map scale) for 1:1000 Scale Mapping in Hong Kong

Source of Discrepancy	Best Case (mm)	Worst Case (mm)
Control survey	0.001	0.02
Detail survey	0.005	0.02
Plotting control	0.2	1.0
Plotting detail	0.2	2.0
Fair drawing	0.06	0.18
Generalization	0.0	1.0
Map reproduction	0.1	0.2
Digitization	0.25	0.25
Total discrepancy (mm)	**0.395**	**2.477**

Source: Adapted from Law, J.S.Y. (1994) Data conversion, updating and integration of LIS data in a CAD system. Unpublished MSc. dissertation. Hong Kong Polytechnic University.

Strategies for managing or reducing slivers from overlay must be able to first determine those polygons that are truly spurious. The most popular approach relies on an epsilon band (ε) that is a buffer zone used to represent the possible error around a point or line (Figure 8.3). The concept was introduced into GIS by Chrisman (1982, 1983b) and Blakemore (1984) based on a formulation by Perkal (1956). Smith and Campbell (1989) found that when using ε on two geomorphological factor maps, an average of 30% error could occur in simple area measurements after overlay. Caspary and Scheuring (1993) give further consideration to the shape of the ε band and present a sagging or pinched buffer zone using $\varepsilon/\sqrt{2}$ in the middle of the line merging with a disk of radius ε at the end points of the line. Zhang and Tulip (1990) use ε as a fuzzy tolerance for automatic removal of spurious polygons, while Law and Brimicombe (1994) use ε to classify mismatches when combining primary and secondary data source in GIS and thereby allowing the use of a decision tree to reconcile the two data sets. Problems naturally arise in quantifying ε. Law (1994) provides a summary of sources of error and, assuming independence, the range of ε for 1:1000 scale mapping in Hong Kong (Table 8.2). Many vendors, however, will merely allow users to specify whatever tolerances they feel appropriate for the removal of slivers during the overlay process.

The types of error evident in the results of vector overlay are not always easy to disentangle as they lie along a continuum (Chrisman, 1989). Although Chrisman considers there is no workable theory, he has devised a framework (Figure 8.12), which considers the influence of scale on how errors might be classified. At the extremes of the diagonal, slivers are clearly distinct from attribute error. It is also useful to distinguish between slivers and more serious positional errors. The intermediate ground, however, is less clear and

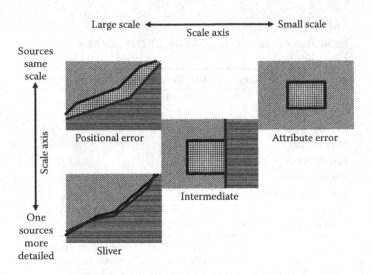

FIGURE 8.12
A framework for distinguishing geometric and attribute errors in vector overlay. (Modified from Chrisman, N.R. (1989) in *The accuracy of spatial databases*, ed. N.F. Goodchild and S. Gopal. Taylor & Francis, London.)

persists as a sizeable grey area in vector overlay. Overlay operations on raster data assume that the data layers are registered to the same grid. This, in many ways, avoids the geometric problems of vector overlay and accounts for why raster is considered the easier of the data models for modeling error propagation (Goodchild, 1990b). Where positional error and attribute error occur during compilation, these tend to manifest themselves in a raster data layer as attribute errors (i.e., a grid cell assigned the wrong attribute). These are then difficult to distinguish. In synthetic tests by Arbia et al. (1998), positional errors consequent on, say, geo-rectification of satellite imagery, vector to raster conversion and resampling to new cell size or cell orientation were found to play a prominent role in accounting for a third of the propagated error. Where arithmetic operators are used in map algebra, any uncertainty for the user may also focus on the weightings and the actual operators used to combine the layers. Decisions on weightings and arithmetic operators are made external to GIS and are more a matter of professional competence.

Turning then to the modeling of categorical error in the overlay process, we have already seen from the studies in the previous section—Early Warnings—that a modified rule of thumb developed: The product of overlay may be less accurate than the individual layers, but in any case could not be more accurate than the least accurate layer used. This rule of thumb assumes that all map combinations employ the INTERSECT or Boolean AND with composite map accuracy thus calculated as the product of individual layer accuracy as in Formula (8.3) below. Veregin (1989a), however, points out that this is not the only way selection can be carried out in map overlay where Boolean OR and

NOT are also commonly used. While an AND (intersection) operation combines the individual probabilities and tends to reduce composite map *accuracy* with the number of layers used (Figure 8.13(a)), an OR (union) operation is unity minus the combined probabilities of the *errors* and acts to increase composite map accuracy (Formula (8.4), Figure 8.13(b)). Furthermore, by de Morgan's law, the NOT of an AND can be taken as an OR and the NOT of an OR as an AND, so only two formulae are required. The implication is that in an analysis involving several stages of union and intersection, the accuracy of the products of each stage could have progressively enhanced or deteriorating accuracy. Rules of thumb regarding the accuracy of final outcomes cannot be relied upon unless only one type of Boolean operator is in use. Lanter (1991), Lanter and Veregin (1992), and Veregin (1994) pursued the idea that propagation of error would have to be traced through the analysis using an internal record of its lineage (the sequence of operators) within GIS from which the accuracy of the global product might be known. An example of such error tracking is given in Figure 8.14 where PCC is propagated through an analysis. This has been modeled on the earlier analysis of potentially unstable slopes illustrated in Chapter 5, Figure 5.10 with an additional consideration of volcanic soils. The "select" command allows the selection of a particular class from within a layer resulting in the use of the PCC for that class from the confusion matrix rather than the global PCC for the layer. Intersect and union propagate as discussed above. The technique assumes that the confusion matrix is known or can be estimated for each map layer. The outcome of the analysis presented here would suggest a final map accuracy for villages at risk from unstable slopes to be 75%. The technique provides a ready means of arriving at a global estimate of accuracy for an analytical product. However, we should not forget the criticisms leveled at both the use of confusion matrices and global indices in GIS context as discussed in the previous section.

Boolean AND:

$$P[\bar{E}_c] = \prod_{i=1}^{n} P[\bar{E}_i] \tag{8.3}$$

Boolean OR:

$$P[\bar{E}_c] = 1 - \prod_{i=1}^{n} P[E_i] \tag{8.4}$$

where $P[\bar{E}_c]$ = composite map accuracy, $P[\bar{E}_i]$ = the accuracy of a given layer as a proportion correctly classified, $P[E_i]$ = the error of a given layer such that $P[\bar{E}_i] = 1 - P[E_i]$. For a recent discussion of propagation of probabilities

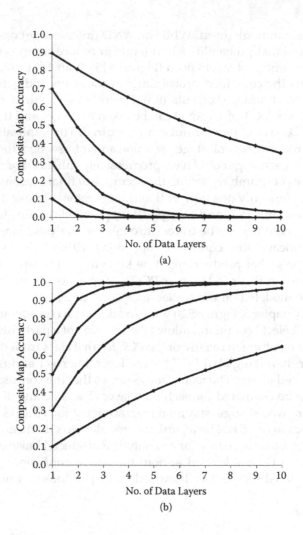

FIGURE 8.13
Composite map accuracy: (a) for the AND operator, (b) for the OR operator. (Adapted from Veregin, H. (1989a) in *The accuracy of spatial databases*, ed. N.F. Goodchild and S. Gopal. Taylor & Francis, London, pp. 3–18.)

in flood simulation studies as a means of assessing overall uncertainty, see Golding (2009).

Interpolation

So far we have looked at categorical data (nominal or ordinal) in polygon or raster coverage. Another common form of data is quantitative attributes that often start out as point data sets, but are transformed into fields through interpolation functionality. This poses a different set of problems for error

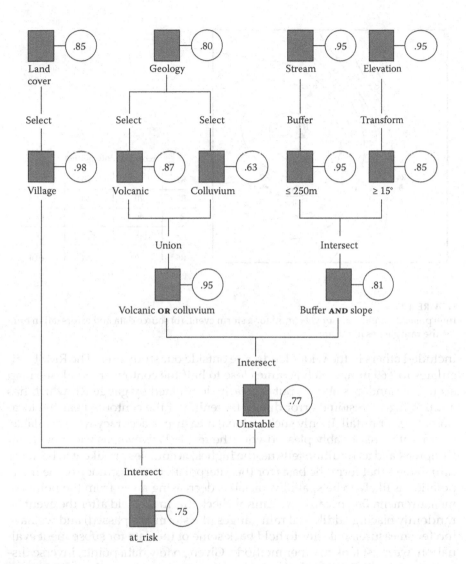

FIGURE 8.14
Propagation of PCC through an analysis using cartographic processing.

propagation. As we have already seen in Chapter 4 and will revisit again in Chapter 9, interpolation is not an exact science, but is itself model-based and can introduce operational uncertainty. Let us consider the isohyet map for our example landscape on which the rainfall stations have now been placed (Figure 8.15(a)). There is one rain gauge in each village and one each near where the main roads cross the rivers (where there are also flow gauges). The table in Figure 8.15(b) gives the recorded rainfall, that which is indicated by the interpolation and the residual error (*observed – expected*). The interpolation was not carried out solely from the five rain gauges that we can see, but

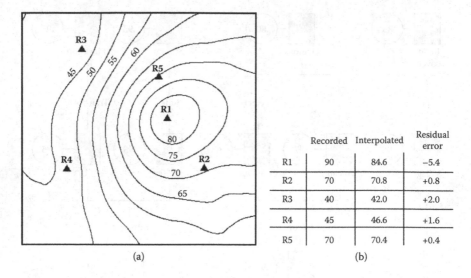

	Recorded	Interpolated	Residual error
R1	90	84.6	−5.4
R2	70	70.8	+0.8
R3	40	42.0	+2.0
R4	45	46.6	+1.6
R5	70	70.4	+0.4

(a)   (b)

FIGURE 8.15
Interpolated rainfall: (a) isohyets (mm) for a storm event, (b) source data and errors (all in mm) for the rain gauges in the area.

included others in the wider landscape outside our study area. The RMSE calculates to 2.69 mm, which is quite close to half the contour interval (see map accuracy standards above), but is clearly dominated by gauge R1, which has a much higher residual error due to flattening of the contours near the location of peak rainfall. If only such a global measure of accuracy were available, we might be reasonably pleased with the result. However, as we know from Chapter 4 and as intuition tells us, the highest accuracies are likely to be at the rain gauges that form the basis for the interpolation. So, accuracy of the interpolation is likely to be spatially variable, decreasing away from the points of measurement. But, because we cannot check this in the field after the event by randomly placing additional rain gauges (the storm has passed) and we have too few measurement sites to hold back some of the data for subsequent evaluation, we must look to other methods. Given so few data points, inverse distance weighted (IDW) interpolation was not used and instead a geostatistical technique called *kriging* was employed. One of the many advantages of kriging is that it can give an indication of its accuracy. Before going farther then, we must look at the details of this particular computational model, which, while being an important spatial modeling technique, is only just beginning to find its way into mainstream GIS software.

Kriging

Named after D.G. Krige (1966), a mining engineer, kriging is a geostatistical method of interpolation, which divides spatial variation into three components (Figure 8.16(a)):

1. A structural component that represents deterministic average behavior or a trend in the behavior.

2. A regionalized component of complex, but spatially correlated variation.

3. A random component of Gaussian noise that represents the residual error term.

In Chapter 4, we saw the important role that spatial dependence (correlation) plays in many geographical phenomena. This is used to effect in kriging. If you imagine a phenomenon represented by a gridded data set, such as the DEM in Figure 4.8, it would be apparent that adjacent values in the grid would be more likely to be similar than grid values farther away. If we were to take the first column of values and calculate the correlation with the adjacent column, we would expect the correlation to be high. With each column farther away, or *lag*, we would expect the correlation to be progressively lower. This is illustrated in Figure 8.16(b), which shows the correlation for successive lags from the first column of the DEM. As expected, the correlation progressively falls with increasing lag. However, because of the periodicy

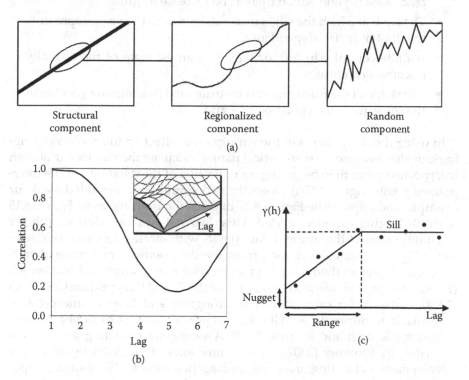

FIGURE 8.16

An example of kriging: (a) components of spatial variation, (b) declining correlation with lag, (c) the variogram.

evident in our example landscape, eventually the correlation begins to rise again, but the point is made—there is a distance within the landscape at which dependence reaches a minimum. Now, imagine carrying this exercise out for every known point in the data set; you would need a computational model to do so. In kriging, this exercise results in a *semivariogram* or often referred to simply as a variogram. The semivariance or γ (gamma) is an estimate of the autocovariance at a particular lag $\mathbf{h}$, thus:

$$\gamma(\mathbf{h}) = \frac{1}{2n} \sum_{i=1}^{n} \{z(x_i) - z(x_i + \mathbf{h})\}^2 \tag{8.5}$$

where $\gamma(\mathbf{h})$ = semivariance at lag $\mathbf{h}$, $z(x_i)$ = value of data point x_i, $z(x_i + \mathbf{h})$ = value of data point at lag $\mathbf{h}$ from x_i, n = the number of data pairs at lag $\mathbf{h}$.

The resulting variogram has a number of components that are evident once a best-fit curve is established (Figure 8.16(c)):

- $\gamma(\mathbf{h})$ increases with lag (varying inversely with autocorrelation) to reach a *sill* beyond which there is no increase in $\gamma(\mathbf{h})$.
- The lag $\mathbf{h}$ at which the sill is reached, known as the *range*, represents the limit of spatial dependence.
- An intercept of $\gamma(\mathbf{h}) > 0$, or *nugget*, is an estimate of the spatially uncorrelated noise.
- The shape of the fitted curve is instrumental in assigning weights in the subsequent interpolation of a grid.

In using the components of the variogram to effect an interpolation, kriging can also, because of its statistical nature, calculate the variance σ^2 at each interpolated point thereby giving an indication of the reliability of the interpolated points. Figure 8.17(a) shows the variogram for the rainfall data in our example landscape while Figure 8.17(b) gives the isohyets, as in Figure 8.15 with the kriging variances added. This clearly shows low variance (higher reliability) around the original data points with increasing variance (lower reliability) as distance increases from the data points. For further details on geostatistical methods, the reader can refer to Burrough and McDonnell (1998), Isaaks and Srivastava (1989), Cressie (1993), and Kanevski and Maignan (2004). Software for calculating the variogram and kriging interpolation is available in Surfer, in ArcGIS, and as part of the GSLIB public domain software (Deutsch and Journel, 1992). A variogram modeling java class is described by Faulkner (2002), while Brimicombe et al. (2009) report on the development and testing of a variogram agent for use in distributed component GIS. A further use of the variogram can be to guide sampling schemes so that the distance between samples will provide adequate support while minimizing spatial dependence between samples.

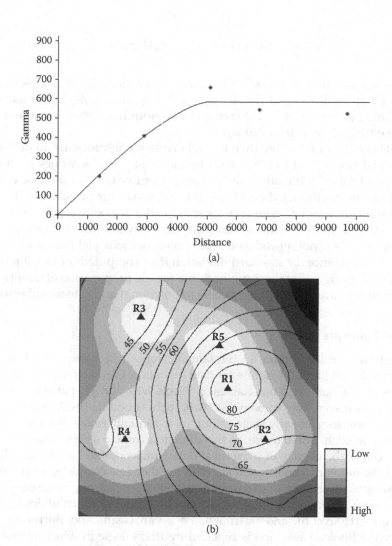

FIGURE 8.17

(a) Variogram and (b) kriging variance for the rainfall interpolation in the example landscape.

If through kriging a data layer representing reliability is generated for each interpolation, then the possibility must exist of propagating such error through analyses that have more than one interpolated layer. This forms the core of Heuvelink's (1998) book. If the input data were without error, then the kriging variance would reflect only the operational uncertainty. However, as we have discussed above, it is unlikely that any input data are free of uncertainty. So, the contribution of any such errors to the kriging needs to be modeled and, if the results of one interpolation forms the input to a subsequent GIS operation, then the uncertainties will be propagated on. Using Heuvelink's formulation, this can be expressed as:

$$U(x) = g(A_1(b_1(x) + V_1(x)), \ldots A_m(b_m(x) + V_m(x)))$$ (8.6)

where x = location of a grid cell where $x \in D$ (domain of interest), $U(\cdot)$ = output map containing all $x \in D$, $g(\cdot)$ = a GIS operation, $A(\cdot)$ = an input map containing all $x \in D$, $b(\cdot)$ = value of x in the input map, $V(\cdot)$ = a random field representing error or uncertainty.

If $g(\cdot)$ is a linear function, then it is relatively straightforward to derive the mean and variance of $U(\cdot)$ that describe the most probable values in $U(\cdot)$ and their reliability. On the other hand, if $g(\cdot)$ is nonlinear (as in the case of most interpolation methods), then Heuvelink suggests four possible methods of estimation: first-order Taylor series, second-order Taylor series, Rosenbleuth's method, and Monte Carlo simulation, all of which are computationally intensive and are not reproduced here. These methods yield maps giving the mean and variance (or standard deviation if so computed) of $U(\cdot)$. It must be reiterated, though, that these methods are for the propagation of quantitative attribute errors and are not applicable to positional and categorical errors.

Fuzzy Concepts in GIS

As we have seen above, there is a fundamental tension between natural variation in the real world and the dominant data models of GIS, which focus on mutually exclusive, homogeneous classes of objects with abrupt spatial boundaries. We have also seen how some researchers such as Bouille (1982) and Burrough (1986a) were from an early stage calling for essentially fuzzy phenomena to be modeled as such. Initial moves in this direction were to use probabilities as we have seen in the overlay of categorical data. But, as pointed out by Ehrliholzer (1995), the more interpretive and complex the data in a coverage, the more suitable are *qualitative* methods of assessment likely to be. Starting in the early 1990s, there was a move toward researching the treatment of probabilities as fuzzy measures (Heuvelink and Burrough, 1993; van Gaans and Burrough, 1993) where the Boolean selection is replaced by fuzzy logic in which intersection (AND) and union (OR) are instead based on MIN and MAX functions, respectively. The resulting probabilities thus reflect the degree of class membership in the final product. By the late 1990s, a considerable body of literature had developed in which fuzzy concepts (fuzzy sets, fuzzy logic, and fuzzy numbers) had been used as a means of accounting for variability and as a means of propagating that variability and any uncertainty to the analytical products.

Zadeh (1965) first introduced, as a concept, fuzzy sets and their associated logic. Whereas traditional mathematics and logic have assumed precise symbols with equally precise meanings, fuzzy sets are used to describe classes of inexact objects. Thus, though Boolean logic relies on a binary (0, 1) (termed *crisp*), fuzzy sets have a continuum of membership. Because imprecisely defined classes are an important element in human thinking, fuzzy sets have found early application in knowledge engineering (Kaufmann, 1975b; Graham and

Jones, 1988), cognitive psychology (Nowakowska, 1986), linguistics (Zadeh, 1972; Lakoff, 1973; Kaufmann, 1975a; Gupta et al., 1979), engineering (Blockley, 1979; Brown and Yao, 1983), and the environment (Ayyub and McCuen, 1987). General texts on fuzzy sets include Kaufmann (1973); Dubois and Prade (1980); Klir and Folger (1988); Harris and Stocker (1998); and Buckley and Eslami (2002).

Theory of Fuzzy Sets

A fuzzy set assigns levels of membership μ in a range [0, 1] for each element of x in a set **A** in a universe **U**:

$$\forall x \in \mathbf{U}, \{x|\mu_A\} \quad 0 \geq \mu_A(x) \leq 1 \tag{8.7}$$

Hence, for intervals of x of 0.1 in the range [0, 1], set **A** is characterized by:

$$\mathbf{A} = \{0|\mu_0, 0.1|\mu_{0.1}, 0.2|\mu_{0.2}, 0.3|\mu_{0.3}, \ldots 0.8|\mu_{0.8}, 0.9|\mu_{0.9}, 1|\mu_1\} \tag{8.8}$$

where μ_0 indicates level of zero level of support for membership of **A** and μ_1 indicates complete support for membership in **A**. Thus, the traditional binary (0, 1) can be viewed as crisp sets in the form:

$$\neg\mathbf{A} = \{0|\mu_1\} \quad \mathbf{A} = \{1|\mu_1\} \tag{8.9}$$

(where ¬ is Boolean NOT) and, thus, can be viewed as a special case of fuzzy set. Crisp and fuzzy sets are illustrated graphically in Figure 8.18.

Fuzzy sets can be combined in Boolean operations and, in general, can be handled in a simpler way than probabilities:

Intersection (∩) **A** AND **B**:

$$\forall x \in \mathbf{U}, \quad \mu_{A \cap B} = \mathrm{MIN}(\mu_A(x), \mu_B(x)) \tag{8.10}$$

Union (∪) **A** OR **B**:

$$\forall x \in \mathbf{U}, \quad \mu_{A \cup B} = \mathrm{MAX}(\mu_A(x), \mu_B(x)) \tag{8.11}$$

In other words, for an intersection AND the minimum membership of all elements x are taken and in the union OR the maximum membership of all elements x are taken (examples are given below). This has important parallels with Formula (8.3) and Formula (8.4). Thus, fuzzy sets can be propagated through analyses typical of those carried out in GIS where overlay is combined with Boolean selection. Fuzzy set operations and the use of fuzzy sets in a geographical context have been reviewed by Macmillan (1995). The term

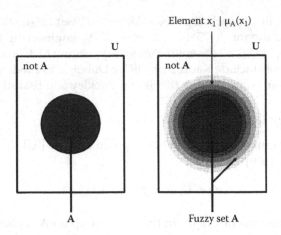

FIGURE 8.18
An illustration of crisp and fuzzy sets.

fuzzy has been introduced into GIS for handling uncertainty, though for the most part it has been loosely applied to any nonbinary treatment of data, such as probabilities. However, there are important differences between probabilities and fuzzy sets. First, probabilities are still crisp numbers in the way they are formally defined. Second, probability of **A** and ¬**A** must sum to unity, but not necessarily so for fuzzy sets where there can be some unknown or unquantified residual. Thus, fuzzy sets, from this perspective, are easier to use than probabilities. Nevertheless, the use of fuzzy set theory proper has thus far been quite restricted (Unwin, 1995) and is reviewed in a spatial analysis context by Altman (1994). One area of application has been the "fuzzification" of data, database queries, and classification schemes through the use of fuzzy membership functions, as a means of overcoming the uncertainty implicit in the binary handling of data (Kollias and Voliotis, 1991; Burrough et al., 1992; Guesgen and Albrecht, 2000). Another area of application has been to quantify verbal assessments of data quality from image interpreters and as a consequence of expert evaluations (Hadipriono et al., 1991; Gopal and Woodcock, 1994; Brimicombe, 1997; 2000a). Use of fuzzy numbers for recording and propagating geometric uncertainty is given in Brimicombe (1993; 1998). However, we are still some way off from seeing fuzzy concepts as part of mainstream GIS software. The theoretical dryness of fuzzy sets (above) can be brought to life and illustrated through a GIS example.

Example of Fuzzy Sets in GIS

In Figure 8.4, I illustrated the problems of interpreting natural variation into discrete classes. Suppose we could express our certainty of class membership linguistically, somehow store that against the appropriate polygon in GIS, and propagate that linguistically expressed uncertainty to any analytical products.

TABLE 8.3

Linguistic Hedges Suggested for Use in Aerial
Photographic Interpretation

	Qualifying Term	Definition
Features	Certain	Well defined, identifiable
	Reliable	Poorly defined, identifiable
	Unreliable	Deduced
Boundaries	Well defined	Full boundary distinct
	Poorly defined	Boundary mainly distinct
	Partly defined	Boundary mainly inferred
	Estimated	Boundary inferred

Source: Based on Edwards, R.J.W. et al., (1982) *Quarterly Journal of Engineering Geology* 15: 265–316.

Nice? Well, entirely feasible. Before we look at a GIS example, first we need to consider how fuzzy sets are used to describe linguistic terms. Verbal assessments or linguistic hedges are a common qualitative indicator of data accuracy and reliability. For example, a set of linguistic hedges (certain, reliable, well-defined, poorly defined) for features and boundaries have been defined and encouraged for use in API for terrain evaluation by the Geological Society Working Party on Land Surface Evaluation for Engineering Practice (Edwards et al., 1982). The problem that arises, however, is that these types of "standard" linguistic hedges (Table 8.3) are defined in terms of yet other hedges that, to each individual, may have different nuances and interpretations. When more than one language is considered, the problem of "meaning" of linguistic hedges is compounded apparently to the point of impossibility.

One of the earliest and main applications of fuzzy sets has been to represent qualifying adjectives such as "tall" or "short." Empirical studies of fuzzy set equivalents of linguistic hedges (Zadeh, 1972; Lakoff, 1973; Kaufmann, 1975a) have shown a general pattern of reduced spread in the fuzzy sets as they tend toward the more definite boundaries of 0 and 1. Examples of such fuzzy set representation of linguistic hedges (Ayyub and McCuen, 1987), and illustrated graphically in Figure 8.19, are:

Small, low, short, or poor:

$$\mathbf{A} = \{0|1, 0.1|0.9, 0.2|0.5\} \tag{8.12}$$

Medium or fair:

$$\mathbf{A} = \{0.3|0.2, 0.4|0.8, 0.5|1, 0.6|0.8, 0.7|0.2\} \tag{8.13}$$

Large, high, long, or good:

$$\mathbf{A} = \{0.8|0.5, 0.9|0.9, 1|1\} \tag{8.14}$$

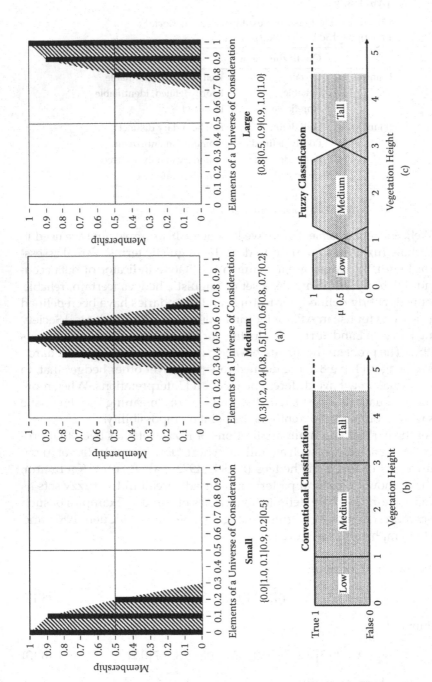

FIGURE 8.19

Illustrative linguistic hedges as fuzzy sets: (a) for formula (8.12) to formula (8.14), (b) an example for classifying vegetation height in conventional crisp sets, (c) as might be done using fuzzy sets.

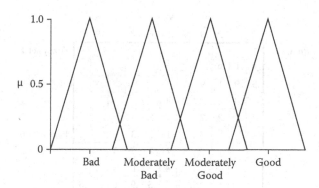

FIGURE 8.20
An example of arbitrary definition of linguistic hedges as fuzzy sets.

In a GIS and RS context, fuzzy sets have been used to represent linguistic hedges and other qualifying adjectives by Hadipriono et al. (1991) and Gopal and Woodcock (1994). However, the choice of linguistic hedge and their "translation" into a fuzzy set space within the literature often appears arbitrary. More worrying are schemas that give linguistic hedges equal spacing and spread of their fuzzy sets (Figure 8.20), which do not accord with the generally accepted use of fuzzy sets to represent linguistic hedges as discussed above (compare with Figure 8.19(a)).

Other difficulties arise in the use of fuzzy sets. While the definition of a linguistic hedge as a fuzzy set can be a fairly straightforward process, the same cannot necessarily be said for the reverse. Faced with a fuzzy set that does not fit any predefined terms, it can be very difficult to interpret as an appropriate linguistic. Furthermore, fuzzy sets are cumbersome to store in a database. Not only is the notation difficult to encode, but there are 39,916,789 useful combinations of fuzzy sets in the range [0,1] for an interval of $x_i = 0.1$. These are serious problems in the use of fuzzy sets and while they may at first appear to be an attractive solution (commented on by many authors), it is problems such as these that have detracted from more widespread use in GIS. To overcome these problems, a sort of "universal translator" is required, which, using a limited number of fuzzy sets, could be used in the two-way translation of linguistic hedges and fuzzy sets. Figure 8.21 gives a series of 11 stylized fuzzy sets in the range [0,1] that have a number of attributes that make them particularly useful as linguistic building blocks:

- The stylized set begins and ends with binary (crisp) numbers 0 = {0|1} and 1 = {1|1}.

- The nine intermediate fuzzy sets are spaced with their maximum membership of $\mu_A(x) = 1$ stepped across the range x_i at 0.1 interval. Thus, each fuzzy set has its maximum membership uniquely placed within the [0, 1] range.

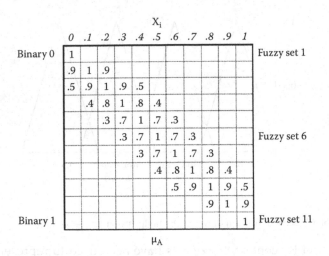

FIGURE 8.21
A graph that shows a set of 11 stylized fuzzy sets unambiguously partitioning the range [0, 1].

- The roughly triangular form of the fuzzy sets spreads toward the center of the range of fuzzy sets. Thus, where there is greatest uncertainty (midway in the [0, 1] range), the fuzzy sets are most spread to reflect higher levels of uncertainty. The reduced spread in moving toward 0 or 1 accords with the empirical evidence for fuzzy set representations of linguistic hedges.

- The Hamming, or orthogonal, distance between each fuzzy set and its immediate neighbor is constant at 2.00 indicating that the fuzzy sets unambiguously partition up the space over the range [0,1]. The orthogonal distance between two fuzzy sets is calculated by:

$$d(\mathbf{A}, \mathbf{B}) = \sum_{i=1}^{n} |\mu_A(x_i) - \mu_B(x_i)| \qquad (8.15)$$

A common form of hedge, other than purely linguistic, is intuitive (subjective) probabilities that individuals use in making judgments under uncertainty (Tversky and Kahneman, 1974). Thus, an individual may say, "I'm 80% sure." Since both types of hedges—linguistic and intuitive probability—are frequently used, it is possible for an individual to make an equivalence between the two. Thus, "I'm reasonably sure" may be, for an individual, an equivalence to "I'm 80% sure." This would be for each individual to define in his or her own language. Given the way the stylized fuzzy sets in Figure 8.21 step across the range [0, 1], each stylized fuzzy set can be "labeled" or identified by the x_i where $\mathbf{A}(x_i) = 1$. These labels at 0.1 intervals in the range [0, 1] provide an intuitive probability-like metric that we will refer to as *fuzzy*

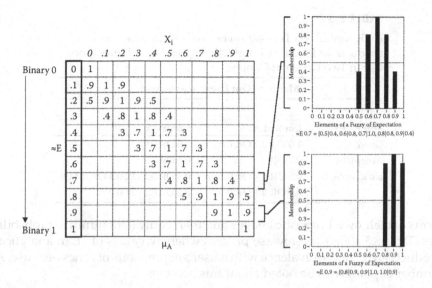

FIGURE 8.22
Illustration of fuzzy expectation ($\approx E$) as an intuitive probability with underlying fuzzy sets.

expectation ($\approx E$). Thus, the choice of a value of $\approx E$ as an intuitive probability gives an underlying stylized fuzzy set (Figure 8.22).

Values of $\approx E$ are both the building blocks for fuzzy set representations of linguistic hedges and the means for "translating" fuzzy sets into qualifying statements of fitness-for-use. So, for example, suppose we have an observer who is collecting data by API in which the observer is able to qualify each polygon as it is delimited and assigned to a class, the range of linguistic hedges used in such a task may be those listed in Table 8.4, which express the observer's degree of certainty. The observer also defines these hedges in terms of $\approx E$. To do so, the observer has to match only intuitive probabilities or sets of intuitive probabilities to the linguistic hedges in use; the underlying stylized fuzzy set or sets are substituted automatically by the GIS software. Users of the system do not have to think in terms of fuzzy sets, only in

TABLE 8.4

An Example of a Translation
between an Observer's Linguistic
Hedges of Certainty and $\approx E$

Mapping in Observer's Uncertainty	
Linguistic	$\approx E$
Certain	1.0 OR 0.9
Reasonably certain	0.8
Moderately certain	0.7 OR 0.6
Not terribly certain	0.6 OR 0.5

TABLE 8.5

An Example of a Translation between ≈E in Analytical
Products and a User's Linguistic Expression of Fitness-for-
Use in Two Different Contexts

Linguistic	Mapping Out Users' Assessment ≈ E	
	Context A	Context B
Good	1.0 OR 0.9 OR 0.8	1.0 OR 0.9
Acceptable	0.7 OR 0.6	0.8 OR 0.7
Unacceptable	0.5 OR 0.4 OR 0.3 or 0.2 OR 0.1 OR 0.0	0.6 OR 0.5 OR 0.4 OR 0.3 OR 0.2 OR 0.1 OR 0.0

terms of their own linguistic hedges and their equivalent intuitive probabilities. Table 8.5 shows the reverse process whereby values of ≈E in analytical products are given equivalence with a user's expressions of fitness-for-use. A number of points can be noted about this process:

- A linguistic hedge may be equivalent to one or more values of ≈E.
- These values would normally be adjacent in the series (logically), but need not necessarily be so.
- Linguistic hedges can overlap in their ≈E equivalence showing that two linguistic hedges may be close in meaning (and, hence, two different hedges may have the same ≈E equivalence if they express the same degree of certainty).
- Where two or more values of ≈E are used, they are not used singly, but are combined using a Boolean OR prior to propagation through analysis.
- The linguistic hedges need not be limited to English, but can be in any language where the user of that language can define ≈E equivalence.
- The degree of certainty associated with every polygon can be entered into a GIS database using natural language.
- A table giving the linguistic hedges and their ≈E equivalence used by an observer could be stored as metadata for future reference.
- The mapping of ≈E in analytical outputs to a user's linguistic fitness-for-use is achieved by matching the relevant ≈E by taking the stylized fuzzy set that is at a least distance from the output fuzzy set using relative orthogonal distance:

$$\delta(\mathbf{A}, \mathbf{B}) = \sum_{i=1}^{n} \left| \mu_A(x_i) - \mu_B(x_i) \right| \Big/ n \qquad (8.16)$$

where n number of nonzero pairs of x_i in $\mathbf{A}$ and $\mathbf{B}$.

Let us now work through a basic example in Figure 8.23. There are two layers, one for "condition of habitat" and the other for "pressure zones." In the first layer, a polygon has been interpreted as having "poor" condition of habitat and the observer is confident of this interpretation and gives a certainty rating of "good" equivalent to ≈E = 0.9 OR 0.8. In the second layer,

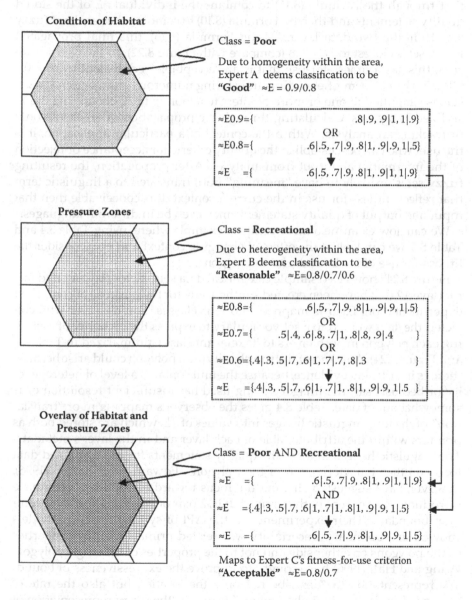

Condition of Habitat

Class = **Poor**

Due to homogeneity within the area, Expert A deems classification to be **"Good"** ≈E = 0.9/0.8

≈E0.9={ .8|.9, .9|1, 1|.9}
OR
≈E0.8={ .6|.5, .7|.9, .8|1, .9|.9, 1|.5}

≈E ={ .6|.5, .7|.9, .8|1, .9|.9, 1|.9}

Pressure Zones

Class = **Recreational**

Due to heterogeneity within the area, Expert B deems classification to be **"Reasonable"** ≈E=0.8/0.7/0.6

≈E0.8={ .6|.5, .7|.9, .8|1, .9|.9, 1|.5}
OR
≈E0.7={ .5|.4, .6|.8, .7|1, .8|.8, .9|.4 }
OR
≈E0.6={.4|.3, .5|.7, .6|1, .7|.7, .8|.3 }

≈E ={.4|.3, .5|.7, .6|1, .7|1, .8|1, .9|.9, 1|.5 }

Overlay of Habitat and Pressure Zones

Class = **Poor** AND **Recreational**

≈E ={ .6|.5, .7|.9, .8|1, .9|1, 1|.9}
AND
≈E ={.4|.3, .5|.7, .6|1, .7|1, .8|1, .9|.9, 1|.5}

≈E ={ .6|.5, .7|.9, .8|1, .9|.9, 1|.5}

Maps to Expert C's fitness-for-use criterion **"Acceptable"** ≈E=0.8/0.7

FIGURE 8.23
A basic example of propagating expert opinion on data quality into a statement of fitness-for-use following an overlay operation.

a similar process has been carried out for a polygon deemed "recreational" with "reasonable" certainty. When the two polygons are overlaid to give poor condition of habitat AND recreational pressure zone (i.e., intersection of both layers), $\approx E$ propagates from the two data layers. The topological overlay is performed as in any standard GIS software, but $\approx E$ is also calculated first through the Formula (8.11) to combine the individual $\approx E$ of the stored quality statements and then by Formula (8.10) to combine the resultant fuzzy sets from the two data layers. Using Formula (8.16), this final propagated fuzzy set is closest to $\approx E = 0.8$ (compare with Figure 8.22). In the user's criterion, this has a fitness-for-use that is an "acceptable" equivalent to $\approx E = 0.8$ OR 0.7. The problem of encoding and storing numerous cumbersome fuzzy sets is simplified to one or more pointers to a lookup table containing the 11 stylized fuzzy sets. Calculating through the propagation of $\approx E$ occurs only in tracking an analysis. Within the context of a particular application, it is the user who can best establish the quality criteria for acceptance or rejection of the informational output from analyses. After propagation, the resulting fuzzy set is unlikely to have "meaning" until translated to a linguistic term that reflects fitness-for-use in the current context. It is conceivable then that input and output of quality statements may even be in different languages.

We can now examine a more complex example where, using Table 8.4 and Table 8.5, we can look at the effect of changing context and also consider the fitness-for-use of the boundaries to polygons.

Figure 8.24 shows two sample categorical data sets, factor (layer) 1 and factor (layer) 2 that have been created synthetically for the purpose of controlled experimentation. Each is mapped into four classes. The observer who collected the data worked to a set vocabulary to express the certainty of assigning each polygon in both layers to its relevant class (summarized in Table 8.4 and Figure 8.24). Any uncertainty in classifying a polygon could arise because there is insufficient evidence, because the unit contains a level of heterogeneity or because, say, the imagery being used has insufficient resolution or is somewhat out of date. Table 8.4 gives the observer's mapping in or "translation" of the four linguistic hedges into values of $\approx E$, which are stored both as pointers within the attribute table of each layer and in the layer's metadata. The linguistic hedges that are the quality statements for the simulated data focus on the characterization of the polygons. Edwards and Lowell (1996), however, have suggested shifting the focus toward boundaries. They have introduced the concept of the *twain*, that is, a pair of polygons and their common boundary. Their experiments on the API of synthetic texture models showed that boundary uncertainty was related primarily to the properties of the polygons on either side and not to the properties of the single polygon. Wang and Hall (1996) have sought to improve the expressiveness of boundary representation to "describe not only the location but also the rate of change of environmental phenomena" through "the fuzzy representation of fuzzy boundaries." For polygon maps of nominal properties, the sharpness of a boundary is determined by the purity of the polygons on either side. The

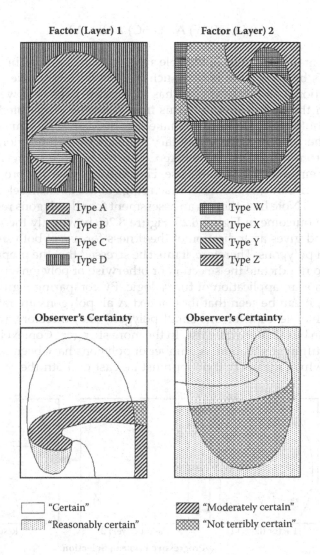

FIGURE 8.24
Two categorical layers with their respective stated certainty according to the observer's terminology (Table 8.4).

linguistic hedge is a statement of conviction of the nominal class of a polygon and differences in such statements on either side of a boundary would influence the characterization of that boundary. Thus, given an appropriate technique, an indication of boundary uncertainty could be derived from the polygon uncertainties, hence, the initial concentration on linguistic hedges of polygons only.

The two categorical maps, layer 1 and layer 2, are combined in an overlay operation (Figure 8.25, top) followed by progressive Boolean selection in the form:

$$(((A \text{ OR } W) \text{ AND } \neg C) \text{ AND } \neg Y)$$

to differentiate suitable and unsuitable areas as a conventional binary output (Figure 8.25, bottom). This is how such a selection might take place using any conventional GIS software that has topological overlay. If we now want to ascertain the fitness-for-use of this map from the propagated $\approx E$ stored in the attribute table, we need to make it context specific. This is given in Table 8.5 where for the same vocabulary there are two "translations" or mappings out from propagated $\approx E$ where Context B is more stringent in terms of its requirements for fitness-for-use. Figure 8.26 shows the propagation of uncertainty in Context A and B for each stage of the Boolean selection given in Figure 8.25. Note how there is an assessment for all polygons regardless of the Boolean outcome in Figure 8.25. Figure 8.26 shows only the propagated certainty and gives an indication of the fitness-for-use of both selected and not selected polygons. Therefore, it must be stressed that the propagated values of $\approx E$ do not dictate the selection or otherwise of polygons in the analysis; this is not an application of fuzzy logic. By comparing Figure 8.25 and Figure 8.26, it can be seen that for Context A all polygons are rated "good" or "acceptable" (except for one small polygon) and, therefore, the Boolean selection can be taken as fit for use. In the more stringent Context B, however, we have a different situation. A number of polygons have been rated "unacceptable," which means that doubt must be cast on both the selection and

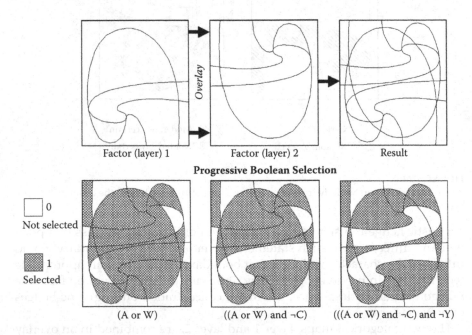

FIGURE 8.25
Topological overlay and Boolean selection of $(((A \text{ OR } W) \text{ AND } \neg C) \text{ AND } \neg Y)$.

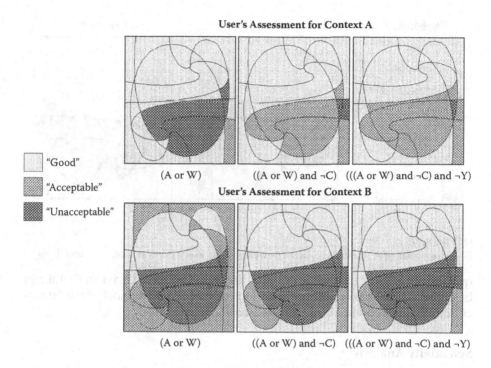

FIGURE 8.26

Propagation of ≈E to map fitness for use of Figure 8.25 in terms of user's Context A and Context B (Table 8.5).

the nonselection of the relevant polygons through the Boolean logic. We also now can turn to the fitness-for-use of the polygon boundaries (Figure 8.27). The method here has been to calculate, using ≈E, an area-weighted average fuzzy set from the polygons on either side of a boundary with a mapping to a linguistic hedge in Table 8.5, again reflecting the particular context of an application. An "unacceptable" boundary in Figure 8.27 is one where its location is too uncertain on the evidence of the uncertainty of the polygons on either side.

As with any technique, some difficulties arise. One specifically is that in propagating fuzzy sets through a union overlay (Boolean AND), it is possible in some instances to arrive at a null fuzzy set (i.e., for all x_i, $\mu_A(x_i) = 0$), which then cannot be resolved further. In such instances, the fitness-for-use defaults to ≈E = 0. Also, where a fuzzy set is equidistant from two adjacent ≈E, the lower one is taken as the match. But, there are also further advantages. A record of the steps in the analysis (its lineage) permits sensitivity analysis (SA). For example, recomputing the fitness-for-use ≈E scores for experimentally altered levels of observer's ≈E can identify what increases in certainty are required to improve the acceptability of the end result, say in Context B. Thus, for Context B, in Figure 8.24, type W in layer 2, which has a

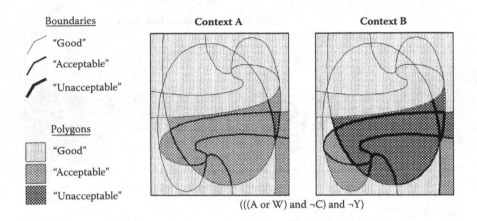

FIGURE 8.27
Boundary fitness-for-use derived from the twain ≈E for user's Context A and Context B.

quality rating of "not terribly certain" and must be resurveyed so that it can be recorded as "reasonably certain" or better in order for most of the "unacceptable" polygons in Figure 8.26 to become "acceptable."

Sensitivity Analysis

In this section on modeling error and uncertainty in GIS, we have looked at ways in which known levels of error or even estimated levels of uncertainty can be propagated through analyses so that a judgment about the fitness-for-use of the analytical products can be made. We even saw in the last example how it would be feasible to backtrack and analyze the sensitivity of some level of uncertainty in order to focus data upgrade efforts to meet minimum criteria of fitness-for-use. As already noted, none of the main GIS vendors provide this type of functionality as standard. In as much as we will look at some technological solutions to this in the last part of Chapter 9, it is as well to consider here what one might do in practical terms having carried out an analysis using GIS. This revolves around the idea of sensitivity analysis (SA). While SA in GIS is by no means new (e.g., Lodwick et al., 1990), there has recently been renewed interest (e.g., Li et al., 2000; Crosetto and Tarantola, 2001; Crosetto et al., 2002; Lilburne and Tarantola, 2009) that appears to have been prompted at least in part by the work of Saltelli et al. (2000). It is often advantageous to differentiate error propagation or *uncertainty analysis* (UA) that seeks to model output uncertainty as propagating from input uncertainties, and SA which looks to apportion relatively among the inputs their degree of influence on the uncertainty of outputs. These then would form a complementary two-stage process. Often the term *sensitivity analysis* is used to cover both UA and SA as defined here. The issue being raised in this section is a global one: in the absence of measures of input uncertainty (data reliability) and/or the degree of added operational uncertainty, how to evaluate fitness-for-use. Figure 8.14 provides a way of quickly arriving at global estimates at a project planning stage and even in helping to define certain targets in data reliability when purchasing or compiling data. However, it is likely to be a ballpark estimate and global to entire layers. Another way was used in Chapter 6 in the reservoir slope instability problem where data perturbations in a *Monte Carlo* (MC)-type simulation were used to study the sensitivity of the output decision to changes in the inputs. The MC process in the context being considered here is quite straightforward:

- Select a *probability density function* (PDF) that reflects the likely pattern of uncertainty within the data (e.g., uniformly, normally, or Poisson distributed).
- Select a sampling scheme (e.g., random, stratified) by which to obtain values from the PDF.
- Carry out the sampling for t number of trials and use these samples to perturb the original data that are then used to replicate the GIS-based analysis.
- Evaluate the t outputs of the GIS-based analysis (e.g., calculate mean and variance, scatterplots, boxplots).

A good introduction to MC analysis can be found in Mooney (1997) with more details in Saltelli et al. (2000). The logic is that given enough trials the mean result will approach the true result, or put another way, "a properly calibrated deterministic model should give a result that is equivalent to the mean value of the output of the equivalent stochastic model" (Burrough, 1997). The simulation can be carried out globally on all inputs simultaneously to identify the overall sensitivity of the output or to achieve a mean result. Alternatively, inputs can be perturbed *one-at-a-time* (OAT) in order to test output sensitivity to the individual input.

We can test this, for example, on the IDW interpolation in Chapter 4, Figure 4.12 and Figure 4.13. In Figure 4.13, we have already systematically tested the sensitivity of the output to the parameter r (distance decay) and we have also looked in Figure 4.12 at what a difference the sampling scheme makes. We would normally evaluate the RMSE of the interpolated DEM against a proportion of the original data held back for such a purpose. From this we can establish the mean and standard deviation of the residual errors ($RMSE$ = standard deviation when the mean = 0) and assuming normally distributed errors gives us the relevant PDF. This PDF can be randomly sampled and used to perturb the survey points. Figure 8.28(a/b) shows the starting point that has a random set of observation errors added to the points used in the interpolation. With an $RMSE$ of ±29.1, it is considerably less accurate than that achieved with our "true" sample points ($RMSE$ ±19.5). The data is perturbed 20 times and IDW interpolation carried out for each. Figure 8.28(c) represents the mean of these 20 perturbed outputs (achieved

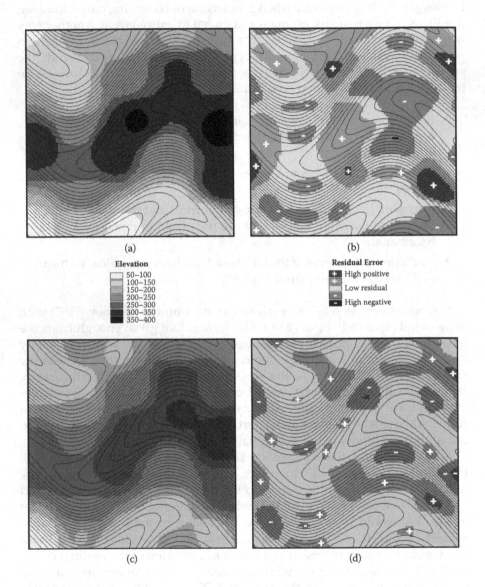

FIGURE 8.28
Monte Carlo analysis: (a) result of IDW interpolation from data points having random observation errors, (b) residual errors of (a), (c) mean surface after 20 trials, (d) residual errors (compare with Figure 4.12(d)).

using map algebra) with Figure 8.28(d) showing the residual errors (*RMSE* ±20.1), which when compared with Figure 4.12(d) gives a very similar result. For more complex models, the number of trials needed will be much higher before a good statistical average can be obtained. Between this example and that given in Chapter 6, we can gain an appreciation of the usefulness of MC-based SA in evaluating fitness-for-use of analytical products for decision making.

Managing Fitness-for-Use

Management refers to the strategies and methods adopted to mitigate against uncertainty in spatial databases and to reduce the uncertainty absorption required of the user (Bedard, 1986; Hunter and Goodchild, 1993; Frank, 2008). Without an underlying conceptual model for handling uncertainty, such strategies may be difficult to develop resulting in a series of loosely organized actions that may not achieve the desired goals. Any strategy needs to anticipate the entire GIS process from the input of data to the output of analytical products and is a quality assurance process.

Beginning with identification of suitable data sets for a project, there is a need from the outset to assess suitability and reliability. *Metadata* are "data about data" (Medyckyj-Scott et al., 1991). They provide further descriptions pertaining to the objects in the database and ideally consist of a series of standardized attributes (Canadian General Standards Board, 1991). Coming under this umbrella then, are definitions of entities and attributes, measurement and coding practice, rules used for spatial delimitation, data sources, and data quality. This gives rise to the notion of spatial data audit (Cornelius, 1991). Thus, the theoretical purpose of metadata is to allow a user to know the nature of the data and its compilation, in particular, the physical and conceptual compatibility of the data for integration and use with other data sets (Hootsman and van der Wel, 1993). Their value and reliability can be judged. Not surprising then, the main consideration of metadata has been in the context of data transfer standards. For example, the U.S. *Spatial Data Transfer Standard* (SDTS) (National Institute of Standards and Technology, 1992) requires a data quality report that specifically requires information on lineage, positional accuracy, attribute accuracy, logical consistency, and completeness. The use of all these data quality modules in a data transfer is mandatory. Key standards for metadata are: ISO/TC211 Geographical Information Metadata Standard, FGDC-STD-001-1998 Content Standard for Digital Geospatial Data, and ISO 15836 Dublin Core Metadata Element Set. While metadata will allow a user to assess the reliability of the data for use in a particular application and compatibility for integration with other data

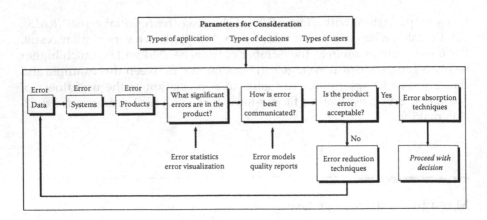

FIGURE 8.29
A strategy for managing uncertainty in spatial databases. (Adapted from Hunter, G.J., and Goodchild, M.F. (1993) *URISA Journal* 5 (2): 56–62.)

sets, they may not be sufficient to assist users in assessing fitness-for-use after propagation of uncertainty during analyses, particularly because such an assessment is context-specific. It is in response to issues such as these that Hunter and Goodchild (1993) have developed an overall strategy for managing uncertainty in spatial databases (Figure 8.29). Data and the system (hardware and software) are first evaluated for error separately and then evaluated again when combined to form an analytical product. These errors then need to be judged in context and communicated so that choices can be made by the user between error reduction (data or system upgrading) and error absorption (accepting the risk in some way).

This type of model, while structurally useful, leaves the specific methodology to the user. What is needed is a conceptual framework that can act both as a strategy and direct specific methodology. The framework presented in Figure 8.30 is based on a communications model (Shannon, 1948; Bedard, 1986) in which there must be sufficient flow of information in order to reduce uncertainty. In this case, not only does data about the real world need to be converted and communicated as information (by means of GIS) to a user/ decision maker, but there must also be sufficient communication about the quality of that information in order to reduce uncertainty in evaluating its fitness-for-use. The initial focus of the framework is on "context zero," the original context in which data are collected, presumably in response to and as specified for a specific or related range of uses. In surveying the real world (or perceived reality), the observers are expected to record or generate measures of positional, thematic, and temporal uncertainty of their data in ways appropriate to the nature of the data being collected and the technology in use. Observers are likely to have their own professionally/culturally conditioned view of the real world and may well be distinctly different from the eventual users of the data. Truth in quality reporting at the highest

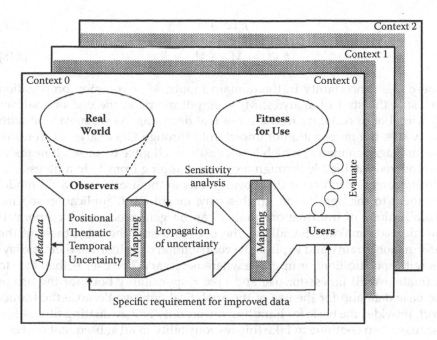

FIGURE 8.30
A communications-based framework for handling uncertainty and fitness-for-use in GIS.

possible resolution and the recording of metadata are, thus, important elements in judging reliability. Aggregate measures of quality may be produced and, where recorded in the metadata, will need to be referred to by the user. Nevertheless, these may not be sufficiently discriminating to give an indication of the spatial variability in quality. For extensive thematic layers compiled possibly by several observers or on different occasions, such variability is likely to be an important component in judging reliability and ultimate evaluation of fitness-for-use of GIS products. Therefore, observers should preferably record quality measures pertaining to individual objects or entities within a thematic layer, such as in the fuzzy set example above.

Where a number of thematic layers are to be used in an analysis, it may be necessary to integrate a range of quality measures in the propagation of uncertainty. Current research has focused on propagating a specific individual quality measure (such as PCC, variances, or probabilities) common to all the thematic layers. Instead, the framework provides for a mapping, in the mathematical sense of $f: X \rightarrow Y$, from a range of quality measures into a common propagation metric M. A prime candidate for M are fuzzy measures (fuzzy sets, fuzzy numbers) into which measures of possibility, plausibility, belief, certainty, and probability can be transformed (Graham and Jones, 1988). The use of M and accepted mappings into it would provide observers with greater flexibility in their choice of appropriate quality measures. Thus, the overall mapping from a domain of inputs to a set of real decisions becomes

$$f: \Omega \rightarrow \Re \qquad (8.17)$$

$$f: \Omega_u \rightarrow M_s \rightarrow M_o \rightarrow \Re_u \qquad (8.18)$$

where Ω_u = uncertainty in the domain inputs, M_s = common propagation metric at the start of analysis, M_o = output metric at the end of analysis, R_u = level of uncertainty in the set of real decisions. As we have seen with fuzzy sets, the propagation of uncertainty through GIS analyses can result in a propagation metric, which is not easily intelligible to a user. It needs to be rendered intelligible through a second mapping from M to a fitness-for-use statement or indicator that corresponds to the user's real world model pertinent to the application. In this way, meaningful, application-specific visualizations of information quality can be generated and incorporated into decision making. SA allows the user to assess the robustness of the information quality and explore the contribution of individual thematic layers with specification for upgrade where necessary. The user is then able to evaluate overall fitness-for-use and take responsibility both for the use of the base data and for the use of the analytical outputs. Because the framework provides the basis for handling uncertainty and evaluating fitness-for-use, users can continue to take this responsibility in all subsequent contexts where the base data are made available for use. Data can have a long shelf life and evaluations of their fitness-for-use in applications should be made possible over their entire life. This framework is intended for implementation with existing data structures in GIS software. For software developers, it provides a framework on which to develop suitable functionality for the handling of uncertainty.

9

Modeling Issues

In Chapter 8, we looked at a range of issues that arise from uncertainty in spatial data, the impact these can have on the analytical products of geographic information systems (GIS) and by implication on the accuracy of environmental simulations. We also saw how these issues might be managed. Sources of uncertainty can be numerous (Figure 8.6) and can be difficult to disentangle. In Chapter 8, the influence of operational uncertainty was played down to concentrate on data issues. In a coupling of GIS and environmental simulation modeling (Chapter 7), operational uncertainty can derive from the nature of the algorithms in GIS and from the operation of the simulation model: choice of spatial discretization and time increment, fixing of parameters, algorithm choice, and calibration. These issues together can be classed as model-induced uncertainty. According to Burrough et al. (1996):

$$Information = Conceptual\ models + data \tag{9.1}$$

and that the link between quality of information, models, and data can be expressed as:

$$Quality\ [information] = f\{quality\ [model],\ quality\ [data]\} \tag{9.2}$$

where "model" encompasses both GIS and environmental simulation model. The issue should not be underestimated. We have already seen the tension that exists between natural variation in the real world and the data models that we use in GIS and simulation models. Thus, data sets are only approximations of reality. Simulation models are also approximations as "very few Earth science processes are understood well enough to permit the application of deterministic models" (Isaaks and Srivastava, 1989). And yet even if we wish to strive toward perfect models, how would we know when we have one because "verification and validation of numerical models of natural systems is impossible" (Oreskes et al., 1994). The answer to this dilemma does not necessarily lie in stochastic models because there are consequent problems in identifying appropriate probability distributions for all the parameters and then there is still the chestnut of verification. However, if we are modeling in the search for engineering solutions rather than purely for the pursuit of science, then we should take a sufficing approach in which the quality of the information need only be dependable in that it contributes toward an effective solution to some problem (Chapter 8). Thus, there is another variable to Equation (9.2)—professional judgment. But, this cuts two ways: first, in interpreting the significance of the analytical products that are the outputs to the environmental modeling process and, second, at an earlier stage in the very choice of data, data processing algorithms, type of simulation model, setting of parameters, achieving an acceptable calibration, and so on. Equation (9.2) appears too deterministic as a good quality model, and good quality data in the hands of an inexperienced modeler may not give dependable results. Therefore, I would propose that:

$$Dependable \text{ [information]} = f\{quality \text{ [model]}, quality \text{ [data]},$$
$$experience \text{ [professional]}\} + \varepsilon \qquad (9.3)$$

where ε = residual uncertainty (but, not necessarily random error).

In this chapter, we will be looking at Equation (9.3) from the perspective of the right-hand side of the equation. Given the plethora of environmental simulation models used by a wide range of disciplines in a large number of situations, I would not be so presumptuous as to evaluate their performance and arrive at a critique of their quality. I leave that to each discipline to establish the means of consensus on the usefulness and applicability of its models. Users should make themselves aware of the assumptions and limitations of models before using them in support of decision making. Data quality issues have already been covered in Chapter 8. What does need to be looked at here are issues around models where professionals need to make choices. By the end of the chapter, I will tie this in with the data quality debate. In Chapter 10, then, we will be looking at Equation (9.3) from the perspective of the left-hand side, i.e., making decisions by dealing with any risk in the residual uncertainty.

Issues of Scale

Scale has doggedly pursued us through this book and, in GIS and environmental modeling, you just can't get away from it. In Chapter 2, scale was an issue affecting how we might model or represent features in GIS. Linked to this was the notion of the characteristic scale (in space–time) at which a particular natural phenomenon is manifestly measurable/observable. We also considered what the meaning of traditional cartographic scale might be for a digital data set without the constraints of the fixed dimensions of a piece of paper (Equation (2.2)); here, resolution was a key determinant. In Chapter 4, we identified an envelope of spatio-temporal domains for process modeling (Figure 4.9), which not only pointed to broad links between spatial and temporal resolution in the construction of models, but that one process model cannot fit all scales. In other words, there are theoretical limits to the scalability of models. In Chapter 8, we considered the problems that arise from natural variation in that spatial units are invariably portrayed as being more homogenous than they are in reality. We also saw how the variogram can be used to measure spatial dependence, but also to describe the scales of spatial variation observable within the data (Atkinson and Tate, 2000). We have already seen that in most simulation modeling there is a trade-off between data resolution and computation time (Chapter 5). There are also implications here for cost of data collection (cost-resolution modeling, Chapter 3) and the fitness-for-use of the modeling outputs. Practical decisions need to be made here, and it's not always a case that smaller is better.

Sensitivity to scale and resolution in process models can be viewed slightly differently from the perspective of lumped parameter models and distributed parameter models. In the latter, the problem focuses on appropriate size of the cells within the grid or triangular mesh. In the former where spatial discretization is often implicit, there is nevertheless some spatial extent that is being modeled, which, for example, in hydrology will be catchment units and yet an area is almost infinitely divisible into catchment units. The question of catchment size and grid size and their effect on parameterization and uncertainty in simulation results has been a focus of attention in hydrology for some time, particularly to separate out which environmental controls are scale dependent and which are not (for two reviews, see Wood et al., 1990 and Clifford, 2002). This is not to play down the importance of other aspects of spatial variability, such as in rainfall (e.g., Arnaud et al., 2002), but it is now well established that there is a tendency for topographic variability to dominate predicted spatial patterns of storm runoff, particularly for small catchments. In progressively reducing the size of catchment area by discretizing to smaller subcatchments, the resulting increase in resolution should lead to a reduction in the variance in subcatchment response. This led Wood et al. (1988) to propose the existence of a threshold resolution or *representative elementary area* (REA) as the fundamental building block for catchment modeling. The REA represents the critical resolution below which it is necessary to account for internal heterogeneity. This was found in their studies to be about 1 km^2. For spatial units larger than the REA, only the statistical representations of control variables (e.g., mean values) need to be known. This has implications for distributed parameter models where generally the discretization is less than 1 km^2 and thus will require a fuller specification of the variables.

Zhang and Montgomery (1994) studied the use of a high-resolution digital elevation model (DEM) in two small catchments in order to study the effects of changing grid size on parameter estimation and simulation of hydrographs on very small catchments (1.5 km^2). They found that the grid size of the DEM significantly affected both. They concluded that a 10 m grid size represented "a reasonable compromise between increasing spatial resolution and data handling requirements" for topographically driven models. Although, in

large drainage basins, the hydrograph will be dominated by channel routing, the influence of DEM grid size on the production of predicted runoff
should be an important consideration in interpreting simulations. On the
other hand, small grid sizes will undoubtedly raise problems concerning
the accuracy of the DEM at that resolution with resultant errors in parameter
determination propagating to the simulation. Bruneau et al. (1995) have also
carried out a sensitivity analysis of grid size and time step on runoff simulation for a 12 km^2 catchment. They found that choices of grid size and time
step were not independent and that there is an optimum region of values for
model building. In their study, this was a grid size of less than 50 m and a
time step of 1 to 2 h. Significantly, larger grid sizes with medium time steps
were found to result in some parameter values to be meaningless, thus giving
inconsistent runoff simulations. They also found that degraded outputs were
more sensitive to larger time steps than to larger grid sizes. Molná and Julien
(2002) have tested a distributed parameter model (CASC2D) on two basins of
21 km^2 and 560 km^2 using grid sizes from 127 to 914 m. This becomes a case of
needing less resolution to model larger features, particularly where computation time becomes intractable with smaller grid sizes. For the smaller basin,
grid sizes up to 380 m were acceptable provided calibration was carried out
to upwardly adjust overland and channel roughness coefficients. However,
these grid sizes are approaching the REA (above). Grid size for the larger
basin was found to be critical for shorter rainfall events where equilibrium
conditions are unlikely to be met. This takes us straight back to Chapter 4,
Figure 4.9: we can't disentangle space–time and, by the same token, we can't
treat them in a disjointed way in our process models. Modeling large, rapid
events is likely to require a different type of model.

While the above findings to do with scale have been explored in the context of surface hydrology, the same broad relationships will apply equally to
the modeling of other processes (Li, 2007). In Chapters 5 and 6, we looked at
coastal oil spill modeling, first at the hydrodynamic modeling that distributed the tidal currents across the study area and then at the trajectory of the
floating oil. One constituent of the trajectory model was the spreading action,
mostly a random component, but dominated by the tidal and wind-blown
currents. Modeled over 3 h on a 200 m grid, an oil spill drifted toward the
coast to eventually make landfall on a beach (Figure 9.1(a/f)). Figure 9.1(g/l)
is a further simulation using a 400-m grid. There are noticeable differences,
but at the same time broad similarity—the oil still ends up on the same
beach. The density values of cells are certainly different. The model has lost
its resolving power for the smaller currents; if some oil just moves across a
grid boundary, then it moves the full 400 m of the grid. At the same time, it is
important not to go for a too fine of a grid. The interpolation of currents from
the larger finite element network to the smaller trajectory modeling grid
leads to a reduction in the size of the current component at each grid node
because of the shorter step in the calculations. If the grid size gets too small,
then the random component in the spreading starts to dominate over tide

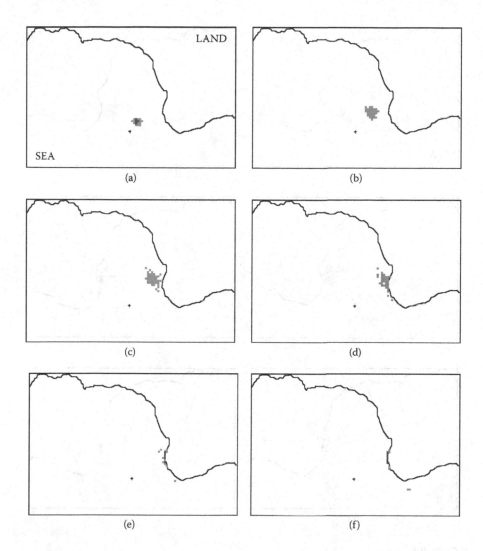

FIGURE 9.1
An example of the consequences of changing grid size on oil spill trajectory modeling: (a) to (f) 200-m grid, (g) to (l) 400-m grid. *Continued*

and wind to produce an unacceptable result. The model becomes unstable. Hence, finer resolution is not necessarily more accurate.

There, of course, may be quite legitimate occasions when a fine resolution is needed, particularly in engineering design. The case study presented by Chen and Brimicombe (1997) and Chen et al. (1998) is of Arha Reservoir in Guizhou Province in southwest China. The capacity of the reservoir is 44.5 million m³ and is the principal source of fresh water for Guiyang City. About 2,000 metric tons of suspended solids are transported into the reservoir from upstream annually, of which some 770 metric tons are Fe (iron)

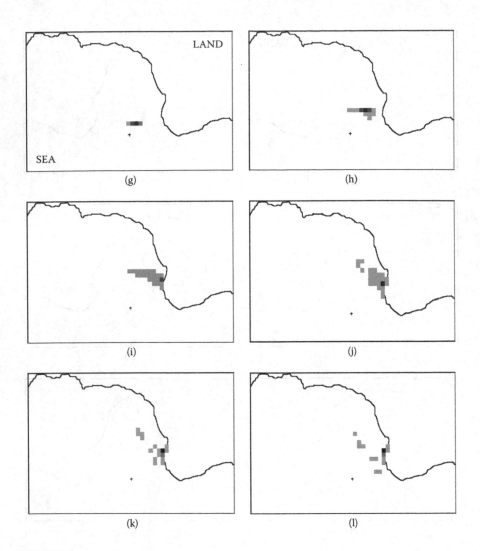

FIGURE 9.1
Continued.

and 50 metric tons are Mn (magnesium). Pollution of Fe and Mn has become a serious problem for the water supply of Guiyang City. The transport of these two heavy metals through the reservoir was modeled using a 3D reservoir quality model on a 50-m grid and with five layers of depth each 4.5 m thick and calibrated against 9 sample points where 22 days of records were collected. The objective was to study the concentrations of heavy metals not only throughout the reservoir, but also at the draw-off outlet from where the city got its water supply. The outlet is itself a relatively small feature in this large reservoir and would require a higher resolution of detail in order to adequately evaluate the problem. A 2-m grid would have meant several orders of larger magnitude model to run and, in any case, would have become unstable. The solution was to run the 50-m model and use the results to construct a variogram for each metal in each of the reservoir layers around the outlet. This was then used to krige a 25-m grid that could then be checked by running the model again on a 25-m grid. With a close fit between the kriged and modeled results at 25 m, the kriging was used to interpolate a 2-m grid around the outlet in order to study the concentrations of the heavy metals and their dynamics. This showed the outlet to be in an area of high Fe and Mn concentration, which could be ameliorated by extending the outlet farther into the reservoir and drawing off from the surface layer. This is an example of how GIS-based postprocessing of simulations can enhance the information content for decision making.

Another approach to testing model sensitivity to scale effects of discretization is through *fractals*. A fractal is defined as a fractional dimension. We are well familiar with the definitions of a point, line, polygon, and solid as being 0-, 1-, 2-, and 3-dimensions, respectively. But, suppose you were to draw a 1D line in a zigzag (ИИ) so close together that its final appearance was a 2D shape (■). At what point would the 1D line suddenly metamorphose itself into a 2D shape? Or can we work with the idea that as the 1D line gets increasingly sinuous and space-filling, so its dimension increases as a fraction from 1.0 (a perfectly straight line) to an upper limit of 1.99, that is, just short of being a 2D shape. The same idea can be envisaged for a 2D surface being folded into an almost 3D shape. The concept of a fractal dimension was first introduced by Mandelbrot (1967; 1983) who argued that our traditional Euclidean view of dimensions was, in fact, a special case, just as we saw in Chapter 8 that a Boolean (0, 1) turns out to be a special case of fuzzy sets. Fractals have found application, for example, in the study of land form (Goodchild, 1980; Lam and Quattrochi, 1992) and urban form (Batty and Longley, 1994). Fractals can be measured by the Hausdorff dimension (Harris and Stocker, 1998):

$$D = \frac{\log N}{\log s} \qquad (9.4)$$

where D = Hausdorff dimension, N = number of segments, s = the scaling factor by which the number of segments increase.

This can be illustrated by Koch's curve (Helge von Koch, a Swedish mathematician, first described the properties of this curve in 1904), which is a self-similar decomposition (Figure 9.2) in which progressively the middle third of every segment is removed and replaced by two new segments that form an angle of 60°. The dimension D is independent of the number of segments N and the scaling factor s. Thus, for the three decompositions in Figure 9.2, D remains 1.262. In other words, the self-similarity of the decomposition maintains the same fractal dimension of the line, although, as can be calculated, the length of the line increases by a factor of 4/3 with each successive

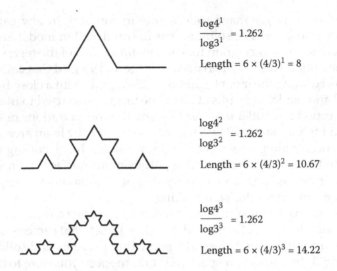

$$\frac{\log 4^1}{\log 3^1} = 1.262$$

Length $= 6 \times (4/3)^1 = 8$

$$\frac{\log 4^2}{\log 3^2} = 1.262$$

Length $= 6 \times (4/3)^2 = 10.67$

$$\frac{\log 4^3}{\log 3^3} = 1.262$$

Length $= 6 \times (4/3)^3 = 14.22$

FIGURE 9.2
Koch's curve with self-similar decomposition.

decomposition, while the Euclidean distance between the start and the end of the line (distance A, B) remains the same. With sufficient decomposition, the distance along Koch's curve from A to B could approach infinity. Another self-similar pattern is the Peano scan in Chapter 2, Figure 2.12(b), which, because it is systematically space-filling of a 2D surface and contains self-similar recursions, has a D of log 15/log 4 = 1.953; it's a line, but nearly a surface.

The decomposition of Koch's curve is like discretization at increasing resolution. This concept has been used by Li et al. (2000) to study the effect of shoreline resolution and roughness on the hydrodynamic simulation of tidal currents. The importance of this lies in accurately simulating the trajectory of oil onto a coastline while not over-discretizing with obvious cost implications for survey, nor inducing instability in the model. Thus, the exercise can be seen as cost-resolution modeling as well. Koch's curve is too geometrically regular to simulate a coastline, so a variant on the principle was used to produce a *fractional Brownian motion* (fBm) for a section of concave and convex coastline (Figure 9.3). If a white noise is a completely random series of frequencies f such that the mean square fluctuations are $1/f^0$, then the most common type of noise found in nature is $1/f$ and a Brownian motion is defined as a $1/f^2$ noise (Voss, 1988). An fBm can be viewed as a self-similar decomposition, but with a random element, a sort of random walk along a fractal dimension. Thus, from an initial straight coastline of 10 km, a complex, rough coastline can be calculated in a systematic way that would be representative of increased resolution of the sampling. The effect on the hydrodynamic modeling is given in Figure 9.4. Remember from Chapter 6 that U is the northerly component of the computed tidal current, V the easterly

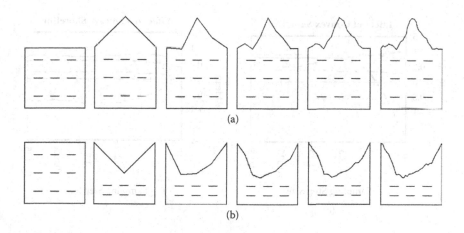

FIGURE 9.3
Synthetic convex and concave shorelines generated using fractional Brownian motion (fBm). (From Li, Y. (2001) Spatial data quality analysis in the environmental modelling. Unpublished PhD dissertation. University of East London, U.K. With permission.)

component, a the current speed, and g the direction. Given that our coastline is in the north of the model, Vg will be most affected because this component running along the shoreline is the most sensitive to the changing resolution. From the graphs in Figure 9.4, when the simulated sampling step-size is less than 2.5 km, the statistical properties (mean and standard deviation) of all the current components have stabilized. A shoreline sample interval of 2.5 km, therefore, can be taken as the minimum survey requirement in order to have a stable tidal current simulation.

The final aspect of scale to be considered here is *aggregation*. Aggregation is the act of joining smaller spatial and/or temporal units into larger units of coarser resolution. There may be several reasons for doing this mostly based on the need to reduce complexity. One reason might be to simplify inputs and outputs so as to ease the modeling burden (or even to make modeling feasible) or in "scaling up" models from local to regional levels. Another might be to reduce the level of noise often present at a higher resolution, thus allowing patterns and relationships to "emerge." Yet, a third reason for aggregation is that policy makers would like to see the "broad picture" rather than all the scientific detail. Aggregation, however, is not without its inherent problems, often referred to in geography as the modifiable areal unit problem (MAUP) (Batty, 1974; Openshaw and Taylor, 1981; Fotheringham and Wong, 1991; Amrhein, 1995). The dual of the MAUP is the ecological fallacy (Robinson, 1950). Both of these can best be explained through the use of a simple example. In Figure 9.5(a) and Figure 9.5(c) are mapped the percentage cover of two species of plants that tend to thrive under the same conditions and would be expected to be correlated. The percentage cover is mapped by quadrat and then by different zone aggregations

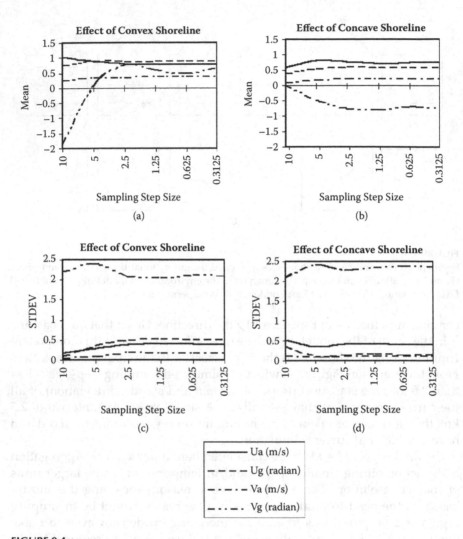

FIGURE 9.4
Effect of increasing shoreline resolution on the mean and standard deviation of simulated tidal components. (From Li, Y. (2001) Spatial data quality analysis in the environmental modelling. Unpublished PhD dissertation. University of East London, U.K. With permission.)

of those quadrats. The most detailed representation is 25 quadrats ($N = 25$), but these have then been aggregated to 7 zones ($N = 7$) and then to 4 zones ($N = 4$). What is more, a choice of two zone arrangements are offered for $N = 7$ and $N = 4$. The series $N = 25$ through $N = 4$ will give scale effects, while differences between the choice of zones at $N = 7$ and $N = 4$ will give zone effects. Figure 9.5(b) and Figure 9.5(d) show summary statistics for all maps. The noticeable trend is that the mean value percentage covers for both species stays roughly the same, whereas the variance s^2 generally declines as the number of zones decreases. This is the scale effect. The difference in variance that occurs between the two maps for $N = 7$ and $N = 4$ are influenced by the specific arrangement of zones in relation to the original data. If we now look at the relationship between the two plant species, we can create regression models and calculate correlations to study the effect on these of scale and zoning (Figure 9.5(e)). Here the *map cross-correlation* (MCC) has been used:

$$MCC = \frac{\sum_{i=1}^{n}(A_i - \bar{A})(B_i - \bar{B})}{\sqrt{\sum_{i=1}^{n}(A_i - \bar{A})^2(B_i - \bar{B})^2}} \quad (9.5)$$

where A_i and B_i = values of spatial elements i in maps A and B, $\bar{A}$ and $\bar{B}$ = average values for maps A and B.

Here again we can see scale and zone effects primarily due to a reduction in variance. In general, with aggregation, both correlation and R^2 will tend to rise. However, as can be seen from the zone effect, the scale effect is not exactly predictable. Looking at Figure 9.5 from $N = 4$ back to $N = 25$ reveals the ecological fallacy; patterns and relationships at a higher level of aggregation cannot be directly inferred on a lower level of aggregation nor attributed to an individual. Looking again at Figure 9.5(a) and Figure 9.5(c), very few of the values that appear in $N = 7$ or $N = 4$ appear in $N = 25$, just as you can't find an average family with 2.2 children. This has some potentially serious impacts on the way in which we manipulate scale and zone data. For example, empirical regression models can be made to have a better fit (R^2) through aggregation. Clearly, as a general rule, analyses and models should be built from appropriate resolution data but, as already stated, there may be very valid reasons for having to aggregate. Bian and Butler (1999) have studied the effects of aggregating data necessary for moving from local models to regional or even global models. In such "scaling up," the original spatial data needs to be reduced to a smaller number of units each covering a larger area. The output of such models may be adversely affected by the altered statistical and spatial characteristics of the data. They studied the effects of three aggregation modes: using the average, the median, and the value of the pixel or cell central to the aggregated unit. The average and median methods were able to maintain the mean and median of the original data across the full hierarchy of aggregation, although the variance in the data was significantly reduced. This basically conforms to our example above. The central pixel method, however, performed poorly in maintaining the central tendency of the data, while increasing somewhat the level of variance. Whereas the mean and median methods tended toward increasingly homogeneous data, the central pixel method was better able to maintain basic spatial patterns within

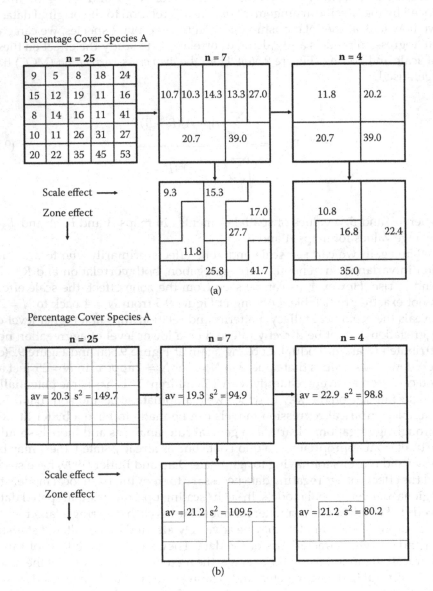

FIGURE 9.5
Illustrative example of the modifiable areal unit problem; see text for details: (a) to (e).

Continued

Percentage Cover Species B

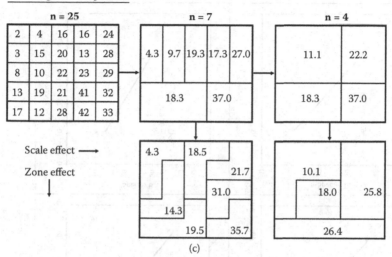

(c)

Percentage Cover Species B

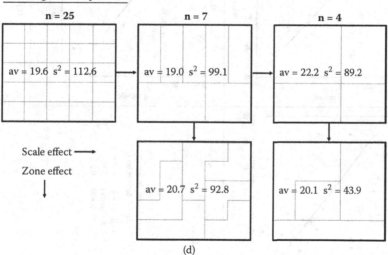

(d)

FIGURE 9.5
Continued.

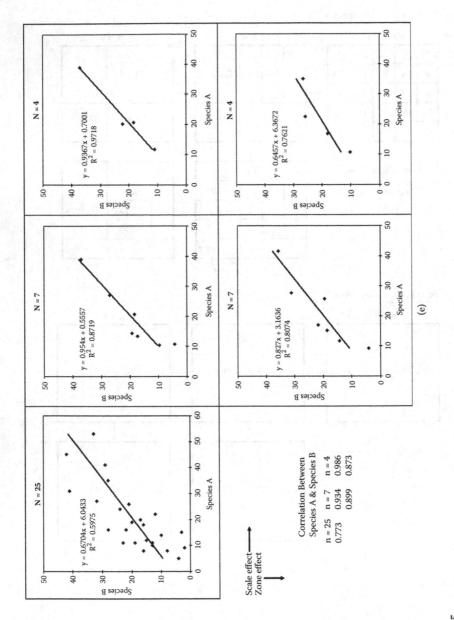

FIGURE 9.5
Continued.

the data. Once again, the variogram of the original data can inform appropriate aggregation levels so that they fall within the autocorrelation range in an attempt to confine the magnitude of uncertainty that might result. Van Beurden and Douven (1999) have studied the problem of aggregating for policy making and political decisions about the regulation of pesticides. They considered two approaches:

1. *Input aggregation*: Local data are first aggregated and then the model is computed at a national level.
2. *Output aggregation*: The data at the higher resolution are first used in the model and the results are subsequently aggregated.

This implies that a scalable model is being used. In this case, a pesticide leaching model is used to calculate the ratio of *predicted environmental concentrations* (PECs) to *no effect concentration* (NEC). A PEC/NEC ratio of greater than one indicates hazardous areas. The specifics of the aggregation were from a 2.5-km grid through to municipalities and provinces in The Netherlands. Two methods of aggregation were used: aggregation by the mean and aggregation by the worst case. The results are shown in Figure 9.6(a/b). Aggregation by worst case gave consistent results for both input and output aggregation, but over exaggerated the level of hazard at municipal and provincial levels. Aggregation by mean gave contrasting results for input and output aggregation with input aggregation resulting in an under exaggeration of the hazard. Certainly in all cases valuable information on spatial variability is lost. Since the results of any similar study are likely to be sensitive to model type, the spatial data model and the aggregation method, van Beurden and Douven recommend that the aggregation sensitivities be tried out before handing results to policy makers.

Issues of Algorithm

Any user of GIS and environmental models often has a bewildering choice of algorithms to choose from. Of course, not all competing algorithms are implemented in every software package and, for some proprietary software, information on the exact algorithms being used may be sparse, sometimes no better than a vague hint. Nevertheless, algorithm choice does often present itself and making the right choice can be important for successful modeling. As we saw in Chapter 4, the success or otherwise in using inverse distance weighted (IDW) interpolation (as one algorithm) rested on choice of parameters (such as r, Figure 4.13) and even the data sampling pattern (Figure 4.12), so choice between competing algorithms is not a clear-cut case, as it also

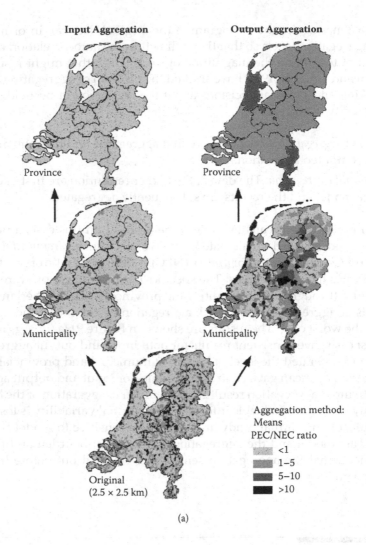

FIGURE 9.6
Comparison of input and output aggregation strategies: (a) aggregation by mean, (b) aggregation by worst case. (Adapted from van Beurden, A.U.C.J., and Douven, W.J.A.M. (1999). *International Journal of Geographical Information Science* 13: 513–527.)   *Continued*

depends on how you use them in relation to the application and the available data. Nevertheless, professionals should acquaint themselves not only with the range of algorithms that are available together with their advantages and disadvantages for use in particular situations, but should also be able to understand at least the broad consequences of fitness-for-use. Staying with the interpolation example, a popular alternative to IDW is triangulated irregular network (TIN; Chapter 2, Figure 2.11) and, in Chapter 8, we used kriging on the rainfall data. We can briefly compare the performance of these

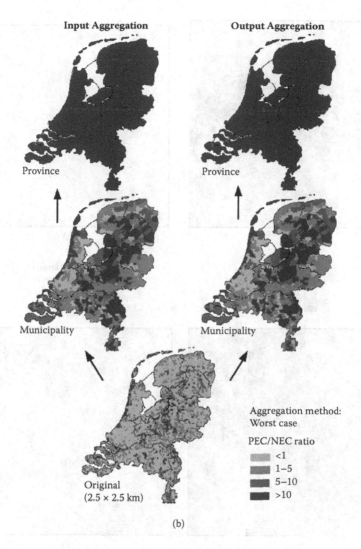

Input Aggregation   **Output Aggregation**

Province   Province

Municipality   Municipality

Aggregation method:
Worst case

PEC/NEC ratio

<1
1–5
5–10
>10

Original
(2.5 × 2.5 km)

(b)

FIGURE 9.6
Continued.

three data sets. To recall, Figure 9.7(a) shows the mathematically computed topography with 10-m contours and 50-m shading. Figure 9.7(b) shows the purposive sampling of 25 points plus the four corner points to bring interpolation to the edge of the study area without extrapolation outside the convex hull subtended by the sample points. Figure 9.7(c/d) shows the results we previously gained for IDW where $r = 4$ (shaded every 50 m with original contours superimposed for comparison) together with the distribution of residual errors giving a *root mean square error* (RMSE) of ±19.5 m. Figure 9.7(e) shows the results of an interpolation using TIN (again shaded every 50 m

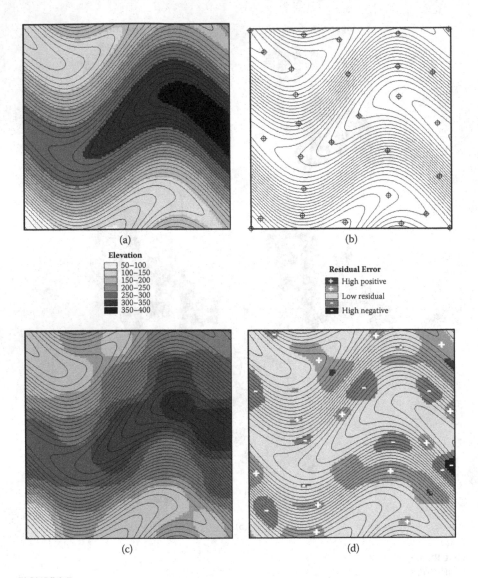

FIGURE 9.7
Comparison of IDW, TIN, and kriging methods of interpolation: (a) mathematically computed, (b) sample, (c) IDW with (d) residual errors, (e–f) TIN and residual errors, (g–h) kriging and residual errors.   *Continued*

with original contours superimposed for comparison) with residual errors in Figure 9.7(f) giving an *RMSE* of ±26.9 m. For this purposive sample where points have been placed to pick out as best as possible local maxima and minima on the surface, TIN for the most part is quite good. The problem that has arisen and is contributing to the high RMSE is with very slim triangles being formed along the boundary truncating the ridgeline. For this reason,

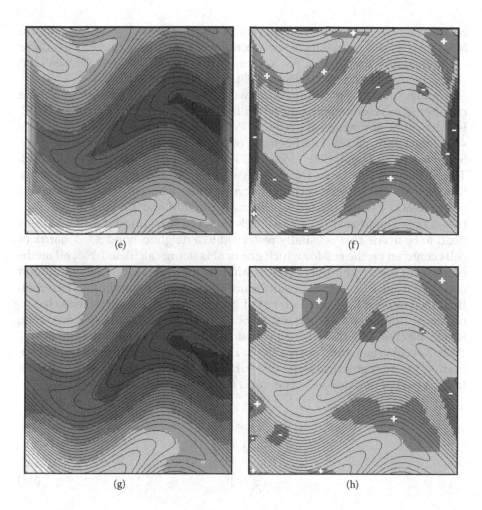

FIGURE 9.7
Continued.

the convex hull resulting from the triangulation of sample points should preferably fall well outside the study area boundary so that this effect can be ignored. TIN tends to produce rather rectilinear-looking results because each triangle is treated as a plane. For this reason, TIN is not very good on sparse data sets. Figure 9.7(g) shows the results of kriging using a spherical variogram model with residual errors in Figure 9.7(h) giving an RMSE of ±15.7 m. Although still not a very good result in relation to the original map, kriging has given the best result of the three and providing that the variogram can be properly constructed and fitted, this geostatistical approach is finding increasing favor among modelers. The point to be made, however, is that all interpolations are approximations and depending on the nature of the sample data and the phenomenon to be modeled by interpolation, the modeler needs to exercise considerable skill in achieving a best fit particularly where (and unlike the above examples) no "truth" model is available for immediate comparison. For a review of interpolation methods, see Lam (1983), Burrough and McDonnell (1998), or de Smith et al. (2007).

Another area of algorithm choice worth looking at in relation to quite a number of environmental simulation models that use routing over a DEM, is calculation of maximum gradient. Within the mapping sciences and GIS, such algorithm choice has been an ongoing debate (e.g., Skidmore, 1989; Srinivasan and Engel, 1991; Hodgson, 1995; Jones, 1998; Schmidt et al., 2003). The problem is that there are so many variant algorithms—at least 12 to my knowledge—that objective comparison becomes a lengthy task. Calculation of gradient is usually carried out from a grid DEM and, because it cannot be based on a single grid cell value of elevation, a surrounding group of cells need to be used. This is usually restricted to a neighborhood 3 × 3 matrix of cells centered on the cell for which gradient is being calculated. Not all methods use all the neighboring cells within this matrix. A larger neighborhood could be used, but this might introduce unacceptable smoothing. Based on this 3 × 3 matrix, there are two broad groups of approaches: (1) those based on finite difference techniques (e.g., Sharpnack and Akin, 1969; Horn, 1981; Ritter, 1987) and (2) those based on calculating a best fit surface (e.g., Evans, 1980; Zevenbergen and Thorne, 1987). These are different approaches to estimating the mathematical slope of a surface G that can be calculated by finding the partial derivatives in the x and y direction as follows:

$$\frac{\partial z}{\partial x} = \beta_1 \tag{9.6}$$

$$\frac{\partial z}{\partial y} = \beta_2 \tag{9.7}$$

$$\tan G = \sqrt{[\beta_1^2 + \beta_2^2]} \tag{9.8}$$

To illustrate some of the finite difference solutions, consider the 3 × 3 matrix of cells in Table 9.1 in which elevations are coded $z0$ through $z8$.

TABLE 9.1

Cell Coding for Finite Difference Calculation of Gradient

	$z5$	$z2$	$z6$
↑ Y	$z1$	$z0$	$z3$
	$z8$	$z4$	$z7$

X →

A second-order finite difference method utilizes only four cells adjacent to the target cell (Ritter, 1987):

$$\frac{\partial z}{\partial x} = [z1 - z3]/2\Delta x \tag{9.9}$$

$$\frac{\partial z}{\partial y} = [z2 - z4]/2\Delta y \tag{9.10}$$

A third-order finite difference method uses all the 3×3 neighbors (Sharpnack and Akin, 1969):

$$\frac{\partial z}{\partial x} = [(z8 + z1 + z5) - (z7 + z3 + z6)]/6\Delta x \tag{9.11}$$

$$\frac{\partial z}{\partial y} = [(z7 + z4 + z8) - (z6 + z2 + z5)]/6\Delta y \tag{9.12}$$

A variation on this method proposed by Horn (1981) and commonly implemented in GIS software, additionally weights the second-order neighbors thus:

$$\frac{\partial z}{\partial x} = [(z8 + 2 \cdot z1 + z5) - (z7 + 2 \cdot z3 + z6)]/8\Delta x \tag{9.13}$$

$$\frac{\partial z}{\partial y} = [(z7 + 2 \cdot z4 + z8) - (z6 + 2 \cdot z2 + z5)]/8\Delta y \tag{9.14}$$

An example of surface fitting to all nine elements in the 3×3 matrix is to fit a least-squares plane (first-order polynomial) in the form:

$$z = \beta_0 + \beta_1 x + \beta_2 y + \varepsilon \tag{9.15}$$

such that if the error term ε is ignored, β_1 and β_2 can be directly substituted into Equation (9.8). Quadratic fitted surfaces (Evans, 1980) take the form:

$$z = \beta_0 + \beta_1 x + \beta_2 y + \beta_3 x^2 + \beta_4 y^2 + \beta_5 xy + \varepsilon \tag{9.16}$$

$$\frac{\partial z}{\partial x} = \beta_1 + \beta_3 2x + \beta_5 y \tag{9.17}$$

$$\frac{\partial z}{\partial y} = \beta_2 + \beta_4 2y + \beta_5 x \qquad (9.18)$$

when the error term ε is ignored. This is only a selection of methods, and there are other variants. Skidmore (1989) tested six methods including all of the above. Srinivasan and Engel (1991) tested four methods including Horn's, best-fit plane, and quadratic surface. Jones (1998) tested eight methods including all of the above except the best-fit plane. Skidmore, and Srinivasan and Engel both use 30 m DEM of natural terrain. Skidmore checked computed gradients against hand calculations from map contours. Srinivasan and Engel evaluated their computed gradients through its use to calculate the length of slope and steepness of slope factors in the Universal Soil Loss Equation for which observed values were available. Jones used a synthetic surface (similar to our example topography, but generated from a 49-term polynomial) and different relative grid sizes for which true values could be known by numerical methods. The results on the synthetic surface for Formula (9.11) to Formula (9.18) above showed that the second-order finite difference (Ritter) performed best overall followed by Horn and Sharpnack and Akin, and then the quadratic surface. Increasing accuracy followed decreasing grid size. When the synthetic surface was rotated, Ritter's showed greater sensitivity to rotation angle than the other methods, which becomes more pronounced as grid size is reduced. Nevertheless, the RMSE remained lower at all times for Ritter's than for the other methods. Skidmore found a statistically significant correlation between all methods used and the "true" gradient. However, the finite difference approaches yielded lower correlations with Ritter and Sharpnack and Akin giving the worst results. Skidmore also noted that spurious gradients could be calculated regardless of method in areas of flat or near flat terrain. This has important implications, for example, for routing in floodplain areas. Finally, Srinivasan and Engel found that Horn's finite difference approach was more accurate on flatter slopes than on steeper slopes where the 3×3 matrix covered too great an area in relation to the length of slope on steeper sections, and was nevertheless the most accurate overall. Differences in algorithm can make a difference of up to 286% on the calculation of the length of slope and steepness of slope factors with considerable variability all around. They conclude that "careful selection of slope prediction method is recommended." These three studies from a comparative perspective have some contrasting results on which method might be better, but from another aspect there is a close agreement, like the interpolation algorithms, each method of calculating gradient will yield a different result. For some simulation models the effect may be negligible, for others it may be more important. Where there is likely to be model sensitivity, that sensitivity should be tested.

A further example of how algorithm choice can be problematic occurs when preparing data for use in an artificial neural network (ANN). Usually the data are transformed in the range [0, 1] or [1, 1] and there are a number of methods of doing this that fall under three broad categories: linear transformation, statistical standardization, and mathematical functions. Linear transformation is simpler and appears to be most frequently used. There is often little guidance as to which method would best suit the type of data being used and little in the literature to suggest that it matters. However, the results of Shi (2000) show that it does matter with improvements in model performance by up to 50% for synthetic trials and up to 13% on a real case study depending on the type of transformation used.

Issues of Model Structure

When working either in a geocomputational mode or in complex modeling situations, it is unlikely that the scientist or professional will be working with just a single tool. It is likely instead that several different tools are used. The way these are configured and pass data may have important influences on the outcome. In Chapter 6 and subsequently, we have seen how in coastal oil spill modeling there are, in fact, three models that cascade. Initially there is the hydrodynamic model that from bathymetric, shoreline, and tidal data calculates the tidal current over the study area using finite element method (FEM). In the next stage, these tidal currents together with other data, such as wind and the properties of the particular type of oil, are used in the oil spill trajectory model. The trajectory model is a routing model requiring only arithmetic calculation and, therefore, is carried out on a grid. However, this requires a reinterpolation of the tidal current from a triangular network to a grid. As we have noted above, not only is there algorithm choice for reinterpolation, but that there is likely to be some level of corruption of the output data from the hydrodynamic modeling as it is transformed to a grid by the chosen interpolation algorithm. Experiments by Li (2001) have shown that errors in the bathymetry and tidal data will be propagated and amplified through the hydrodynamic modeling and affect the computed currents. Interpolation of those currents to a grid further degrades the data by increasing the amount of variance by about 10%. This is then propagated through the next stage of modeling. These are inbuilt operational errors that are a function of overall model structure. Because different components or modules within the overall modeling environment work in very different ways such that they cannot be fully integrated but remain instead tightly coupled, then resampling or reinterpolation becomes necessary. Both model designers and model users, however, should be more aware of and try to limit the effects.

Another aspect of model structure that is hard to guard against is inadvertent misuse. While blunders, such as typographic errors in setting parameters, are sure to occur from time to time and will normally manifest themselves in nonintuitive outputs, there are subtle mistakes that result in believable but wrong outputs. In hydrodynamic modeling, for example, the forced tidal movement at the open boundary requires a minimum number of iterations in order for its effect to be properly calculated throughout the study area. For a large network with many thousands of elements in the triangular mesh, this may take many iterations at each time step. The model usually requests of the modeler the number of iterations that should be carried out; too many can be time consuming for a model with many thousands of time steps, but too few can give false results. Figure 9.8(a/b) shows the results of hydrodynamic modeling for an adequate number of iterations at 0 h and at 2 h. The tide is initially coming in and then starts to turn on the eastern side of the study area. In Figure 9.8(c/d), the exact same modeling has been given an insufficient number of iterations at each time step to give the correct answer.

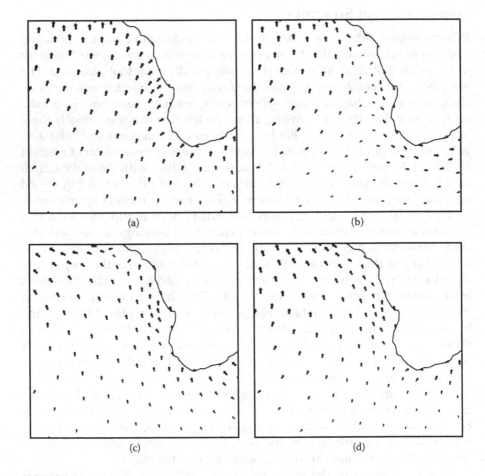

FIGURE 9.8
An illustration of the consequences of allowing insufficient iteration in hydrodynamic modeling. (See text for explanation.)

After 2 h, the tide continues to flow in and has not turned. The result of this error on the oil spill trajectory modeling can be seen in Figure 9.9. This can be compared for half-hourly intervals against Figure 9.1(a/f), which uses the correctly simulated tidal currents. With an insufficient number of iterations, the oil spill ends up in quite a different place and may adversely affect decision making.

In agent-based modeling there is a different, but by no means less complex set of issues in assessing the validity and usefulness of the results (Batty and Torrens, 2005; Li et al., 2008). In Chapter 5, we identified how agent-based

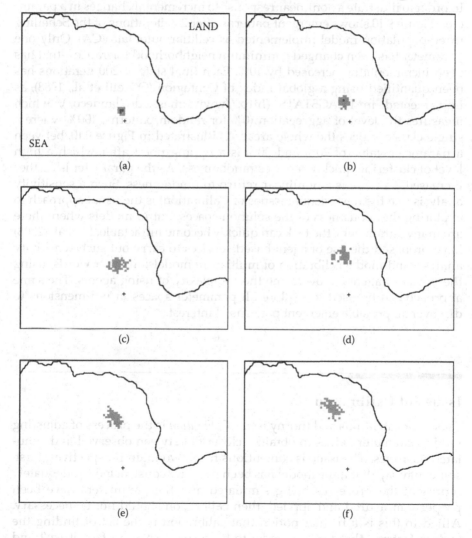

FIGURE 9.9
Knock-on effect of Figure 9.8 on oil spill trajectory modeling for half-hourly intervals up to 3 h. (Compare with Figure 9.1(a/f)).

models could have many thousands of agents all programmed with micro-level behaviors in order to study the macro patterns that emerge over time from these behaviors (Figure 5.15). It is this level of geographical complexity and emergent behavior that raises methodological challenges in validating them (Amblard et al., 2005; Manson, 2007). Aspects for consideration include the stability or robustness of the emergent patterns, possible equifinality of emergent patterns from different initial states, boundary conditions and parameter values, different emergent patterns depending on parameter values, nonlinear responses to parameter change, and the propagation of error. In order to illustrate a nonlinear response to incremental changes in a parameter, Figure 9.10 shows emergent patterns after 200 iterations of the Schelling three-population model implemented as cellular automata (CA). Only one parameter has been changed—minimum neighborhood tolerance—that has been incrementally increased by 10%. Each final state at 200 iterations has been quantified using a global Index of Contagion (O'Neill et al., 1988), as implemented in FRAGSTATS (http://www.umass.edu/landeco/), which measures the level of aggregation (0% for random patterns, 100% where a single class occupies the whole area). As illustrated in Figure 9.10, between a parameter value of 20% and 30% is a tipping point after which a high level of clustering quickly replaces randomness. As the parameter is further increased, so there is a nonlinear return to randomness. Such a sensitivity analysis (see the next section: Issues of Calibration) is the usual approach to exploring the robustness of the solution spaces, but in models where there are many parameters, the task can quickly become intractable. Li et al. (2008) have proposed the use of agent-based services to carry out such sensitivity analysis and model calibration of multiagent models; in other words, using the power of agents to overcome the complexity of using agents. The same approach can be used to explore all parameter spaces in n-dimensions to discover all possible emergent patterns of interest.

Issues of Calibration

This is a contentious and thorny issue. *Calibration* is the process of adjusting model parameter values to obtain a closer fit between observed and simulated variables. The issue is contentious from two main perspectives. First, some may say that if the model has been properly constructed to adequately represent the processes being simulated and the parameters have been properly measured and applied, then calibration should not be necessary. Allied to this is a further notion that calibration is the act of finding the "fiddle factors" that force a model to be correct when, in fact, it isn't and perhaps either the wrong model was chosen for the circumstances or the model needs instead to be rethought. The issue is thorny because there are

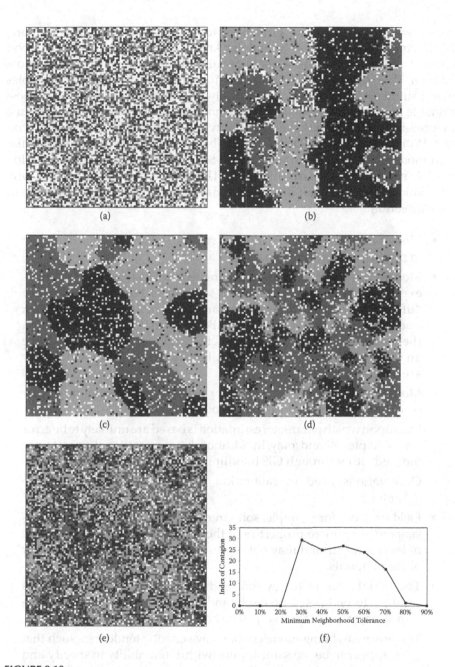

FIGURE 9.10
Sensitivity analysis of emergent patterns (after 200 iterations) for a three-population Schelling model implemented as CA, based on incremental changes to minimum neighborhood tolerance: (a) to (e) emergent patterns at parameter values of 20, 30, 50, 70, and 80%; (f) the effect measured using Index of Contagion at 10% increments. (Based on Li, Y., Brimicombe, A.J., and Li, C. (2008) *Computers, Environment and Urban Systems* 32: 464–473.)

very few industry-standard guidelines for model calibration (Whittemore, 2001). And then, how does a modeler know when the calibration is good enough and the model fit-for-use? For calibration to be possible, there must be data. This is often part of the problem as the observed data for variables being simulated may be sparse (spatially and temporally), may have questionable accuracy (interpolated?), may not be continuous, may even be just inaccessible (bureaucracy, confidentiality), or deemed too expensive to collect. Without adequate data on which to base a calibration, applying simulation models can become an act of faith. So, even the simplest of simulation models need some form of calibration. There are a number of reasons why simulation outputs do not fit observed data and, hence, the justifiable need for calibration:

- Models are only approximations of reality and are therefore unlikely to produce a perfect fit (Chapters 4 and 5).

- Models are often expensive to create and test, new software is expensive to buy and maintain, so models have a considerable "life span" and get reused on many studies where conditions, due to natural variation, are unlikely to match those for which the model was originally constructed and tested; users also have an inertia in changing from a familiar model to one in which they are less familiar.

- Models may get reused at scales for which they were not originally constructed and tested (see issues of scale above).

- Data upon which parameter estimation is based are unlikely to be error free (Chapter 8) and may in addition have operational uncertainty induced either through GIS handling and/or in the simulation.

- Observations used in calibration are unlikely to be error free (Chapter 8).

- Field data on, for example, soil strengths or infiltration rates represent only the micro properties of the materials measured over areas of less than 1 m^2 and may not represent the meso or macro behavior of the materials.

- The initial state of the system and boundary conditions are thus unlikely to be known precisely and even the most carefully determined parameters are likely to be best estimates.

- The processes being modeled may have chaotic tendencies such that what appear to be the same inputs (within our ability to specify and measure) may not consistently produce the same response (but, this would be an indication that a stochastic rather than a deterministic model might be preferable).

- The processes may exhibit *nonstationarity* where the relationship over time between inputs and outputs may change in response to other evolving changes in the system; in Chapter 4, Figure 4.4, for example, we looked at the reduction in the factor of safety (FoS) of a slope over time in response to weathering such that the reduction in factor the FoS in response to a storm event will change as the soil gets progressively weaker.

Taking all into consideration, the odds are that a simulation will initially not fit the observed data well; in fact, so much so that a good fit would indeed be a surprise that might even turn to suspicion.

So, why is the act of calibration so problematic? As we shall see, good model calibration is as much an art as it is a science, perhaps more so since you need a "feel" for the model, the processes being modeled, and the situation in which it is being applied—a "feel" that is both intuitive and inspirational. However, first there needs to be some measure of fit between simulated and observed (such as the RMSE in Chapter 4) and these are usually based on the empirical variance of residual errors:

$$\sigma_e^2 = \sum_{i=1}^{n} (\tilde{v}_i - v_i)^2 \Big/ n-1 \tag{9.19}$$

where σ_e^2 = variance of the residual errors, $\tilde{v}_i$ = observed value of the simulated variable, v_i = expected value of the variable, n = number of observations or time steps.

One easily applied goodness-of-fit measure based on the variance of residual error is called modeling efficiency E, as defined by Nash and Sutcliffe (1970):

$$E = \left[1 - \frac{\sigma_e^2}{\sigma_o^2} \right] -1 \leq E \leq 1 \tag{9.20}$$

where σ_o^2 = variance of the observations.

Similar to a correlation coefficient, $E = 1$ indicates a perfect fit. This may not be the most suitable way of calculating goodness-of-fit for some types of models where small time lags can dramatically increase σ_e^2 for oscillating variables or where the residuals are strongly autocorrelated in space and/or time (for a comparison of a range of goodness-of-fit statistics, see, for example, Fotheringham and Knudsen, 1987). Then a choice needs to be made about which parameters might usefully be varied to produce the desired changes toward a better goodness-of-fit. A local sensitivity analysis, such as OAT (one-at-a-time) might allow the sensitivity of the model to individual parameters to be quantified and ranked and also allow an observation to be made as to how each parameter influences model behavior. One measure of sensitivity S_i is (Saltelli et al., 2000):

$$S_i = \frac{\ln v - \ln v_b}{\ln x_i - \ln x_{ib}} \qquad (9.21)$$

where x_i = value of a parameter i, x_{ib} = baseline value of parameter i, v = value of the simulated variable, v_b = baseline value of the simulated variable.

If all parameters are varied by the same amount (e.g., ±5%), then the relative importance of parameters can be evaluated. But what sounds simple quickly gets quite complicated. First of all, assessing what contribution a particular changed parameter has had on the overall outcome of the simulation can be difficult. This is because many environmental simulation models are nonlinear. If they were linear, then inputs would have linear relationship to outputs. But, this is rarely the case. In hydrology, for example, antecedent conditions have an important role in determining the relationship between inputs and outputs for any one storm. This produces a nonlinear relationship between inputs and outputs and because antecedent conditions vary (and may not be well measured), a unit of input may not consistently produce the same unit of output for all time steps. Another cause of nonlinearity is nonstationarity, which was already mentioned above. Second, when GIS are coupled with simulation modeling, there is the added consideration of what preprocessing took place to establish parameter values and what postprocessing might have taken place to reach the calibration stage (as, for example, in the basin management planning example in Chapter 6). If there were algorithm choices (see above), then for a proper evaluation, some different approaches may need to be tried and then tested for sensitivity as in Equation (9.21). Third, such a strategy may be tractable for a small number of parameters, but where there are a large numbers of parameters such sensitivity analyses could take weeks or even months. Such a strategy is based on the premise that there is a single global optimum in the parameter space that can be found by varying the parameters for which the model is most sensitive such that a best fit with observations can be found. If this is the case and the number of parameters are few, then all well and good. But, unfortunately, as summarized by Beven (2001), there may not be a single global optimum, but a series of local optima instead. This would mean having to accept that there is *equifinality* in the solution, that is, there may be a number of model states that are acceptably consistent with the observed behavior of the processes being modeled. Therefore, it may be better to think of *corroboration* in which there is a noncontradiction between the output of a model and the evidence from reality (Oreskes et al., 1994).

In the absence of a single global optimum and without an *a priori* knowledge of the parameter space (for multiple interacting parameters), the safest approach, though often time consuming, is to fall back on techniques based on Monte Carlo (MC) simulation. The theoretical basis of MC has already been discussed in Chapter 8 with illustrations there and in Chapter 6. One such tool is the "generalized likelihood uncertainty estimator" (GLUE) of

Beven and Binley (1992). In GLUE, random parameter sets are generated from a prior distribution of parameter values. Each of these is input to the model and the results compared with the calibration data. Outcomes that produce a better fit against the calibration are weighted by a likelihood measure (probability, belief) and all outcomes with a likelihood measure of greater than zero contribute toward a cumulative weighted distribution of outcomes. From this, upper and lower bounds to the model outcomes can be established as well as quantiles. While versatile in the type of likelihood function used, GLUE is computationally intensive. It is nevertheless effective in establishing prediction limits on the basis of the calibration data and the parameter ranges modeled.

Bringing Data Issues and Modeling Issues Together

Digital spatial data sets have grown rapidly in scope, coverage, and volume over the last decade. We have moved from data-poverty to data-richness. On the other hand, environmental models have steadily grown more complex, are more frequently used, are expected to deal with larger data volumes, and give better predictions over a wider range of issues from local to global scales. The abundance of digital data is leading to its own set of problems in identifying and locating relevant data and in evaluating choices of resolution, coverage, provenance, cost, and conformance with the models to be used. Furthermore, for environmental models it may not be so much the characteristics of the raw data that are the most critical, but their characteristics once converted, aggregated, and implemented in the model. Given that a modeling task may access data from multiple sources, there is the added difficulty of assessing combined performance in relation to the implementation of the simulation such that outputs have fitness-for-use. Then there are the modeling issues of choosing appropriate space–time discretization, data transformations, algorithms where choice presents itself. Finally, as we have seen, there are difficulties in achieving an adequate calibration. Clearly, tools are required in order to help resolve the data and modeling issues discussed over the last two chapters. Should such tools be part of GIS or built into the environmental simulation model? Given the trajectory that we are on toward tool coupling strategies where the network is the core technology (Chapter 7, Figure 7.5 and Figure 7.10), neither approach need be the solution. Instead, in order to meet the diversity of requirements just listed, with sufficient flexibility, a wide range of functionality from different sources could be tightly coupled to form a *quality analysis engine* (QAE), as suggested by Li et al. (2000) and as an agent-based implementation by Li (2006). There now exists a richness of public domain and proprietary software, which can be used for various aspects of quality analysis. Varekamp et al. (1996), for

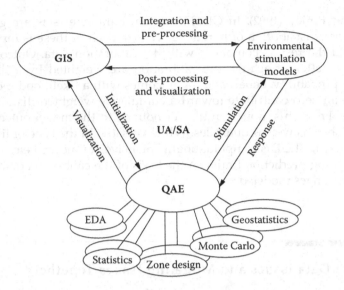

FIGURE 9.11
Process architecture for a quality analysis engine (QAE). (Based on Li, Y., Brimicombe, A.J., and Ralphs, M.P. (2000) *Computers, Environmental and Urban Systems* 24: 95–108.)

example, have already demonstrated the availability and efficacy of public domain geostatistical software, such as GSLIB (Deutsch and Journel, 1992), PCRaster (http://pcraster.geog.uu.nl/), and GEO-EAS (ftp://eliot.unil.ch), to which could be added GSTAT (http://www.gstat.org); further tools, such as GLUE (www.es.lancs.ac.uk/Freeware/Freeware.html), discussed above, and spatial analysis tools, such as GeoDa (http://geoda.uiuc.edu/). A wealth of proprietary software is also available. Given the speed, sophistication, and growing interoperability of these tools there is little reason to invest in reimplementing these tools within GIS or environmental models, but instead to couple them within a process architecture, as suggested in Figure 9.11. Here GIS and environmental models are portrayed largely in their *de facto* relationship, whereby GIS integrate and preprocess spatial data inputs for simulation models and postprocess and present for visualization the simulation outputs. The QAE is a series of tightly coupled tools, which depending on the type of modeling being undertaken might include exploratory data analysis, statistics, geostatistics, interpolators, zone designers, cluster detectors, MC analysis, and tools for simulating synthetic data sets and error surfaces. GIS would have a role in initializing the QAE with spatial data and assisting in the visualization of results. Much of the interaction is then between the tools of the QAE and the environmental simulation model in a stimulation and response mode in order to carry out both uncertainty analysis and sensitivity analysis of both data and model components. By using simulated synthetic data, for example, key issues around data quality, discretization, and model performance can be studied and understood at project inception stage, as illustrated above in establishing minimum sampling requirements using fractalized shorelines (see Figure 9.3). Using the tools of the QAE, the effects of algorithm choice could also be explored as well as analysis in support of calibration. In as much as research in this area is ongoing, there are already sufficient QAE components on the market and in the public domain together with programming languages like Visual Basic that allow the creation of wrappers, for environmental modeling and GIS professionals to proceed with their own implementations in support of their specific requirements.

10

Decision Making under Uncertainty

In Chapter 9, Formula (9.3) expressed fitness-for-use of analytical outputs as a function of model, data, and the professional where each of these is a set of entities. Throughout this book, we have been progressively looking at the mapping:

$$f : \Omega \rightarrow \Re \qquad (10.1)$$

which in this last chapter can be expressed as:

$$f : \Omega_u \rightarrow \Re_u \qquad (10.2)$$

where Ω = set of domain inputs, $\Re$ = set of real decisions, u = uncertainty.

In Chapter 8, we looked at the issues surrounding data uncertainties and the evolving strategies for knowing and reducing the level of uncertainty in spatial data, the analytical products of GIS, and inputs to environmental simulation models. However, as Frank (2008) has observed, it is "difficult to observe directly the effect of data quality on decisions." While good data = good decisions is a common-sense belief, there are, as we have seen, so many intermediate steps between data collection and model output—the transformation of *data* into *information* and *understanding*—that better data do not necessarily lead to better decisions. In Chapter 9, we considered a range of issues in model uncertainty and, again, the evolving strategies for knowing and reducing the level of uncertainty in the outputs of GIS and environmental simulation modeling. Here again, higher resolution, more detailed models are not necessarily the path to better decisions. In any case, as we saw from Equation (9.3), there will inevitably be some residual uncertainty. However, having gotten to this stage, a decision needs to be made by somebody: Is there a problem or isn't there, what are the risks, should something be done about them and if so what, is it technically the most appropriate solution, will the majority agree with it, how much is it going to cost, can we afford it, should we afford it, and does it represent value for money? This typifies the *decision space* that needs to be explored and navigated. Many GIS analysts and professional modelers may well say it is not their decision, they just ascertain and present the facts as they see them. But as we discussed in Chapter 5 (Figure 5.7), there has to be communication with the policy

FIGURE 10.1
The above graph depicts certainty about risk and policy decisions. (Adapted from Rejeski, D. (1993) in *Environmental modeling with GIS.* ed. M.F. Goodchild, B.P. Parks, and L.T. Steyaert. Oxford University Press, New York.)

makers and the public in an iterative process that should ideally bring about an informed consensus. Bellamy et al. (1999), for example, have emphasized the political and social context of environmental decision making requiring an inclusive process of collaboration and participation of scientists, professionals, and stakeholders. The reduction of risks is not just a matter of science and engineering (Tansel, 2005), but needs to include people's perceptions of hazard and how to deal with them. Social and economic factors are key in distinguishing disasters from ordinary events and, therefore, are crucial in assessing disasters (Wisner et al., 2004). Those affected by disasters tend to be people who are geographically marginalized to hazard-prone areas, socially marginalized because they suffer poverty and other inequalities, and politically marginalized because their voice is disregarded (Gaillard et al., 2007). Environmental decision making is inevitably negotiated in an arena of power relations where some actors have more power, resources, and better tactics with which to be heard (Few, 2002; Haque, 2003; Mercer et al., 2008).

If the level of propagated uncertainty to $\Re$ is high, one policy option is to continue improving the level of knowledge about risks so that a policy decision on whether or not to intervene eventually can be taken (Figure 10.1). This means that GIS analysts and professional modelers should be honest in giving information about the limitations and uncertainties of their analytical products (Rejeski, 1993). That having been said, in the arena of environmental policy making, particularly at national level, the degree of uncertainty can have a perverse effect when coupled with irreversibility (Arrow and Fisher, 1974; Saphores, 2004). Where there are high risks of irreversible effects but low certainty, then an optimal solution can be to act immediately. To some extent, we have seen this with Hurricane Katrina that devastated New Orleans. On the one hand, despite the low level of certainty in predicting when the next hurricane of this magnitude might strike, or the effect of climate change on increasing the incidence and severity of hurricanes, and, on the other, the human, economic, and political cost of that hurricane alongside what appears to be irreversible damage to some residential neighborhoods and potentially irreversible effects to the city as a whole (and other cities and communities in the region) from similar or more severe events in the future, did appear to trigger national policy changes toward climate change.

In the rest of this chapter, we will be looking at spatial decision support systems (SDSSs) as a means of exploring the decision space as part of the strategy for managing risks through risk reduction. We will then briefly look at the communication of spatial concepts as part of the iteration of decision making before moving on to a consideration of Web-based GIS as part of participatory approaches to decision making.

Exploring the Decision Space: Spatial Decision Support Systems

In Chapter 5, Figure 5.6 showed how there is a range of alternative, even complementary strategies that could be put forward to reduce the level of risk depending on the degree of prevention or preparedness that society is willing to pay for and accept. We also noted that mitigation in terms of prevention or preparedness was a spectrum from maximum risk reduction usually at a high financial cost to much smaller levels of risk reduction at a much reduced financial cost. What is more, they are not mutually exclusive. Given the magnitude and frequency relationship of natural hazards, it may be prohibitively expensive or technically impossible, for example, to prevent loss from higher magnitude events, in which case some form of preparedness through zoning or early warning may be prudent. Lower magnitude events that happen more frequently, on the other hand, may indeed be preventable at an economic cost and allow land to be kept in productive use. Such decisions may well change from one area to another depending on the local level of risk. Decisions on where and when to employ structural or nonstructural measures of mitigation and the mix that may represent a good, affordable strategy need to be based on an assessment of realistic alternative scenarios. SDSSs allow such assessments to be structured.

Decision support systems (DSS) began their technical development in the 1960s at the Massachusetts Institute of Technology (MIT) as computer-based solutions to support complex decision making and problem solving. The classic DSS tool comprised of the following components:

- Database management system for accessing internal and external data, information, and knowledge.
- Modeling functions.
- User interface designs for interactive queries, graphical display, and reporting.

The main focus of DSS research has been on how information technology (IT) can improve both the efficiency with which decisions are made and the effectiveness of those decisions (Shim et al., 2002). A *structured* problem is one that is fully specified in terms of the objective of the decision and the nature of the problem to be solved. They, therefore, tend to be problems that are routine, repetitive, and easily solved. Structured problems clearly don't need the aid of IT to effect a solution. The bath is overflowing, so turn off the tap. But, there are a whole range of problems that are *ill-structured* in which the nature of the problem itself may not be known or cannot be fully and coherently specified, the objective of the decision may not be clear. These may be new, novel problems or ones that are difficult to solve. A heightened incidence of lung cancer is identified on the south side of the town. Sure, cancers are a problem, but are they a consequence of some hidden cause in that part of town (asbestos in the building materials?), caused nearby and wafted in (that new incinerator?) or caused entirely elsewhere (they all work in the same factory in the next town?). What is the real problem, how do we find it, and when we've found it, how are we going to solve it? *Semi-structured* problems would have a mix of characteristics of structured and ill-structured problems. DSS are designed to assist in solving semi- and ill-structured problems. The decision process that DSS try to mimic is based on intelligence, design, and choice (Simon, 1960) where intelligence is the means by which we search for and clarify problems; design involves the development of alternatives and choice consists in analyzing the alternatives and choosing one for implementation. Another way of putting this is illustrated in Figure 10.2 and concerns methods and goals. DSS provide structure and assistance in moving through the contingency table in a counterclockwise direction (and never the other way), first in defining goals through a proper definition of the problem to be solved and then in defining methods first through the development of candidate solutions and then through the design and delivery of the solution to be implemented.

SDSS are not fundamentally different from DSS other than a focus toward the solution of complex spatial problems (Densham and Goodchild, 1989; Densham, 1991). Because of the spatial dimension, SDSS will need to have additional capabilities that:

- allow handling of spatial data;
- allow representation of spatial relations (e.g., topology);
- include spatial analysis techniques (e.g., buffering, overlay);
- provide visualization of spatial data.

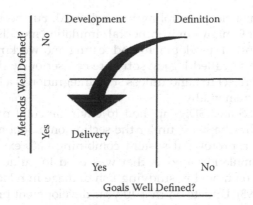

FIGURE 10.2
Approach to solving semi- and ill-structured problems.

Of course, most of what is required in terms of these additional capabilities can be provided by GIS, though GIS on their own do not constitute SDSS, as most semi- and ill-structured problems of a spatial nature have a complexity that cannot be solved purely by query and recombination of geographic data alone. Moreover, GIS only have a narrow set of modeling capabilities falling short of what would be desirable in SDSS. From the same perspective, environmental simulation models are not DSS either (spatial or otherwise). Bring the two together and you start to have the minimum configuration for an SDSS, the ability to carry out "what if"-type analyses that provide multiple outcomes that characterize the decision space being explored for implementable solutions. The architecture then of an SDSS is not dissimilar to those discussed in Chapter 7 for the coupling of technologies where one would expect at least a loose coupling for multiple "what ifs" to be tractable. While the ideal for an SDSS would be a tool coupling (Figure 7.10) to form a modeling framework with subsystems for data management, spatial data processing, model building and management, model execution, quality management and visualization, this is not always feasible. SDSS are often implemented for specific problem domains that, in a spatial context, often means within a finite geographical area. The up-front cost and time to implement a tool coupling may not be justified and instead a lower order of integration may be chosen. Also, if each of the basic tools being brought together are in themselves sophisticated with extensive user-interface and a range of industry "standard" input and output formats, then there is often little incentive in working toward a fuller integration, especially when GIS analyst and professional modeler are separate individuals working extensively with their own tools. In the Hong Kong basin management planning example given in Chapter 6, the initial phase was characterized by a loose coupling with GIS and hydraulic modeling activities being carried out in different locations (university and consultants offices) with a two-way data transfer every few days. It was only in subsequent phases of the project (it ran its course over several years) that a closer coupling was achieved. Furthermore, as we discussed in Chapter 9, many environmental simulation models have a long life span due to the cost of development and testing and working toward closer integration of the so-called legacy software can seriously detract from getting on with the project in hand unless such integration is a longer-term goal to be achieved incrementally.

Examples of DSS and SDSS applied to environmental problems abound in the literature. In Chapter 7, under the section on model management, we briefly looked at a prototypal system combining GIS, expert system, and environmental simulation models that was used to guide the appropriate choice of simulation model in studying fish damage in relation to pH levels in lakes (Lam, 1993). Update on that 10-year development program to create RAISON (Regional Analysis by Intelligent Systems ON microcomputers), an environmental DSS, is provided by Booty et al. (2001). RAISON DSS is seamlessly linked within the MSWindows® environment as a hierarchy of tools (Figure 10.3(a)). The interconnectivity of tools has been achieved through the use of Windows interoperability features (clipboard, OLE, DDE) as well as purpose-built linkages. The architecture is illustrated in Figure 10.3(b). Three separate databases are used to store different types of data, though all were built using Microsoft Access. RAISON DSS has mass balance models (for toxic chemicals) and trajectory models (air and water) converted to Visual Basic. The artificial neural network (ANN, back-propagation) is used predominantly for filling data gaps in input data (such as meteorological inputs to the air trajectory model) and for data classification. The expert system is rule-based with fuzzy logic and, while also used for data classification, has an important function on advising on scientific processes and knowledge limitations. The GIS component is used for the creation and editing of spatial objects, general spatial data handling, and for some aspects of visualization. Visualization also includes graphing and animation. Booty et al. report that creation of a system for a lake from within the RAISON DSS shell is relatively straightforward. The most time-consuming task is the collection of all the georeferenced and nongeoreferenced data from the various sources. The most technically difficult task is the selection of the most appropriate models for the system and becoming knowledgeable in their use. They consider the key benefit of the DSS is being able to bring together environmental science and economics for decision makers in meeting mandated elimination of pollutants.

An SDSS for the purpose of multiple-criteria site analysis is reported by Jun (2000). Here, DSS become the means for handling multiple socioeconomic criteria while considering physical suitability. By coupling GIS with expert systems and a tool for multiple-criteria methods, decisions can be coupled with prioritization. Within the loose coupling that transfers criteria and rules using text files, the expert system provides rules that advise on physical suitability constraints for generating suitability maps using GIS, while the multiple-criteria decision tool accommodates trade-offs in the multiple and conflicting decisions in determining a preferred location. Feedback

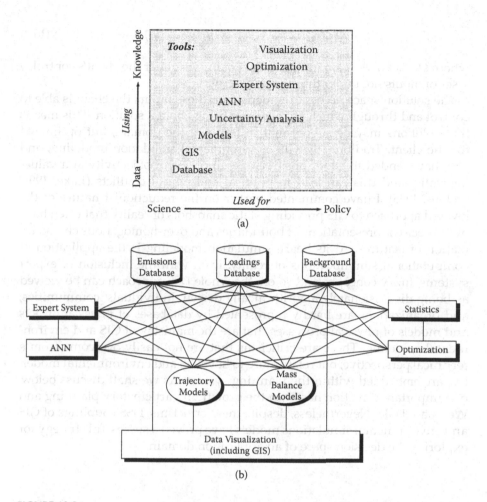

FIGURE 10.3
The RAISON DSS: (a) hierarchy of tools, (b) system architecture. (Based on Booty, W.G. et al. (2001) *Environmental Modelling & Software* 16: 453–458.)

loops allow the consequences of successive decisions to be assessed, thus allowing confirmation or rejection of earlier decisions and a gradual narrowing down of the solution space. The SDSS set-up in Hong Kong by coupling GIS and hydraulic modeling (Chapter 6, Figure 6.10) allowed multiple "what if" scenarios of different land use developments (prepared using GIS) against different return periods of rainstorms (hydraulic modeling) for which appropriate engineering solutions for mitigation and their associated costs could be identified. On the basis of these multiple outcomes, a preferred development scenario and a mix of structural and preparedness measures against flooding could be identified. In DSS applications, it is common to split the set of inputs into two components: those that are under the control of the client or stakeholders and those that are not:

$$\Omega : (x_1, x_2, x_3, \ldots, x_n; y_1, y_2, y_3, \ldots, y_n) \qquad (10.3)$$

where Ω = set of domain inputs, x = set of inputs under the client's control, y = set of inputs not under the client's control.

The solution space necessarily focuses on those inputs the client is able to control and through which changes can bring about a solution. This means that solutions may be suboptimal in terms of the domain, but optimized for the client. Traditionally, GIS, environmental simulation modeling, and DSS have tended to adopt positivist assumptions of objectivity and value-neutrality and this can lead to political and ethical conflicts (Lake, 1993; Pickles, 1995). I have commented earlier on the reductionist nature of the layered approach to GIS providing static snapshots of reality that often have over-precise representations of boundaries and over-homogenous character-ization of features. At its heart, simulation modeling is the application of computational simplifications of reality. Even with the inclusion of expert systems, fuzzy concepts, and so on, the whole DSS approach can be viewed as being dispassionately algorithmically driven. Individuals, communities, and organizations are hardly represented by databases of physical objects and models of physical processes that are the mainstay of GIS and environ-mental modeling. These are criticisms that are not easily overcome from a technical perspective, but relate more to how GIS and environmental model-ing are embedded within the planning system. As we shall discuss below, one important direction has been the growth in participatory planning and Web-based GIS. Nevertheless, despite these criticisms, DSS couplings of GIS and environmental simulations models have proved a successful strategy for exploring the decision space of an application domain.

Communication of Spatial Concepts

An important function of GIS when coupled with environmental modeling is postprocessing and visualization. The postprocessing has already been discussed in previous chapters. Here we will give some consideration to the visualization and communication of spatial information within the general context of this book. A good piece of modeling, with uncertainty reduced to a minimum, will fall flat if the results cannot be adequately communi-cated. This comes down to good cartographic skills with quantitative and qualitative data. It must also be recognized that not everybody can instantly recognize and feel comfortable with plan representations of multidimen-sional phenomena (Keates, 1982) and, therefore, the means of communica-tion and its effectiveness need to be carefully considered. It is often a case of having to supplement maps with other representations, such as tables, graphs, and images. The Web is providing new tools in this regard (Kraak and Brown, 2001).

The principle of communication through maps can be summed up as: "How do I say what to whom, and is it effective?" (Kraak, 2001). This concept that underpins cartographic design which aims to create maps that can be understood effectively by users. The "what" aspect concerns the information content to be presented on a map—the data model. Such content needs to be analyzed and determined in relation to the objective of presenting such content. The aspect of "how" is the means by which the information is represented cartographically. In achieving the creation of a good map, the basic concept of the cartographic theory developed by Bertin (1967) can be regarded as key guidelines (see below). Another aspect is how a user reads the map, which leads to "whom": the user. Users should always be considered in the process of creating maps with either an emphasis toward a particular group or more toward individual users, their expected background, likely level of understanding of the concepts being mapped, etc. The "effective" aspect demonstrates the amount of information extracted and understood by users. The information intended to be communicated through a map and the information retrieved by a user will rarely achieve an exact match, which can be viewed as the different levels of effectiveness. There can be information loss or, as in Chapter 8, Figure 8.6, levels of uncertainty in use can arise through confusion, ambiguity, or misrepresentation of the information to be imparted. In deciding an appropriate cartographic design, it is important then to analyze the characteristics of the information that is to be visualized and understood.

Bertin (1967) has distinguished six visual variables that are important in thematic map construction and relate mostly to the symbology used to represent features. The variables are size (e.g., thickness of lines, size of points), shape (e.g., circles versus triangles, dashed lines), orientation (say, of labeling), color (scheme, range), value (e.g., the degree of contrast over the color range), and texture (e.g., smooth versus coarse patterning). These all need careful consideration and even some experimentation and testing on colleagues for legibility. Tufte (1983) has further laid down some principles of graphical excellence, which can be well applied to thematic mapping. The cornerstone of graphical excellence is interesting data well presented; in other words, there must be something worth communicating. A well-designed presentation becomes a matter of substance, statistics, and design; it consists of complex ideas communicated with clarity, precision, and efficiency. What this means is that the graphics (map or other visualization) should give the viewer "the greatest number of ideas, in the shortest time, with the least ink in the smallest space" (Tufte, 1983). Above all, it requires telling the truth about the data. In thematic maps of quantitative data, the main challenge in telling that truth often boils down to the number of class intervals, the identification of class boundaries, and what symbology to use for each class. Most GIS packages provide default approaches to help the user through this task, usually

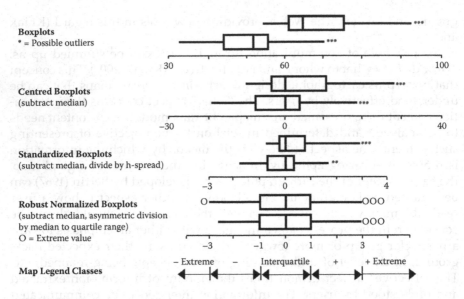

FIGURE 10.4
An illustration of the principle of robust normalization.

giving a choice of equal interval, equal area, natural breaks, standard deviations and quantiles, and inevitably in shades of red. One problem is the over-
use of a particular default by individuals or within an organization without
much thought given to the nature of the data or the nature of the message,
results in throw-away graphics. The issues of good thematic map design are
still being debated in the search for good representations that provide more
objectively comparable maps (e.g., Brewer and Pickle, 2002). One key here is
the normalization of data prior to map creation. A common means of normalization is the z-score:

$$Z = \frac{x - \bar{x}}{\sigma} \qquad (10.4)$$

where $\bar{x}$ = the mean value, σ = the standard deviation.

Unfortunately, unless the data are normally distributed (which very often
they are not), this can lead to bias. Recently an alternative to the z-score,
robust normalization, has been introduced (Brimicombe 1999b, 2000b). The
process of robust normalization can be visualized in Figure 10.4 where there
are initially two sets of data with contrasting distributions. The first stage
of normalization is the centering of the boxplots through subtraction of the
median. Sibley (1987) suggests division by the interquartile range to give
standardized boxplots, but this gives values that can be difficult to interpret
in terms of core and extreme values. Instead, robust normalization uses an
asymmetric division in the form:

$$RN = (x - \text{median})/(\text{median} - \text{lower quartile}) \quad \text{for } x < \text{median} \quad (10.5)$$

$$RN = (x - \text{median})/(\text{upper quartile} - \text{median}) \quad \text{for } x > \text{median} \quad (10.6)$$

$$RN = 0 \qquad\qquad\qquad\qquad\qquad\qquad \text{for } x = \text{median} \quad (10.7)$$

This ensures that all data are transformed to have a median of zero and an interquartile range of [-1, 1]. Extreme values can be taken as values of $RN < -3$ or $RN > 3$ (equivalent to 2 standard deviations of a normal distribution). This also leads to an intuitive set of class intervals, which make maps more objectively comparable. Figure 10.5 gives a robust normalized mapping of the pipe burst case study. The eye can quickly pick out the one location where there is both a high extreme occurrence of cases as represented by density and a high extreme occurrence of risk. That would be the first area to prioritize for remedial action.

MacEachren and Kraak (1997) have identified different purposes for map visualizations depending on the intended audience, the degree of interaction with the viewer, and the degree of unknowns still to be resolved. Thus, maps can be used to explore data and issues, analyze them, provide synthesis, and for presentation. Further advances in data exploration useful to GIS and environmental modeling are interaction with the graphics through the use of "brushing" techniques with linked, alternative views of the data (Dykes, 1997). We are also now seeing the incorporation of images, panoramas, and video alongside maps in the exploration of real world phenomena (Dykes, 2000). Through the Web, we have seen the growth of clickable maps in a multimedia setting that facilitate pathways to data and information. This underscores the next issue we need to consider.

Participatory Planning and the Web-Based GIS

Two important, synergistic trends occurred during the 1990s that have changed the face of decision making. The first was the established ubiquity of the Internet and the Web over a highly interconnected telecommunications environment, which has not only facilitated a network-oriented approach to GIS and environmental modeling (and most other IT applications of significance), but has promoted almost instantaneous communication worldwide. The second is the high level of access to PCs in the industrialized nations promoting a user-centered approach to information gathering and participation socially and politically as actors over the network. Surfing the Web has become an important part of people's social and business lives. This has been mirrored in the migration of complex decisions from single individuals to large groups of individuals as governments strive to promote inclusiveness

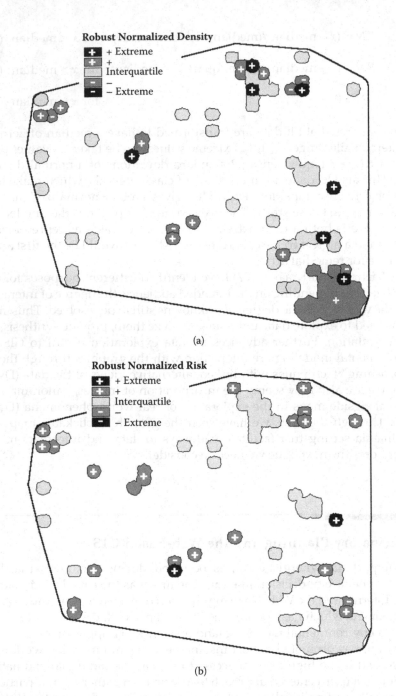

FIGURE 10.5
A robust normalized mapping of the pipe burst cluster detection (a) density, (b) risk.

to offset the growing socio-economic rifts and marginalization of the poor that has also grown over the same period. Such notions are also at the heart of online "community mapping" initiatives in which residents collaboratively produce maps of their locale that feature local knowledge and resources (Parker, 2006). Examples can be found at http://www.greenmap.org. Such projects aim to promote inclusion, transparency, and empowerment. IT is seen on both sides of the Atlantic as transforming lives (e.g., PITAC, 1998), but not all is rosy with the emergence of a "digital divide" (NTIA, 2000; Hull, 2003) where segments of society are nevertheless slipping into the margins of the informational economy (Castells, 1998). The term *e-government* has been added to our expanded vocabulary of "e" words. We are now in an age of growing participatory or collaborative decision making at the local level, including environmental and planning issues, in which the Web is the major vehicle for access.

Spatial planning has always been a complex process in which the recommendations emanating from a narrowed decision space as an outcome of, say, GIS and environmental modeling within an SDSS framework, needs to be mediated by the social and political reality (highly subjective and selective of perceived issues that they are) into a consensus for action. The environmental awareness that is now ingrained has made the process more complex and yet at the same time there has been a growing distrust or skepticism that scientists and experts know the right answers. Ferrand (1996) identifies three main elements: (1) the environment (the territory and related elements), (2) the proposed project, and (3) a "set of embedded actors" who will usually be a large, diverse group. Reality will inevitably be a social construction as perceived by the planners, the policy makers, and the embedded actors who "emphasize the contingency of information." Howard (1999) has detailed a number of advantages of Web-based approaches to participation in the planning process:

- Participation does not have to be restricted to attendance at meetings in specific geographical locations.
- Access to pertinent information can be from anywhere that has Web access.
- Access to information is not subject to "opening hours," but is available 24 hours a day, 7 days a week.
- Views can be expressed relatively anonymously and in a nonconfrontational situation.

Web-based participatory GIS (PGIS) provides an important adjunct to participatory planning as a means of facilitating and structuring access to data and information in map or mixed-media formats. This can be achieved in a number of ways (Harder, 1998):

- Static, noninteractive maps that are the digital equivalent of the paper map.
- Time-slice, noninteractive maps that are changed periodically, such as weather maps.
- Static, interactive maps (HTML image map) that provide a cartographic gateway to more detailed maps and/or other forms of data and information by clicking on the relevant location.
- Static, interactive maps that can be zoomed and panned by clicking relevant on-screen icons or scale bars.
- Dynamic, interactive maps with regular refresh that show changing situations, which are clickable for more information locally, such as traffic congestion maps.
- Static, user-defined maps that are created online via an interactive menu of options.
- User-defined, interactive maps that allow the user to undertake online spatial analysis using data layers and allowing subsequent clickable queries of the map.

These are all useful configurations for participatory planning and the first three in the above list require only simple architecture. The last four would normally require a vendor's Internet server version of their GIS software. A general overview of PGIS is given by Weiner and Harris (2008) and their use in relation to decision making by Jankowski and Nyerges (2008).

Kingston et al. (2000) describe the use of Web-based PGIS as part of a participatory planning initiative that took place in a village called Slaithwaite in West Yorkshire, United Kingdom (http://www.ccg.leeds.ac.uk/projects/slaithwaite/). The particular planning issues concerned the reopening of a canal through the village center, commercial traffic access to industrial sites, and many old buildings in disrepair. A Web-based GIS was considered appropriate as the whole community would benefit from being able to provide input to decisions about difficult problems. The design of the system centered on a Java map application called GeoTools, which allows pan, zoom, simple spatial query, and attribute input through a text box. If on navigating the map, a user wished to comment on a building or particular feature, it could be selected using the mouse and a free-form text box would pop up into which comments could be typed. The database would be immediately updated with the comment for collation and analysis at a later stage. Each user was requested to fill in a form that could later be used to profile respondents. The system was found to offer a high degree of flexibility with continual update of the database with comments. Any corrections/changes that needed to be made to the base data as a result of a user's local knowledge could be carried out quickly. Kingston et al. observed a high level of proficiency in map usage and where a feature could not readily be found, users would navigate from recognized, known objects. Most problematic for the older generation was the use of a mouse to select features. All users seemed to appreciate the opportunity to write as much as they wished about any issue. Within the respondents was a strong gender bias toward male users and there was an occupational weighting toward professional/managerial and educational employment. The dominant age group of respondents, however, was school-age children. One significant problem recognized by Kingston et al. centered on making all the data and information available to users in relation to who controls and owns them. Contracts for use of data are often quite restrictive and for Web-based media, owners want to charge on a per-view basis, which could make participatory planning very expensive with resources tied up in complex copyright and legal issues. Nevertheless, the basic principles underscoring participatory processes incorporating Web-based PGIS are:

- All sectors of the community should have equal access to data and information.
- The community should be empowered through the provision of data and information that maps onto the communities' needs.
- The legitimacy and accountability of the process needs to be under-scored by a high degree of trust and transparency.

Mercer et al. (2008) have studied the methodological advantages, limitations, and ethical issues of participatory techniques in the context of disaster risk reduction. While the advantages outweigh the limitations resulting in broadening the capacity for dialog between communities likely to be impacted by disasters and the relevant stakeholders with effective knowledge sharing and transfer, it was found that groups can inhibit individual voice, the process can be time consuming, and unequal power levels between the participants influence the interactions. Ethical issues focus on unduly raising expectations: Who is really being empowered, are participants being exploited, and where does ownership reside? Nevertheless, PGIS is becoming far more sophisticated (e.g., http://www.miltonkeynes-windfarm.com) and seen as an increasingly convenient and cost-effective means of carrying through mandated public consultation around planning decisions.

All's Well That Ends Well?

Burrough et al. (1996) have given consideration to the advantages and disadvantages of computerized environmental modeling:

Advantages

- Models can accommodate many variables and complex interactions.
- Data storage for a model is standardized, while simulation provides an organized approach.
- Modeling requires interdisciplinary cooperation in its development.
- Simulation/modeling languages can be a means of communicating ideas.
- Simulation is a fast and inexpensive method of investigation.
- Using simulation, it is possible to compress decades and centuries into minutes.
- Models can be evolved as knowledge and understanding of processes improves.
- The ability to experiment with models is greater than it is in the real world.
- Models permit the extrapolation of known results to new locations.
- The addition of GIS allows data to reflect a range of spatial and temporal situations.

Disadvantages

- Inexperienced users may uncritically accept results and assume that models are working correctly.
- Even experts may accept modeling results without validation of fitness-for-use.
- Mechanistic use of simulation can be detrimental to rational decision making.
- Simulation does not replace the need for fieldwork.
- Over-emphasis on models may overshadow important field experience on the significance of site-specific factors.
- Careful calibration of models is required and there are few guidelines.
- Adequate calibration may be difficult or even impossible for some models.
- The use of computational models requires accurate data at appropriate spatial and temporal resolutions.
- The development of complex models is expensive.
- Correct results do not guarantee a correct model.

The majority of the disadvantages cited here could be put down to the effects of inexperience. This is not to belittle students and younger professionals and it has been the purpose of this book to promote a healthy criticality in the use of GIS and simulation models, particularly when they are used together. The advantages are all for the taking and although the number of disadvantages equals the number of advantages, the disadvantages are to a greater or lesser extent avoidable, so there is, in fact, everything to be gained. But, does all this effort influence, for example, policy toward hazard mitigation and risk reduction? Is it making the world a better place? That's a tough call, and I don't think there can be any easy measure. Certainly a greater understanding of the processes underlying hazards and their mitigation has been achieved through the use of models. Most of our knowledge is bound up in models. However, having mapped Ω (the set of domain inputs) into $\mathfrak{R}$ (the set of real decisions), having evaluated the uncertainties along the way such that the risks are known and any residual risks in making a decision have been minimized, then what is left is all down to politics.

References

Achinstein, P. (1964) Models, analogies and theories. *Philosophy of Science* 31: 328–350.

Ackoff, R.L. (1964) General systems theory and systems research: contrasting conceptions of systems science. In *Views on general systems theory*, ed. M.D. Mesarovic. New York: John Wiley & Sons.

Albrecht, J.H. (1996a) Universal GIS-operations: A task-oriented systematization of data structure-independent GIS functionality leading towards a geographic modelling language. Institut für Strukturforschung und Planung in agrarischen Intensivegebieten, Vechta.

Albrecht, J.H. (1996b) Universal GIS operations for environmental modeling. In *NCGIA: Proceedings of the Third International Conference/Workshop on Integrating GIS and Environmental Modeling*, Santa Fe, NM, on CD. joche.santafe.html

Albrecht, J.H. (2005) A new age for geosimulation. *Transactions in GIS* 9: 451–454.

Aleotti, P., and Chowdhury, R. (1999) Landslide hazard assessment: Summary review and new prospects. *Bulletin of Engineering Geology and the Environment* 58: 21–44.

Almeida, C.M., Gleriani, J.M., Castejon, E.F., and Soares-Filho, B.S. (2008) Using neural networks and cellular automata for modeling intra-urban land-use dynamics. *International Journal of Geographical Information Science* 22: 943–963.

Altman, D. (1994) Fuzzy set theoretic approaches for handling imprecision in spatial analysis. *International Journal of Geographical Information Systems* 8: 271–289.

Amblard, F., Rouchier, J., and Bommel, P. (2005) Evaluation et validation de modèles multi-agents. In *Modélisation et simulation multi-agents*, ed. F. Amblard and D. Phan, 103–140. Paris: Lavoisier.

American Society of Civil Engineers (1983) *Maps uses, scale and accuracies for engineering and associated purpose.* New York: ASCE.

American Society of Photogrammetry and Remote Sensing (1985) Accuracy specifications for large scale maps. *Photogrammetric Engineering and Remote Sensing* 51: 195–199.

Amrhein, C.G. (1995) Searching for the elusive aggregation effect: Evidence from statistical simulations. *Environment & Planning* A 27: 105–119.

Amrhein, C.G., and Griffith, D.A. (1991) A model for statistical quality control of spatial data in a GIS. *Proceedings GIS/SIG'91*, Ottawa, Canada, pp. 91–103.

Anderson, J.M., and Mikhail, E.M. (1998) *Surveying theory and practice.* 7th ed. New York: McGraw-Hill.

Anselin, L. (1995) Local indicators of spatial association—LISA. *Geographical Analysis* 27: 93–155.

Anselin, L., Dodson, R.F., and Hudak, S. (1993) Linking GIS and spatial data analysis in practice. *Geographical Systems* 1: 3–23.

Antoniou, G., and van Harmelen, F. (2004) *A semantic Web primer.* Cambridge, MA: The MIT Press.

Arbia, G., Griffith, D., and Haining, R. (1998) Error propagation modelling in raster GIS: Overlay operations. *International Journal of Geographical Information Science* 12: 145–167.

315

Armstrong, M.P. (1991) Knowledge classification and organisation. In *Map generalisation—Making rules for knowledge representation*, ed. B.P. Buttenfield and R.B. McMaster, 86–102. New York: John Wiley & Sons.

Armstrong, M.P. (2000) Geography and computational science. Annals of the *Association of American Geographers* 90: 146–156.

Arnaud, P., Bouvier, C., Cisneros, L., and Dominguez, R. (2002) Influence of rainfall spatial variability on flood prediction. *Journal of Hydrology* 260: 216–230.

Arrow, K.J., and Fisher, A.C. (1974) Environmental preservation, uncertainty and irreversibility. *Quarterly Journal of Economics* 88: 312–319.

ASCE Task Committee in Applications of Artificial Neural Networks in Hydrology (2000a) Artificial neural networks in hydrology I: Preliminary concepts. *Journal of Hydrologic Engineering* 5: 115–123.

ASCE Task Committee in Applications of Artificial Neural Networks in Hydrology (2000b) Artificial neural networks in hydrology I: Hydrologic applications. *Journal of Hydrologic Engineering* 5: 124–137.

ASCE Task Committee on GIS Modules and Distributed Models of the Watershed (1999) *GIS modules and distributed models of the watershed*. Reston, VA: ASCE.

ASCE Task Committee on Modeling of Oil Spill (1996) State-of-the-art review of modelling transport and fate of oil spills. *Journal of Hydraulic Engineering* 11: 594–609.

Aspinall, R., and Pearson, D. (2000) Integrated geographical assessment of environmental condition in water catchments: Linking landscape ecology, environmental modelling and GIS. *Journal of Environmental Management* 59: 299–319.

Atkinson, P.J., and Unwin, D.J. (2002) Density and local attribute estimation of an infectious disease using MapInfo. *Computers & Geosciences* 28: 1095–1105.

Atkinson, P.M., and Tate, N.J. (2000) Spatial scale problems and geostatistical solutions: A review. *Professional Geographer* 52: 607–623.

Augarde, A., ed. (1991) *The Oxford dictionary of modern quotations*. Oxford, U.K.: Oxford University Press.

Ayyub, B.M., and McCuen, R.H. (1987) Quality and uncertainty assessment of wildlife habitat with fuzzy sets. *Journal of Water Resources Planning and Management* 113: 95–109.

Bailey, T.C., and Gatrell, A.C. (1995) *Interactive spatial data analysis*. Harlow, U.K.L Longman.

Batty, M. (1974) Entropy in spatial aggregation. *Geographical Analysis* 8: 1–21.

Batty, M., and Longley, P. (1994) *Fractal cities*. London: Academic Press.

Batty, M., and Torrens, P. (2005) Modelling and prediction in a complex world. *Futures* 37: 745–766.

Beach, E.F. (1957) *Economic models*. New York: John Wiley & Sons.

Beanlands, C. (1988) Scoping methods and baseline studies in EIA. In *Environmental impact assessment: Theory and practice*, ed. P. Wathern, London: Routledge, 33–46.

Beard, K. (1989) Use error: The neglected error component. *Proceedings AutoCarto 9*, April 2–7, Baltimore, MD, pp. 808–817.

Beck, M.B., Jakeman, A.J., and McAleer, M.J. (1993) Construction and evaluation of models of environmental systems. In *Modelling change in environmental systems*, ed. A.J. Jakeman, M.B. Beck, and M.J. McAleer, Chichester U.K.: Iohn Wiley & Sons.

Bedard, Y. (1986) A study of the nature of data using a communication-based conceptual framework for land information systems. *Canadian Surveyor* 40: 449–460.

Bellamy, J.A., McDonald, G.T., Syme, G.J., and Butterworth, J.E. (1999) Evaluating integrated resource management. *Society and Natural Resources* 12: 337–353.

Bennett, D.A. (1997a) Managing geographical models as repositories of scientific knowledge. *Geographical & Environmental Modelling* 1: 115–133.

Bennett, D.A. (1997b) A framework for the integration of geographical information systems and modelbase management. *International Journal of Geographical Information Science* 11: 337–357.

Berry, B.J.L., Griffith, D.A., and Tiefelsdorf, M.R. (2008) From spatial analysis to geospatial science. *Geographical Analysis* 40: 229–238.

Berry, J.K. (1995) *Spatial reasoning for effective GIS*. Fort Collins, CO: GIS World Books.

Bertin, J. (1967) *Sémiologie graphique*. Den Haag, The Netherlands: Mouton.

Besag, J., and Newell, J. (1991) The detection of clusters in rare diseases. *Journal of the Royal Statistical Society* A 154: 143–155.

Beven, K.J. (1996) The limits of splitting: hydrology. *Science of the Total Environment* 183: 89–97.

Beven, K.J. (2001) *Rainfall-runoff modelling: The primer*. Chichester, U.K.: John Wiley & Sons.

Beven, K.J. and Binley, A.M. (1992) The future of distributed models: Model calibration and uncertainty prediction. *Hydrological Processes* 6: 279–298.

Bian, L. (2007) Object-oriented representation of environmental phenomena: Is everything best represented as an object? *Annals of the Association of American Geographers* 97: 267–281.

Bian, L., and Butler, R. (1999) Comparing the effects of aggregation methods on statistical and spatial properties of simulated spatial data. *Photogrammetric Engineering & Remote Sensing* 65: 73–84.

Bian, L., and Hu, S. (2009) Identifying components for interoperable process models using concept lattice and semantic reference system. *International Journal of Geographical Information Science* 23: 1009–1032.

Blakemore, M. (1984) Generalisation and error in spatial data bases. *Cartographica* 21: 131–139.

Block, P. (1981) *Flawless consulting*. Austin, TX: Learning Concepts.

Blockley, D.I. (1979) The calculation of uncertainty in civil engineering. *Proceedings of the Institution of Civil Engineers, Part 2* 67: 313–326.

Blockley, D.I. and Henderson, J.R. (1980) Structural failure and growth of engineering knowledge. *Proceedings of the Institution of Civil Engineers, Part 1* 68: 719–728.

Blockley, D.I. and Robinson, C.I. (1983) An analysis of the characteristics of a good engineer. *Proceedings of the Institution of Civil Engineers, Part 1* 75: 77–94.

Bolstad, P.V., Gessler, P., and Lillesand, T.M. (1990) Positional uncertainty in manually digitized map data. *International Journal of Geographical Information Systems* 4: 399–412.

Booty, W.G., Lam, D.C.L., Wong, I.W.S., and Siconolfi, P. (2001) Design and implementation of an environmental decision support system. *Environmental Modelling & Software* 16: 453–458.

Botkin, D.B. (1990) *Discordant harmonies: A new ecology for the twenty-first century*. Oxford, U.K.: Oxford University Press.

Bouille, F. (1982) Actual tools for cartography today. *Cartographica* 19: 27–32.

Brand, E.W., Premchitt, J., and Phillipson, H.B. (1984) Relationship between rainfall and landslides in Hong Kong. *Proceedings* 4th *International Symposium of Landslides*, Toronto, 1: 377–384.

Brandmeyer, J.E., and Karimi, H.A. (2000) Coupling methodologies for environmental models. *Environmental Modelling & Software* 15: 479–488.

Bregt, A.K., Denneboom, J., Gesink, H.J., and van Randen, Y. (1991) Determination of rasterizing error: A case study with the soil map of The Netherlands. *International Journal of Geographical Information Systems* 5: 361–367.

Brewer, C.A., and Pickle, L. (2002) Evaluation of methods for classifying epidemiological data for choropleth maps in series. *Annals of the Association of American Geographers* 92: 662–681.

Brimicombe, A.J. (1982) Engineering site evaluation form aerial photographs. *Proceedings 7th Southeast Asian Geotechnical Conference*, Hong Kong, 2: 139–148.

Brimicombe, A.J. (1985) The application of geomorphological triangular databases in geotechnical engineering. Unpublished MPhil thesis, University of Hong Kong.

Brimicombe, A.J. (1992) Flood risk assessment using spatial decision support systems. *Simulation* 59: 379–380.

Brimicombe, A.J. (1993) Combining positional and attribute uncertainty using fuzzy expectation in a GIS. *Proceedings GIS/LIS'93*, Minneapolis, MN, 1: 72–81.

Brimicombe, A.J. (1997) A universal translator of linguistic hedges for the handling of uncertainty and fitness-for-use in geographical information systems. In *Innovations in GIS 4*, ed. Z. Kemp, 115–126. London: Taylor & Francis.

Brimicombe, A.J. (1998) A fuzzy co-ordinate system for locational uncertainty in space and time. In *Innovations in GIS 5*, ed. S. Carver, 143–152. London: Taylor & Francis.

Brimicombe, A.J. (1999a) Geographical information systems and environmental modelling for sustainable development. In *Land reform and sustainable development*, ed. R. Dixon-Gough, 77–92. Aldershot, U.K.: Ashgate.

Brimicombe, A.J. (1999b) Small may be beautiful, but is simple sufficient. *Geographical & Environmental Modelling* 3: 9–33.

Brimicombe, A.J. (2000a) Encoding expert opinion in geo-information systems: A fuzzy set solution. *Transactions in International Land Management* 1: 105–121.

Brimicombe, A.J. (2000b) Constructing and evaluating contextual indices using GIS: A case of primary school performance tables. *Environment and Planning A* 22: 1909–1933.

Brimicombe, A.J. (2002) Cluster discovery in spatial data mining: A variable resolution approach. *Proceedings Data Mining 2002*, Bologna, Italy, pp. 625–634.

Brimicombe, A.J. (2006) A dual approach to cluster discovery in point event data sets. *Computers Environment and Urban Systems* 31: 4–18.

Brimicombe, A.J. and Bartlett, J.M. (1993) Spatial decision support in flood hazard and flood risk assessment: A Hong Kong case study. *Proceedings of the 3rd International Workshop on GIS*, Beijing, China, 2: 173–182.

Brimicombe, A.J. and Bartlett, J.M. (1996) Linking geographic information systems with hydraulic simulation modeling for flood risk assessment: The Hong Kong approach. In *GIS and environmental modeling: Progress in research issues*, ed. M.F. Goodchild et al., 165–168. New York: John Wiley & Sons.

Brimicombe, A.J., and Li, C. (2009) *Location-based services and geo-information engineering*. Chichester, U.K.: John Wiley & Sons.

Brimicombe, A.J., Li, Y., Al-Zakwani, A., and Li, C. (2009) Agent-based distributed component services in spatial modeling. In *Computational science and its applications—ICCSA 2009*, ed. O. Gervasi et al., 300–312. Berlin: Springer-Verlag.

Brimicombe, A.J. and Tsui, P. (2000) A variable resolution, geocomputational approach to the analysis of point patterns. *Hydrological Processes* 14: 2143–2155.

Brimicombe, A.J. and Yeung, D. (1995) An object oriented approach to spatially inexact socio-cultural data. *Proceedings 4th International Conference on Computers in Urban Planning & Urban Management*, Melbourne, Australia, 2: 519–530.

Bristow, R. (1989) *Hong Kong's new towns, A selective review*. Hong Kong: Oxford University Press.

Bronstert, A. (1999) Capabilities and limitations of distributed hillslope hydrological modelling. *Hydrological Processes* 13: 21–48.

Brooks, S.M. (2003) Modelling in physical geography. In *The students companion to geography*, ed. A. Rogers and H.A. Viles, 161–166. Oxford, U.K.: Blackwell.

Brown, C.B., and Yao, J.T.P. (1983) Fuzzy sets and structural engineering. *Journal of Structural Engineering* 109: 1211–1225.

Brown, J.D. and Heuvelink, G.B.M. (2008) On the identification of uncertainties in spatial data and their quantification with probability distribution functions. In *The handbook of geographic information science*, ed. J.P. Wilson and A.S. Fotheringham, 94–107, Malden, MA: Blackwell.

Bruneau, P., Gascuel-Odoux, C., Robin, P., Merot, P., and Beven, K. (1995) Sensitivity to space and time resolution of a hydrological model using digital elevation data. *Hydrological Processes* 9: 69–81.

Brunsden, D. (1973) The application of systems theory to the study of mass movement. *Geologia Applicata e Idrogeologia* 7: 185–207.

Brunsden, D. (2002) Geomorphological roulette for engineers and planners: Some insights into an old game. The Fifth Glossop Lecture. *Quarterly Journal of Engineering Geology and Hydrogeology* 35: 101–142.

Buckley, J.J. and Eslami, E. (2002) *An introduction to fuzzy logic and fuzzy sets*. Heidelberg, Germany: Physica-Verlag.

Buehler, K. and McKee, L., eds. (1998) *The OpenGIS guide*. 3rd ed. Wayland, MA.: The Open GIS Consortium Inc.

Bulut, F., Boynukalin, S., Tarhan, F., and Ataoglu, E. (2000) Reliability of landslide isopleth maps. *Bulletin of Engineering Geology and the Environment* 58: 95–98.

Bureau of Budget, U.S. (1947). *National map accuracy standards*. Washington D.C.: U.S. Government Printing Office.

Burrough, P.A. (1986a) *Principles of geographical information systems for land resources asssessment*. Oxford, U.K.: Clarendon Press.

Burrough, P.A. (1986b) Five reasons why geographical information systems are not being used efficiently for land resources assessment. *Proceedings AutoCarto London* 2: 139–147.

Burrough, P.A. (1992) Are GIS data structures too simple minded? *Computers & Geosciences* 18: 395–400.

Burrough, P.A. (1997) Environmental modelling with geographical information systems. In *Innovations in GIS 4*, ed. Z. Kemp, 143–153. London: Taylor & Francis.

Burrough, P.A. (1998) Dynamic modelling and geocomputation. In *Geocomputation: A primer*, ed. P.A. Longley et al., 165–191. Chichester, U.K.: John Wiley & Sons.

Burrough, P.A. (2000) Whither GIS (as systems and as science)? *Computers, Environment & Urban Systems* 24: 1–3.

Burrough, P.A., and Frank, A.U., eds. (1996) *Geographical objects with indeterminate boundaries*. London: Taylor & Francis.

Burrough, P.A., and McDonnell, R.A. (1998) *Principles of geographical information systems*. Oxford, U.K.: Oxford University Press.

Burrough, P.A., MacMillan, R.A., and van Deursen, W. (1992) Fuzzy classification methods for determining land suitability from soil profile observations and topography. *Journal of Soil Science* 43: 193–210.

Burrough, P.A., van Rijn, R., and Rikken, M. (1996) Spatial data quality and error analysis issues: GIS functions and environmental modelling. In *GIS and environmental modeling: Progress in research issues,* ed. M.F. Goodchild et al., 29–34. New York: John Wiley & Sons.

Canadian General Standards Board (1991) *CGSB-171.1 Spatial archive and interchange format (SAIF).* Ottawa, Canada.

Carrara, A., Cardinali, M., and Guzzetti, F. (1992) Uncertainty in assessing landslide hazard and risk. *ITC Journal* 1992-2: 172–183.

Carson, R. (1963) *Silent spring.* London: Hamish Hamilton.

Caspary, W., and Scheuring, R. (1993) Positional accuracy in spatial databases. *Computers, Environment & Urban Systems* 17: 103–110.

Castells, M. (1998) *The information age: Economy, society and culture. Vo. III: End of millennium.* Oxford, U.K.: Blackwell.

Chen, X. (2008) Microsimulation of hurricane evacuation strategies of Galveston Island. *The Professional Geographer* 60: 160–173.

Chen, X.H., and Brimicombe, A.J. (1997) Temporal and spatial variations of Fe and Mn in Arha Reservoir. In *Water pollution IV: Modelling, measuring and prediction,* ed. R. Rajar and C.A. Brebbia, 95–104. Southampton, UK.: Computational Mechanics Publications.

Chen, X.H., Brimicombe, A.J., Whiting, B.M., and Wheeler, C. (1998) Enhancement of three-dimensional reservoir quality modelling by geographical information systems. *Geographical & Environmental Modelling* 2: 125–139.

Chorley, R.J., and Haggett, P. (1967) *Models in geography.* London: Methuen.

Chrisman, N.R. (1982) A theory of cartographic error and its measurement in digital databases. *Proceedings AutoCarto 5,* Crystal City, VA, pp. 159–168.

Chrisman, N.R. (1983a) The role of quality information in the long-term functioning of a geographic information system. *Cartographica* 21: 79–87.

Chrisman, N.R. (1983b) Epsilon filtering: A technique for automated scale changing. Paper presented at the *43rd Annual Meeting of the American Congress on Surveying & Mapping,* Washington D.C., pp. 322–331.

Chrisman, N.R. (1987) The accuracy of map overlays: a reassessment. *Landscape & Urban Planning* 14: 427–439.

Chrisman, N.R. (1989) Modelling error in overlaid categorical maps. In *The accuracy of spatial databases,* ed. M.F. Goodchild and S. Gopal, 21–34. London: Taylor & Francis.

Chrisman, N.R. (1991). The error component of spatial data. In *Geographical information systems principles and practice,* ed. D.J. Maguire, M.F. Goodchild, and D.W. Rhind, 165–174. Harlow, U.K.: Longman.

Chrisman, N.R. (1995) Beyond Stevens: A revised approach to measurement for geographic information. *Proceedings AUTOCARTO 12,* Charlotte, NC, pp. 271–280.

Chrisman, N.R. (1997) *Exploring geographic information systems.* New York: John Wiley & Sons.

Clark, M.J. (1998) Putting water in its place: A perspective on GIS in hydrology and water management. *Hydrological Processes,* 12: 823–834.

Clark, P.J. and Evans, F.C. (1954) Distance to nearest neighbour as a measure of spatial relations in populations. *Ecology* 35: 445–453.

Clarke, K., Parks, B., and Crane, M. (2000) Integrating geographic information systems (GIS) and environmental models. *Journal of Environmental Management* 59: 229–233.

Clayton, A.M.H., and Radcliffe, N.J. (1996) *Sustainability: A systems approach.* London: Earthscan Publications.

Cliff, A.D., and Ord, J.K. (1981) *Spatial processes: Models and applications.* London: Pion.

Clifford, N.J. (2002) Hydrology: The changing paradigm. *Progress in Physical Geography* 26: 290–301.

Coad, P. and Yourdan, E. (1991) *Object-Oriented Analysis.* Prentice Hall International, Englewood Cliffs, NJ.

Cohen, J., and Stewart, I. (1994) *The collapse of chaos, discovering simplicity in a complex world.* London: Penguin Books.

Congalton, R.G., and Mead, R.A. (1983) A quantitative method to test for consistency and correctness in photo-interpretation. *Photogrammetric Engineering and Remote Sensing* 49: 69–74.

Cooke, R.U. (1992) Common ground, shared inheritance: Research imperatives for environmental geography. *Transactions of the Institute of British Geographers* 17: 131–151.

Coppock, J.T., and Rhind, D.W. (1991) History of GIS. In *Geographical information systems principles and practice*, ed. D.J. Maguire, M.F. Goodchild, and D.W. Rhind, 21–43. Harlow, U.K.: Longman.

Cornelius, S. (1991) Spatial data auditing. In *Metadata in geosciences*. ed. D. Medyckyj-Scott, I. Newman, C. Ruggles, and D. Walker, 39–54. Loughborough, U.K.: Group D Publications.

Cotter, D.M., and Campbell, R.K. (1987) Concept for a digital flood hazard data base. *Proceedings of the 25th URISA Annual Conference*, Fort Lauderdale, FL. 2: 156–170.

Couclelis, H. (1998) Geocomputation and space. *Environment and Planning B* (25th anniversary issue) 41–47.

Council for Science and Society (1989) *Benefits and risks of knowledge-based systems.* Oxford, U.K.: Oxford University Press.

Cowen, D.J. (1988) GIS versus CAD versus DBMS: What are the differences? *Photogrammetric Engineering and Remote Sensing* 54: 1551–1555.

Crain, I.K., and Macdonald, C.L. (1984) From land inventory to land management— the evolution of an operational GIS. *Cartographica* 21: 39–46.

Cressie, N.A.C. (1993) *Statistics for spatial data.* New York: John Wiley & Sons.

Crosetto, M., Crosetto, F., and Tarantola, S. (2002) Optimized resource allocation for GIS-based model implementation. *Photogrammetric Engineering & Remote Sensing* 68: 225–232.

Crosetto, M., and Tarantola, S. (2001) Uncertainty and sensitivity analysis: Tools for GIS-based model implementation. *International Journal of Geographical Information Science* 15: 415–437.

Crow, S. (2000) Spatial modelling environments: Integrating GIS and conceptual modeling frameworks. In *Proceedings Third International Conference/Workshop on Integrating GIS and Environmental Modeling*, Santa Fe, NM, National Center for Geographic Information and Analysis (NCGIA); on CD.

Cutter, S.L. (1996) Vulnerability to environmental hazards. *Progress in Human Geography* 20: 529–539.

D'Ambrosio, D., Iovine, G., Spataro, W., and Miyamoto, H. (2007) A macroscopic collisional model for debris-flows simulation. *Environmental Modelling & Software* 22: 1417–1436.

Dacey, M.F. (1960) A note on the derivation of nearest neighbour distances. *Journal of Regional Science* 2: 81–87.

Dangermond, J. (1983) A classification of software components commonly used in geographical information systems. In *Design and implementation of computer-based geographic information systems*, ed. D. Peuquet and J. O'Callaghan, 70–91. New York: IGU Commission of Geographical Data Sensing and Processing.

Davis, F.W., Quattrochi, D.A., Ridd, M.K., Lam N.S-N., Walsh, S.J., Michaelson, J.C., Franklin, J., Snow, D.A., Johannsen, C.J., and Johnston, C.A. (1991) Environmental analysis using integrated GIS and remotely sensed data: Some research needs and priorities. *Photogrammetric Engineering and Remote Sensing* 57: 689–697.

Davis, J.C. (1973) *Statistics and data analysis in geology*. New York: John Wiley & Sons.

de Jongh, P. (1988) Uncertainty in EIA. In *Environmental impact assessment: Theory and practice*, ed. P. Wathern, 62–84. London: Routledge.

de Smith, M.J., Goodchild, M.F., and Longley, P.A. (2007) *Geospatial analysis: A comprehensive guide to principles, techniques and software tools*. Leicester, U.K.: Matador.

Deaton, M.L., and Winebrake, J.L. (2000) *Dynamic modeling of environmental systems*. New York: Springer.

Delcourt, H.R., and Delcourt, P.A. (1988) Quarternary landscape ecology: relevant scales in space and time. *Landscape Ecology* 2: 23–44.

Densham, P.J. (1991) Spatial decision support systems. In *Geographical information systems principles and practice*, ed. D.J. Maguire, M.F. Goodchild, and D.W. Rhind, 403–412. Harlow, U.K.: Longman.

Densham, P.J., and Goodchild, M.F. (1989) Spatial decision support systems: A research agenda. *Proceedings of the GIS/LIS'89*, Orlando, FL, pp. 707–716.

Department of the Environment (1987) *Handling geographic information: The report of the committee of inquiry*. London: Her Majesty's Stationary Office.

Deutsch, C.V., and Journel, A.G. (1992) *GSLIB: Geostatistical library and user's guide*. New York: Oxford University Press.

Diamond, J. (2005) *Collapse: How societies choose to fail or succeed*. New York: Vicking Press.

Drainage Services Department (2008) *Flood prevention*. www.dsd.gov.hk/flood_prevention/ long_term_improvement_measures/dmp/index.htm (accessed June 5, 2009).

Dror, Y. (1963) The planning process: A facet design. *International Review of Administrative Sciences* 29: 46–58.

Drumm, D., Purvis, M., and Zhou, Q. (1999) Spatial ecology and artificial neural networks: Modeling the habitat preference of the sea cucumber (*Holothuria leucospilota*) on Rorotonga, Cook Islands. *Proceedings of the SIRC 99*, Dunedin, New Zealand, pp. 141–150.

Drummond, J.E. (1987) A framework for handling error in geographic data manipulation. *ITC Journal* 1987-1: 73–82.

Dubois, D., and Prade, H. (1980) *Fuzzy sets and systems*. New York: Academic Press.

Dueker, K.J. (1980) An approach to integrated information systems for planning. In *Computers in local government urban and regional planning*, ed. K. Krammer and J. King, J., 1–12. Pennsauken, NJ: Auerbach.

Dueker, K.J., and Kjerne, D. (1989) *Multi-purpose cadastre: Terms and definitions*. Falls Church, VA: American Society of Photogrammetry & Remote Sensing and American Congress of Surveying & Mapping.

Dunn, R., Harrison, A.R., and White, J.C. (1990) Positional accuracy and measurement error in digital databases of land use. *International Journal of Geographical Information Systems* 4: 385–398.

Dykes, J.A. (1997) Exploring spatial data representations with dynamic graphics. *Computers and Geosciences* 23: 345–370.

Dykes, J.A. (2000) An approach to virtual environments for visualisation using linked geo-referenced panoramic imagery. *Computers, Environment and Urban Systems* 24: 127–152.

Edwards, G., and Lowell, K.E. (1996) Modelling uncertainty in photointerpreted boundaries. *Photogrammetric Engineering and Remote Sensing* 62: 377–391.

Edwards, R.J.G., Brunsden, D., Burton, A.N., Dowling, J.W.F., Greenwood, J.G.W., Kelly, J.M.H., King, R.B., Mitchell, C.W., and Sherwood, D.E. (1982) Land surface evaluation for engineering practice. *Quarterly Journal of Engineering Geology* 15: 265–316.

Ehrliholzer, R. (1995) Quality assessment in generalization: Integrating quantitative and qualitative methods. *Proceedings of the 17th International Cartographic Conference*, Barcelona, 2: 2241–2250.

Elith, J., Burgman, M.A., and Regan, H.M. (2002) Mapping epistemic uncertainties and vague concepts in predictions of species distribution. *Ecological Modelling* 157: 313–329.

Emery, F.E., ed. (1969) *Systems hinking*. London: Penguin Books.

Epstein, E.F. (1991) Legal aspects of GIS. In *Geographical information systems principles and practice*, ed. D.J. Maguire, M.F. Goodchild, and D.W. Rhind, 489–503. Harlow, U.K.: Longman.

Estivill-Castro, V., and Lee, I. (2002) Argument free clustering for large spatial point-data sets via boundary extraction from Delaunay diagrams. *Computers, Environment and Urban Systems* 26: 315–334.

Etkin, D.S., and Welch, J. (1997) Oil spill intelligence report, international oil spill database: Trends in oil spill volumes and frequency. *Proceedings of the 1997 International Oil Spill Conference*. Fort Lauderdale: American Petroleum Institute, pp. 949–952.

Evans, I.S. (1980) An integrated systems of terrain analysis and slop mapping. *Zeitschrift fur Geomorphologies*, (Suppl.) 36: 274–295.

Eveleigh, T.J., Mazzuchi, T.A., and Sarkani, S. (2006) Systems engineering design and spatial modeling for improved natural hazard risk assessment. *Disaster Prevention and Management* 15: 636–648.

Ewen, J., and Parkin, G. (1996) Validation of catchment models for predicting land-use and climate change impacts, 1: Method. *Journal of Hydrology* 175: 583–594.

Faulkner, B.R. (2002) Java classes for nonprocedural variogram modeling. *Computers & Geosciences* 28: 387–397.

Fedra, K. (1993) GIS and environmental modeling. In *Environmental modeling with GIS*, ed. M.F. Goodchild, B.O. Parks, and L.T. Steyaert, 35–50. New York: Oxford University Press.

Ferber, J. (2005) Concepts et méthodologies multi-agents. In *Modélisation et simulation multi-agents*, ed. F. Amblard, and D. Phan, D., 23–48. Paris: Lavoisier.

Fernandez, B., and Salas, J.D. (1999) Return period and risk of hydrological events I: Mathematical formulation. *Journal of Hydrological Engineering* 4: 297–307.

Ferrand, N. (1996) Modelling and supporting multi-actor spatial planning using multi-agents systems. In *Proceedings Third International Conference/Workshop on Integrating GIS and Environmental Modeling*, Santa Fe, NM. ed. National Center for Geographic Information and Analysis (NCGIA): /ferrand_nils/santafe.html

Ferris, T. (1988) *Coming of age in the Milky Way*. London: Vintage.

Few, R. (2002) Researching actor power: Analyzing mechanisms of interactions in negotiations over space. *Area* 34: 29–38.

Fisher, P.F. (1991) Spatial data sources and data problems. In *Geographical information systems principles and practice*, ed. D.J. Maguire, M.F. Goodchild, and D.W. Rhind, 175–189. Harlow, U.K.: Longman.

Flemming, P. (1934) *One's company*. London: Jonathan Cape.

Fookes, P.G., Dale, S.G., and Land, J.M. (1991) Some observations on a comparative aerial photography interpretation of a landslide area. *Quarterly Journal of Engineering Geology* 24: 249–265.

Forman, M.G.G., and Walters, R.A. (1990) A finite-element tidal model for the southwest coast of Vancouver Island. *Atmosphere-Ocean* 28: 261–287.

Fotheringham, A. S. (1992) Exploratory spatial data analysis and GIS. *Environment and Planning A* 24: 1675–1678.

Fotheringham, A.S. (1998) Trends in quantitative methods II: Stressing the computational. *Progress in Human Geography* 22: 283–292.

Fotheringham, A.S., Brunsdon, C., and Charlton, M. (2000) *Quantitative geography: Perspectives on spatial data analysis*. London: Sage.

Fotheringham, A.S., and Knudsen, D.C. (1987) *Goodness-of-fit statistics*. Norwich, U.K.: Geo Books.

Fotheringham, A.S., and Wong, D.W.S. (1991) The modifiable areal unit problem in multivariate statistical analysis. *Environment & Planning A* 23: 1025–1044.

Fotheringham, A.S., and Zhan, F.B. (1996) A comparison of three exploratory methods for cluster detection in spatial point patterns. *Geographical Analysis* 28: 200–218.

Frank, A.U. (2000) Geographical information science: New methods and technology. *Journal of Geographical Systems* 2: 99–105.

Frank, A.U. (2008) Analysis of dependence of decision quality on data quality. *Journal of Geographical Systems* 10: 71-88.

Frank, A.U., and Raubal, M. (2001) GIS education today: From GI science to GI engineering. *URISA Journal* 13 (2): 5–10.

Gaillard, J.C., Liamzon, C., and Villanueva, J.D. (2007) 'Natural' disasters? A retrospect into the causes of the late-2004 typhoon disaster in Eastern Luzon, Philippines. *Environmental Hazards* 7: 257–270.

Gatrell, A.C., Bailey, T.C., Diggle, P.J., and Rowlingson, B.S. (1996) Spatial point pattern analysis and its application in geographical epidemiology. *Transactions of the Institute of British Geographers* 21: 256–274.

Gerrard, A.J. (1981) *Soils and landforms*. London: George Allan & Unwin.

Gilbert, N. (2008) *Agent-based models*. Thousand Oaks, CA: Sage Publications.

Gleick, J. (1987) *Chaos, making a new science*. London: Heinemann.

Golding, B.W. (2009) Uncertainty propagation in a London flood simulation. *Journal of Flood Risk Management* 2: 2–15.

Goldsmith, V., McGuire, P.G., Mollenkopf, J.H., and Ross, T.A. (2000) *Analysing crime patterns: Frontiers of ractice*. Thousand Oaks, CA: Sage.

Goodchild, M.F. (1980) Fractals and the accuracy of geographical measures. *Mathematical Geology* 12: 85–98.

Goodchild, M.F. (1986) *Spatial autocorrelation.* Catmog 47. Norwich, U.K.: Geo Books.

Goodchild, M.F. (1990a) Keynote address: Spatial information science. *Proceedings of the 4ᵗʰ International Symposium on Spatial Data Handling,* Zurich, 13–14.

Goodchild, M.F. (1990b) Modeling error in spatial databases. *Proceedings of the GIS/ LIS'90,* Anaheim, CA, 154–162.

Goodchild, M.F. (1992a) Geographical data modelling. *Computers & Geosciences* 18: 401–408.

Goodchild, M.F. (1992b) Geographical information science. *International Journal of Geographical Information Systems* 6: 31–45.

Goodchild, M.F. (1993) The state of GIS for environmental problem-solving. In *Environmental modeling with GIS,* ed. M.F. Goodchild, B.O. Parks, and L.T. Steyaert, 8–15. New York: Oxford University Press.

Goodchild, M.F. (2001) Metric scale in remote sensing and GIS. *International Journal of Applied Earth Observation and Geoinformation* 3: 114–120.

Goodchild, M.F., and Gopal, S., eds. (1989) *The accuracy of spatial databases.* London: Taylor & Francis.

Goodchild, M.F., Parks, B.O., and Steyaert, L.T., eds. (1993) *Environmental modeling with GIS.* New York: Oxford University Press.

Goodchild, M.F., and Proctor, J. (1997) Scale in a digital geographic world. *Geographical & Environmental Modelling* 1: 5–23.

Goodchild, M.F., Steyaert, L.T., Parks, B.O., Johnson, C., Maidment, D., Crane, M., and Glendinning, S., eds. (1996) *GIS and environmental modeling: Progress in research issues.* New York: John Wiley & Sons.

Goodchild, M.F., Yuan, M., and Covas, T.J. (2007) Towards a general theory of geographic representation in GIS. *International Journal of Geographical Information Science* 21: 239–260.

Gopal, S., and Woodcock, C. (1994) Theory and methods of assessment of thematic maps using fuzzy sets. *Photogrammetric Engineering & Remote Sensing* 60: 181–188.

Goudie, A., ed. (1997) *The human impact reader.* Oxford, U.K.: Blackwell.

Graham, I., and Jones, P.L. (1988) *Expert systems: Knowledge, uncertainty and decision.* London: Chapman & Hall.

Grayson, Pr.B., Bloschi, G., Barling, R.D., and Moor, I.D. (1993) Process, scale and constraints to hydrological modelling in GIS. In *Applications of geographical information systems and water resources management,* ed. K. Kovar, and H.P. Nachtnebel, 83–92. Wallingford, U.K.: IAHS Press.

Greenberg, M.R., Anderson, R., and Page G.W. (1978) *Environmental impact statements.* Washington, D.C.: Association of American Geographers.

Greig-Smith, P. (1964) *Quantitative plant ecology.* Oxford, U.K.: Blackwell Scientific.

Grejner-Brzezinska, D.A., Li, R., Haala, N., and Toth, C. (2004) From mobile mapping to telegeoinformatics: Paradigm shift in geospatial data acquisition, processing and management. *Photogrammetric Engineering & Remote Sensing* 70: 197–210.

Griffith, D.A. (1987) *Spatial autocorrelation* : A primer. Washington D.C.: Association of American Geographers.

Griffiths, J.S. (1999) Proving the occurrence and cause of a landslide in a legal context. *Bulletin of Engineering Geology and the Environment* 58: 75–85.

Gronlund, A.G., Xiang, W.N., and Sox, J. (1994) GIS, expert systems technologies improve forest fire management techniques. *GIS World,* (February) 32–36.

Gruber, T.R. (1995) Towards principles of the design of ontologies used for knowledge sharing. *International Journal of Human-Computer Studies* 43: 907–28.

Grunblatt, J. (1987) An MTF analysis of Landsat classification error at field boundaries. *Photogrammetric Engineering & Remote Sensing* 53: 639–643.

Guesgen, H.W., and Albrecht, J. (2000) Imprecise reasoning in geographic information systems. *Fuzzy Sets and Systems* 113: 121–131.

Gupta, M.M., Ragade, R.K., and Yager, R.R. (1979) *Advances in fuzzy set theory and applications.* Amsterdam: North-Holland Publishing Company.

Hadipriono, F.C., Lyon, J.G., and Li, T.W.H. (1991) Expert opinion in satellite data interpretation. *Photogrammetric Engineering & Remote Sensing* 57: 75–78.

Haines-Young, R., Green, D.R., and Cousins, S., eds. (1993) *Landscape ecology and geographic information systems.* London: Taylor & Francis.

Hallett, S.H., Jones, R.J.A., and Keay, C.A. (1996) Environmental information systems development for planning sustainable land use. *International Journal of Geographical Information Systems* 10: 47–64.

Halls, P.J., Bulling, M., White, P.C.L., Garland, L., and Harris S. (2001) Dirichlet neighbours: Revisiting Dirichlet tessellation for neighbourhood analysis. *Computers, Environment and Urban Systems* 25: 105–117.

Hamilton, R.N. (2000) Science and technology for natural disaster reduction. *Natural Hazards Review* 1: 56–60.

Haque, A. (2003) Information technology, GIS and democratic values: Ethical implications for IT professional in public service. *Ethics and Information Technology* 5: 39–48

Harder, C. (1998) *Serving maps on the Internet.* Redlands, CA: ESRI.

Hardisty, J., Taylor, D.M., and Metcalfe, S.E. (1993) *Computerised environmental modelling, a practical introduction using Excel.* Chichester, U.K.: John Wiley & Sons.

Harris, C. (1997) The Sea Empress incident: Overview and response at sea. *Proceedings of the 1997 International Oil Spill Conference,* American Petroleum Institute, Fort Lauderdale, FL, pp. 177–184.

Harris, J.W., and Stocker, H. (1998) *Handbook of mathematics and computational science.* New York: Springer.

Harvey, D.W. (1966) Geographical processes and point patterns: Testing models of diffusion by quadrat sampling. *Transactions of the Institute of British Geographers* 40: 81–95.

Harvey, D.W. (1973) *Explanation in Geography.* Edward Arnold, London, UK.

Heil, R.J. and Brych, S.M. (1978) An approach to consistent topographic representation of varying terrain. *American Society of Photogrammetry DTM Symposium,* St. Louis: 397–411.

Hellweger, F.L., and Maidment, D.R. (1999) Definition and connection of hydrological elements using geographical data. *Journal of Hydrologic Engineering* 4: 10–18.

Heuvelink, G.B.M. (1998) *Error-propagation in environmental modelling.* London: Taylor & Francis.

Heuvelink, G.B.M., and Burrough, P.A. (1993) Error propagation in cartographic modelling using Boolean logic and continuous classification. *International Journal of Geographical Information Systems* 7: 231–246.

Heywood, I., Cornelius, S., and Carver, S. (2006) *An introduction to geographical information systems.* Harlow, U.K.: Longman.

Hirzel, A., and Guisan, A. (2002) Which is the optimal sampling strategy for habitat suitability modelling? *Ecological Modelling* 157: 331–341.

Hodgson, M.E. (1995) What cell size does the computed slope/aspect angle represent? *Photogrammetric Engineering & Remote Sensing* 61: 513–517.

Hoffman, C. (1989) *Geometric and solid modeling*. San Mateo, CA: Morgan Kaufmann.

Holt, A., and Benwell, G.L. (1999) Applying case-based reasoning techniques in GIS. *International Journal of Geographical Information Science* 13: 9–25.

Hootsman, R.M., and van der Wel, F. (1993) Detection and visualization of ambiguity and fuzziness in composite spatial datasets. *Proceedings of the EGIS'93*, Genoa, Italy, pp. 1035–1046.

Hord, M.R., and Brooner, W. (1976) Land-use map accuracy criteria. *Photogrammetric Engineering & Remote Sensing* 42: 671–677.

Horn, B.K.P. (1981) Hill shading and the reflectance map. *Proceedings of the IEEE* 69: 14–47.

Howard, D. (1999) Geographic information technologies and community planning: Spatial empowerment and public participation. In *Empowerment, marginalisation and public participation GIS*, ed. W. Craig, T. Harris, and D. Weiner, 41–43. Santa Barbara, CA: NCGIA.

Huang, B., and Jiang, B. (2002) AVTOP: A full integration of TOPMODEL into GIS. *Environmental Modelling & Software* 17: 261–268.

Hudson, W.D., and Ramm, C.W. (1987) Correct formulation of the kappa coefficient of agreement. *Photogrammetric Engineering & Remote Sensing* 53: 421–422.

Hull, B. (2003) ICT and social exclusion: The role of libraries. *Telematics and Informatics* 20: 131–142.

Hunter, G.J. (1999) New tools for handling spatial data quality: moving from academic concepts to practical reality. *URISA Journal* 11 (2): 25–34.

Hunter, G.J., and Goodchild, M.F. (1993) Managing uncertainty in spatial databases: Putting theory into practice. *URISA Journal* 5 (2): 55–62.

Isaaks, E.H., and Srivastava, R.M. (1989) *An introduction to applied geostatistics*. New York: Oxford University Press.

Jacquez, G.M. (2008) Spatial cluster analysis. In *The handbook of geographic information science*, ed. J.P. Wilson and A.S. Fotherinham, 395–416. Malden, MA: Blackwell.

Janbu, N. (1968) *Slope stability computations*. Soil Mechanics and Foundation Engineering Report, Technical University of Norway, Trondheim, Norway.

Jankowski, P., and Nyerges, T.L. (2008) Geographic information systems and participatory decision making. In *The handbook of geographic information science*, ed. J.P. Wilson and A.S. Fotherinham, 481–493. Malden, MA: Blackwell..

Jennings, N. (2000) On agent-based software engineering. *Artificial Intelligence* 117: 277–296.

Jensen, J.R., Halls, J.N., and Michel, J. (1992) Utilisation of remote sensing and GIS technologies for oil spill planning and response: Case studies in Florida, United Arab Emirates and Saudi Arabia. *Proceedings from the First Thematic Conference on Remote Sensing for Marine and Coastal Environments*, New Orleans, LA.

Jensen, J.R., Halls, J.N., and Michel, J. (1998) A systems approach to environmental sensitivity index (ESI) mapping for oil spill contingency planning and response. *Photogrammetric Engineering & Remote Sensing* 64: 1003–1014.

Jones, D.K.C. (1983) Environments of concern. *Transactions of the Institute of British Geographers* 8: 429–457.

Jones, K.H. (1998) A comparison of algorithms used to compute hill slope as a property of the DEM. *Computers & Geosciences* 24: 315–323.

Jun, C. (2000) Design of an intelligent geographic information system for multi-criteria site analysis. *URISA Journal* 12 (3): 5–17.

Kanevski, M., and Maignan, M. (2004) *Analysis and modelling of spatial environmental data*. Lausanne, Switzerland: EPFL Press.

Karimi, H.A., and Houston, B.H. (1996) Evaluating strategies for integrating environmental models: Current trends and future needs. *Computers, Environment and Urban Systems* 20: 411–425.

Karssenberg, D., and de Jong, K. (2005a) Dynamic environmental modelling in GIS: 1. Modelling in three spatial dimensions. *International Journal of Geographical Information Science* 19: 559–579.

Karssenberg, D., and de Jong, K. (2005b) Dynamic environmental modelling in GIS: 2. Modelling error propagation. *International Journal of Geographical Information Science* 19: 623–637.

Kaufmann, A. (1973) *Introduction a la theorie des sous-ensembles flous*, Tome 1: *Elements theoretiques de base*. Paris: Masson.

Kaufmann, A. (1975a) *Introduction a la theorie des sous-ensembles flous*, Tome 2: *Applications a la linguistique, a la logique et a la sementique*. Paris: Masson.

Kaufmann, A. (1975b) *Introduction a la theorie des sous-ensembles flous*, Tome 3: *Applications a la classification et a la reconnaissance des formes, aux automates et aux systemes aux choix de critieres*. Paris: Masson.

Keates, J.S. (1982) *Understanding maps*. London: Longman.

Keefer, B., Smith, J., and Gregoire, T. (1988) Simulating manual digitizing error with statistical models. *Proceedings of the GIS/LIS'88*, San Antonio, TX, pp. 475–483.

Kingston, R., Carver, S., Evans, A., and Turton, I. (2000) Web-based public participation geographical information systems: An aid to local environmental decision-making. *Computers, Environment and Urban Systems* 24: 109–125.

Kirkby, M.J. (2000) Limits to modelling in the Earth and environmental sciences. In *GeoComputation*, ed. S. Openshaw and R.J. Abrahart, London: Taylor & Francis.

Kirkby, M.J., Naden, P.S., Burt, T.P., and Butcher, D.P. (1987) *Computer Simulation in Physical Geography*, Chichester, U.K.: John Wiley & Sons.

Klir, G.J., and Folger, T.A. (1988) *Fuzzy sets, uncertainty and information*. Upper Saddle River, NJ: Prentice Hall.

Knox, E.G. (1964) The detection of space-time interactions. *Applied Statistics* 66: 13: 25–29.

Kollias, V.J., and Voliotis, A. (1991) Fuzzy reasoning in the development of geographical information systems. *International Journal of Geographical Information Systems* 5: 209–223.

Kong, W.K. (2002) Risk assessment of slopes. *Quarterly Journal of Engineering Geology and Hydrogeology* 35: 213–222.

Kraak, M-J. (2001) Cartographic principles. In *Web cartography: Developments and prospects*, ed. M-J. Kraak and A. Brown, 53–72. London: Taylor & Francis.

Kraak, M-J. and Brown, A., eds. (2001) *Web cartography: Developments and prospects*. London: Taylor & Francis.

Krige, D.G. (1966) Two-dimensional weighted moving average trend surfaces for ore valuation. *Journal of South African Institute of Mining & Metallurgy* 13–38

Lake, R.W. (1993) Planning and applied geography: positivism, ethics and geographic information systems. *Progress in Human Geography* 17: 404–413.

Lakoff, G. (1973) Hegdes: A study in meaning criteria and the logic of fuzzy concepts. *Journal of Philosophical Logic* 2: 458–508.

Lam, D.C.L. (1993) Combining ecological modeling, GIS and expert systems: A case study of regional fish species richness model. In *Environmental modeling with GIS*, ed. M.F. Goodchild, B.O. Parks, and L.T. Steyaert, 270–275. New York: Oxford University Press.

Lam, N.S-N. (1983) Spatial interpolation methods: A review. *The American Cartographer* 10: 129–149.

Lam, N.S-N., and Quattrochi, D.A. (1992) On the issues of scale, resolution and fractal analysis in the mapping sciences. *Professional Geographer* 44: 88–98.

Lanter, D.P. (1991) Design of a lineage-based meta-data base for GIS. *Cartography and Geographic Information Systems* 18: 255–261.

Lanter, D.P., and Veregin, H. (1992) A research paradigm for propagating error in layer-based GIS. *Photogrammetric Engineering & Remote Sensing* 58: 825–833.

Laurini, R., Servigne, S., and Tanzi, T. (2001) A primer on TeleGeoProcessing and TeleGeoMonitoring. *Computers, Environment & Urban Systems* 25: 248–265.

Lave, C.A., and March, J.G. (1993) *An introduction to models in the social sciences*. New York: University Press of America.

Law, J.S.Y. (1994) Data conversion, updating and integration of LIS data in a CAD system. Unpublished MSc dissertation, Hong Kong Polytechnic University.

Law, J.S.Y., and Brimicombe, A.J. (1994) Integrating primary and secondary data sources in land information systems for recording change. *Proceedings of the XX FIG Congress*, Melbourne, pp. 478–489.

Lawson, A.B. (2001) *Statistical methods in spatial epidemiology*. Chichester, U.K.: John Wiley & Sons.

Lee, J., and Wong, D.W.S. (2001) *Statistical analysis with ArcView GIS*. New York: John Wiley & Sons.

Leopold, L., Clarke, B., Hanshaw, B., and Balsley, J. (1971) *A procedure for evaluating environmental impact*. Washington D.C.: U.S. Geological Survey, Circular 645.

Lewin, R. (1993) *Complexity, life on the edge of chaos*. New York: Macmillan.

Li, C., and Maguire, D.J. (2003) The handheld revolution: Toward ubiquitous GIS. In *Advanced spatial analysis*, ed. P. Longley, and M. Batty. Redlands, CA: ESRI Press.

Li, Y. (2001) Spatial data quality analysis in the environmental modelling. Unpublished PhD dissertation, University of East London.

Li, Y. (2006) Spatial data quality analysis with agent technologies. *Proceedings of the GISRUK 2006*, Nottingham, U.K., 250–254.

Li, Y. (2007) Control of spatial discretisation in coastal oil spill modeling. *International Journal of Applied Earth Observation and Geoinformation* 9: 392–402.

Li, Y., Brimicombe, A.J., and Ralphs, M.P. (1998) Spatial data quality and coastal spill modelling. In *Oil and hydrocarbon spills: Modelling, analysis and control*, ed. R. Garcia-Marinez and C.A. Brebbia, 51–62. Southampton, U.K.: Computational Mechanics Publications.

Li, Y., Brimicombe, A.J., and Ralphs, M.P. (2000) Spatial data quality and sensitivity analysis in GIS and environmental modelling: The case of coastal oil spills. *Computers, Environment and Urban Systems* 24: 95–108.

Li, Y, Brimicombe, A.J., and Li, C. (2008) Agent-based services for the validation and calibration of multi-agent models. *Computers, Environment and Urban Systems* 32: 464–473.

Lilburne, L. (1996) The integration challenge. *Proceedings of the 8th Annual Colloquium of the Spatial Information Research Centre*, Dunedin, New Zealand, 85–94.

Lilburne, L., and Tarantola, S. (2009) Sensitivity analysis of spatial models. *International Journal of Geographical Information Science* 23: 151–168.

Livingstone, D., and Raper, J. (1994) Modelling environmental systems with GIS: Theoretical barriers to progress. In *Innovations in GIS 1*, ed. M.F. Worboys, 229–240. London: Taylor & Francis.

Lodwick, W.A., Monson, W., and Svoboda, L. (1990) Attribute error and sensitivity analysis of map operations in geographical information systems: suitability analysis. *International Journal of Geographical Information Systems* 4: 413–428.

Lofgren, B.M. (1995) Surface albedo-climate feedback simulated using two-way coupling. *Journal of Climate* 8: 2543–2562.

Long, M.E. (2002) America's nuclear waste. *National Geographic* 202 (1): 2–33.

Longley, P.A. (1998) Editorial: developments in geocomputation. *Computers, Environment and Urban Systems* 22: 81–83.

Longley, P.A., Brooks, S.M., McDonnell, R., and Macmillan, W.D., eds. (1998) *Geocomputation: A primer.* Chichester, U.K.: John Wiley & Sons.

Longley, P.A., Goodchild, M., Maguire, D., and Rhind, D. (2005) *Geographical information systems and science.* Chichester, U.K.: John Wiley & Sons.

Lovelock, J. (1988) *The ages of gaia, a bibliography of our living Earth.* Oxford, U.K.: Oxford University Press.

Lowry, J.H., Ramsey, R.D., Stoner, L.L., Kirby, J., and Schulz, K. (2008) An ecological framework for evaluating map errors using fuzzy sets. *Photogrammetric Engineering & Remote Sensing* 74: 1509–1519.

Lunetta, R.S., Congalton, R.G., Fenstermaker, L.K., Jensen, J.R., McGwire, K.C., and Tinney, L.R. (1991) Remote sensing and geographic information systems: Error sources and research issues. *Photogrammetric Engineering & Remote Sensing* 57: 677–687.

MacDougall, E.B. (1975) The accuracy of map overlays. *Landscape Planning* 2: 23–30.

MacEachren, A.M., and Kraak, A.J. (1997) Exploratory cartographic visualization: Advancing the agenda. *Computers & Geosciences* 23: 335–344.

Macmillan, W.D. (1995) Modelling: Fuzziness revisited. *Progress in Human Geography* 19: 404–413.

Macmillan, W.D. (1998) Epiloque. In *Geocomputation: A primer,* ed. P.A. Longley et al., 257–264. Chichester, U.K.: John Wiley & Sons.

Maeda, E.J., Formaggio, A.R., Shimabukuro, Y.E., Arcoverde, G.F.B., and Hansen, M.C. (2009) Predicting forest fire in the Brazilian Amazon using MODIS imagery and artificial neural networks. *International Journal of Applied Earth Observation and Geoinformation* 11: 265–272.

Maffini, G., Arno, M., and Bitterlich, W. (1989) Observations and comments on the generation and treatment of errors in digital GIS data. *Proceedings of the CISM National Conference on GIS*, Ottawa, Canada, 201–219.

Maguire, D.J. (1991) An overview and definition of GIS. In *Geographical information systems principles and practice*, ed. D.J. Maguire, M.F. Goodchild, and D.W. Rhind, 9–20. Harlow, U.K.: Longman.

Maidment, D.R. (1993a) GIS and hydrologic modelling. In *Environmental modeling with GIS*, ed. M.F. Goodchild, B.O. Parks, and L.T. Steyaert, 147–167. New York: Oxford University Press.

Maidment, D.R. (1993b) Environmental modelling within GIS. In *Environmental modeling with GIS*, ed. M.F. Goodchild, B.O. Parks, and L.T. Steyaert, 315–323. New York: Oxford University Press.

Malkina-Pykh, I.G. (2000) From data and theory to environmental models and indices formation. *Ecological Modelling* 130: 67–77.

Mandelbrot, B. (1967) How long is the coast of Britain? Statistical self-similarity and fractal dimension. *Science* 155: 636–638.

Mandelbrot, B. (1983) *The fractal geometry of nature*. New York: Freeman.

Manson, S.M. (2007) Challenges in evaluating models of geographic complexity. *Environment and Planning B: Planning and Design* 34: 245–260.

Mark, D. (1975) Computer analysis of topography: A comparison of terrain storage methods. *Geografiska Annaler* 57a: 179–188.

Marr, A.J., Pascoe, R.T., Benwell, G.L., and Mann, S. (1998) Development of a generic system for modelling spatial processes. *Computers, Environment and Urban Systems* 22: 57–69.

Marsh, G.P. (1864) *Man and nature*. New York: Scibner.

Martin, Y., and Church, M. (2004) Numerical modeling of landscape evolution: Geomorphological perspectives. *Progress in Physical Geography* 28: 317–339.

Martinez-Casasnovas, J.A., and Stuiver, H.J. (1998) Automated delineation of drainage networks and elementary catchments from digital elevation models. *ITC Journal* 1998-3: 198–208.

Martz, L.W., and Garbrecht, J. (1998) The treatment of flat areas and depressions in automated drainage analysis of raster digital elevation models. *Hydrological Processes* 12: 843–855.

Mason, P.J., and Rosenbaum, M.S. (2002) Geohazard mapping for predicting landslides: An example from Langhe Hills in Piemonte, NW Italy. *Quarterly Journal of Engineering Geology and Hydrogeology* 35: 317–326.

Mazzotti, F.J., and Vinci, J.J. (2007) Validation, verification, and calibration: Using standardized terminology when describing ecological models. *Document WEC216*. Wildlife Ecology and Conservation Department, University of Florida, Gainsville.

McDonnell, R.A. (1996) Including the spatial dimension: Using geographical information systems in hydrology. *Progress in Physical Geography* 20: 159–177.

McHaffie, P. (2000) Surfaces: Tacit knowledge, formal language, and metaphor at the Harvard Lab for Computer Graphics and Spatial Analysis. *International Journal of Geographical Information Science* 14: 755–773.

McHarg, I. (1969) *Design with nature*. New York: Doubleday.

McIntosh, B.S. (2003) Qualitative modelling with imprecise ecological knowledge: A framework for simulation. *Environmental Modelling & Software* 18: 295–307.

Medyckyj-Scott, D., Newman, I., Ruggles, C., and Walker, D., eds. (1991) *Metadata in geosciences*. Loughborough, U.K.: Group D Publications.

Mercer, J. Kelman, I., Lloyd, K., and Suchet-Pearson, S. (2008) Reflections on use of participatory research for disaster risk management. *Area* 40: 172–183.

Messerli, B., and Messerli, P. (1978) Wirtschaftliche entwicklung und ökologische belastbarkeit im berggebiet. *Geographica Helvetica* 33: 203–210.

Middelkoop, H. (1990) Uncertainty in GIS: A test of quantifying interpretation input. *ITC Journal* 1990-3: 225–232.

Miles, S.B., and Ho C.L. (1999) Application and issues of GIS as a tool for civil engineering modelling. *Journal of Computing in Civil Engineering* 13: 144-152.

Miller, H.J. and Han, J., eds. (2001) *Geographical data mining and knowledge discovery*. London: Taylor & Francis.

Millsom, R. (1991) Accuracy moves from maps to GIS: Has anything changed? *Proceedings of Symposium on Spatial Database Accuracy*, Melbourne, 17–22.

Milne, P., Milton, S., and Smith, J.L. (1993) Geographical object-orientated databases—A case study. *International Journal of Geographical Information Systems* 7: 39–56.

Molenaar, M. (1998) *An introduction to the theory of spatial object modelling.* London: Taylor & Francis.

Molkethin, F. (1996) Impact of grid modelling on physical results. In *Coastal environment: Environmental problems in coastal regions*, ed. A.J. Ferrante and C. Brebbia, 295–304. Southampton, U.K.: Computational Mechanics Publications.

Molná, D.K., and Julien, P.Y. (2000) Grid-size effects on surface runoff modeling. *Journal of Hydrologic Engineering* 5: 8–16.

Mooney, C.Z. (1997) *Monte Carlo simulation.* Thousand Oaks, CA: Sage Publications,

Morisawa, M., and LaFure, E. (1979) Hydraulic geometry, stream equilibrium and urbanisation. In *Adjustments of the fluvial system*, ed. D. Rhodes and G. Williams, 333–350. London: George Allen & Unwin.

Morita, M. (2008) Flood risk analysis for determining optimal flood protection levels in urban river management. *Journal of Flood Risk Management* 1: 142–149.

Morton, G. (1966) A computer-oriented geodetic data base and a new technique in file sequencing. Unpublished report, IBM Canada Ltd, Ottawa, Canada.

Muller, J-C. (1987) The concept of error in cartography. *Cartographica* 24: 1–15

Munich Reinsurance Company (2001) *Annual review of natural catastrophes 2000.* Munich: Munich Reinsurance Company.

Murray, A.T., and Estivill-Castro, V. (1998) Cluster discovery techniques for exploratory spatial data analysis. *International Journal of Geographical Information Science* 12: 431–443.

Nash, J.E., and Sutcliffe, J.V. (1970) River flow forecasting through conceptual models 1: A discussion of principles. *Journal of Hydrology* 10: 282–290.

National Center for Geographic Information and Analysis (NCGIA) (1996) *Proceedings of the Third International Conference/Workshop on Integrating GIS and Environmental Modeling*, Santa Fe, NM, on CD.

National Center for Geographic Information and Analysis (NCGIA) (2000) *Proceedings of the Fourth International Conference/Workshop on Integrating GIS and Environmental Modeling*, Banff, Alberta, Canada. http://www.colorado.edu/research/cires/banff/ (accessed on June 9, 2009)

National Institute of Standards and Technology (1992) *Federal information processing standard publication 173 (spatial data transfer standard).* Washington, D.C.: U.S. Department of Commerce.

National Telecommunications and Information Administration (NTIA) (2000) *Falling through the net: Towards digital inclusion.* Washington, D.C.: NTIA.

Nazaroff, W., and Alvarez-Cohen, L. (2001) *Environmental engineering science.* New York: John Wiley & Sons.

Nefeslioglu, H.A., Gokceoglu, C., and Sonmez, H. (2008) An assessment of the use of logistic regression and artificial neural networks with different sampling strategies for the preparation of landslide susceptibility maps. *Engineering Geology* 97: 171–191.

Newcomer, J.A., and Szajgin, J. (1984) Accumulation of thematic map errors in digital overlay analysis. *The American Cartographer* 11: 58–62.

Nowakowska, M. (1986) *Cognitive sciences.* New York: Academic Press.

O'Callaghan, J.F., and Mark, D.M. (1984) The extraction of drainage networks from digital elevation data. *Computer Vision, Graphics and Image Processing* 28: 323–344.

O'Neill, R.V., Krummel, J.R., Gardner, R.H., Sugihara, G., Jackson, B., DeAngelis, B.L., Milne, B.T., Turner, M.G., Zygmunt, B., Christensen, S., Dale, V.H., and Graham, R.L. (1988) Indices of landscape pattern. *Landscape Ecology* 1: 153–162.

O'Sullivan, D. (2008) Geographical information science: Agent-based models. *Progress in Human Geography* 32: 541–550.

O'Sullivan, D., and Unwin, D.J. (2003) *Geographical information analysis*. Hoboken, NJ: John Wiley & Sons.

Odum, H.T. (1971) *Environment, power and society*. New York: John Wiley & Sons.

OGC (2002) *Vision, mission & values*. Open Geospatial Consortium, http://www.opengeospatial.org/ogc/vision (accessed June 9, 2009)

Openshaw, S. (1989) Learning to live with errors in spatial databases. In *The accuracy of spatial databases*, ed. M.F. Goodchild and S. Gopal, 263–276. London: Taylor & Francis.

Openshaw, S. (1998) Building automated geographical analysis and explanation machines. In *Geocomputation: A primer*, ed. P.A. Longley et al., 956–115. Chichester, U.K.: John Wiley & Sons.

Openshaw, S., and Abrahart, R.J. (1996) Geocomputation. *Proceedings of the GeoComputation '96*, University of Leeds, U.K., 665–666.

Openshaw, S., Charlton, M.E., Wymer, C., and Craft, A.W. (1987) A mark I geographical analysis machine for the automated analysis of point data sets. *International Journal of Geographical Information Systems* 1: 359–377.

Openshaw, S. and Openshaw, C. (1997) *Artificial intelligence in geography*. Chichester, U.K.: John Wiley & Sons.

Openshaw, S., and Taylor, P.J. (1981) The modifiable areal unit problem. In *Quantitative geography*, ed. N. Wrigley and R.J. Bennett, 60–70. Henley-on-Thames, U.K.: Routledge.

Ord, J.K., and Getis, A. (1995) Local spatial autocorrelation statistics: Distributional issues and an application. *Geographical Analysis* 27: 286–306.

Oreskes, N., Schrader-Frechette, K., and Belitz, K. (1994) Verification, validation and confirmation of numerical models in the Earth sciences. *Science* 263: 641–646.

Owens, S. (1994) Land, limits and sustainability: a conceptual framework and some dilemmas for the planning system. *Transactions of the Institute of British Geographers* 19: 439–456.

Parker, B. (2006) Constructing community through maps? Power and praxis in community maps. *The Professional Geographer* 58: 470–484.

Parkin, G., O'Donnell, G., Ewen, J., Bathurst, J.C., O'Connell, P.E., and Lavabre, J. (1996) Validation of catchment models for predicting land use and climate change impacts, 1: Case study for a Mediterranean catchment. *Journal of Hydrology* 175: 595–613.

Parks, B.O. (1993) The need for integration. In *Environmental modeling with GIS*, ed. M.F. Goodchild, B.O. Parks, and L.T. Steyaert, 31–34. New York: Oxford University Press.

Patterson, D. (1996) *Artificial neural networks*. Singapore: Prentice Hall.

Payne, J.A. (1982) *Introduction to simulation: Programming techniques and methods of analysis*. New York: McGraw-Hill.

Perkal, J. (1956) On the ε-length. *Bulletin de L'Academie Polonaise des Sciences* IV (7): 399–403.

Peuquet, D.J. (1984) A conceptual framework and comparison of spatial data models. *Cartographica* 21(4): 66–113.

Phillips, J. D. (1999) Spatial analysis in physical geography and challenge of deterministic uncertainty. *Geographical Analysis* 31: 359–372.

Pickles, J. (1995) *Ground truth*. New York: Guilford Press.

Pickles, J. (1997) Tool or science? GIS, technoscience and the theoretical turn. *Annals of the Association of American Geographers* 87: 363–372.

Pirozzi, M.A., and Zicarelli, M. (2000) Environmental modelling on massively parallel computers. *Environmental Modelling & Software* 15: 489–496.

Power Technology (ND) Black Point Power Plant, China. http://www.power-technology.com/projects/blackpoint/index.html (accessed June 9, 2009)

Pozo-Vázquez, D., Olmo-Reyes, F.J., and Aldos-Arboledas, L. (1997) A comparative study of algorithms for estimating land surface temperature from AVHRR data. *Remote Sensing of Environment* 62: 215–222.

President's Information Technology Advisory Committee (PITAC) (1998) *Interim report to the president*. Arlington, VA: National Coordination Office for Computing, Information and Communications.

Pullar, D. (1991) Spatial overlay with inexact numerical data. *Proceedings of the AutoCarto 10*, Baltimore, MD, 313–329.

Raetzo, H., Lateltin, O., Bollinger, D., and Tripet, J.P. (2002) Hazard assessment in Switzerland—Codes of practice for mass movements. *Bulletin of Engineering Geology and the Environment* 61: 263–268.

Reed, M., and French, D. (1991) A natural resource damage assessment model and GIS for the Great Lakes. *Environment Canada Chemical Spills 8th Technical Seminar*, Vancouver, 1–9.

Refsgaard, J.C., and Knudsen, J. (1996) Operational validation and intercomparison of different types of hydrological models. *Water Resources Research* 32: 2189–2202.

Refsgaard, J.C., Storm, B., and Abbott, M.B. (1996) Comment on: A discussion of distributed hydrological modeling by K. Beven. In *Distributed hydrological modelling*, ed. M.B. Abbott and J.C. Refsgaard, 279–331. Dordrecht, Germany: Kluwer.

Rejeski, D. (1993) GIS and risk: A three-cultured problem. In *Environmental modeling with GIS*, ed. M.F. Goodchild, B.O. Parks, and L.T. Steyaert, 318–331. New York: Oxford University Press.

Ritter, P. (1987) A vector-based slope and aspect generation algorithm. *Photogrammetric Engineering & Remote Sensing* 53: 1109–1111.

Robinove, C.J. (1981) The logic of multispectral classification and mapping of land. *Remote Sensing of Environment* 11: 231–244.

Robinson, A.H. and Sale, R.D (1969) *Elements of Cartography*, 3rd ed. New York: John Wiley & Sons.

Robinson, V.B., and Frank, A.U. (1985) About different kinds of uncertainly in collections of spatial data. *Proceedings of the AutoCarto 7*, Washington D.C., 440–449.

Robinson, W.S. (1950) Ecological correlations and the behaviour of individuals. *American Sociological Review* 15: 351–357.

Robrigues, A., and Raper, J. (1999) Defining spatial agents. In *Spatial multimedia and virtual reality*, ed. A.S. Camara and J. Raper, 111–129. London: Taylor & Francis.

Rodda, J.C. (1969) The flood hydrograph. In *Introduction to physical hydrology*, ed. R.J. Chorley, 162–175. London: Methuen.

Rosenberg, D.J., and Martin, G. (1988) Human performance evaluation of digitizer pucks for computer input of spatial information. *Human Factors*, 30 (2): 231–235.

Rosenfield, G.H. (1986) Analysis of thematic map classification error matrices. *Photogrammetric Engineering & Remote Sensing* 52: 681–686.

Royse, K.R., Rutter, H.K., and Entwisle, D.C. (2009) Property attribution of 3D geological models in the Thames Gateway, London: New ways of visualizing geoscientific information. *Bulletin of Engineering Geology and the Environment* 68: 1–16.

Russel, B. (1961) *History of western philosophy*. London: George Allen & Unwin.

Rykiel, E.J. (1996) Testing ecological models: The meaning of validation. *Ecological Modeling*, 90: 224–229.

Sadahiro, Y. (1997) Cluster perception in the distribution of point objects. *Cartographica* 34: 49–61.

Sah, N.K., Sheorey, P.R., and Upadhyaya, L.N. (1994) Maximum likelihood estimation of slope stability. *International Journal of Rock Mechanics, Mineral Science and Geomechanics Abstracts* 31: 47–53.

Sakellariou, M.G., and Ferentinou, M.D. (2001) GIS-based estimation of slope stability. *Natural Hazards Review* 2: 12–21.

Salgé, F. (1995) Semantic accuracy. In *Elements of spatial data quality*, ed. S. Guptil and J. Morrison, 139–151. Oxford, U.K.: Elsevier Science.

Saltelli, A., Chan, K., and Scott, E.M., eds. (2000) *Sensitivity analysis*. Chichester, U.K.: John Wiley & Sons.

Samet, H. (1984) The quadtree and related hierarchical data structure. *Computing Surveys* 16: 187–260.

Saphores, J-D.M. (2004) Environmental uncertainty and the timing of environmental policy. *Natural Resource Modeling* 17: 162–189

Schelling, T.S. (1971) Dynamic models of segregation. *Journal of Mathematical Sociology* 1: 143–186.

Schlesinger, J., Ripple, W., and Loveland, T.R. (1979) Land capability studies of the South Dakota automated geographic information system (AGIS). *Harvard Library of Computer Graphics* 4: 105–114.

Schmidt, J., Evans, I.S., and Brinkmann, J. (2003) Comparison of polynomial models for land surface curvature calculation. *International Journal of Geographic Information Science* 17: 797–814.

Schumacher, E.F. (1973) *Small is beautiful*. London: Blond & Briggs.

Schumm, S.A., and Lichty, R.W. (1965) Time, space and causality in geomorphology. *American Journal of Science* 263: 110–119.

Schuster, R.L. (1996) Socioeconomic significance of landslides. In *Landslides: Investigation and mitigation*, ed. A.K. Turner and R.L. Schuster, 12–35. Washington D.C.: Transport Research Board, National Research Council, Special Report 247.

Sebastiao, P., and Soares, C.G. (1995) Modeling the fate of oil spill at sea. *Spill Science and Technology Bulletin* 2: 121–131.

Selman, P. (1992) *Environmental planning*. London: Paul Chapman Publishing.

Sengupta, R.R., and Bennett, D. (2007) Agent-based modelling environment for spatial decision support. *International Journal of Geographic Information Science* 17: 157–180.

Sengupta, R.R., and Sieber, R.E. (2007) Geospatial agents, agents everywhere ... *Transactions in GIS* 11: 483–506.

Shannon, C.E. (1948) A mathematical theory of communications. *Bell Systems Technical Journal* 27: 379–423, 623–656.

Sharpnak, D.A., and Akin, G. (1969) An algorithm for computing slope and aspect from elevations. *Photogrammetric Engineering* 35: 247–248.

Shearer, J.W. (1990) The accuracy of digital terrain models. In *Terrain modelling in surveying and civil engineering*, ed. G. Petrie and T. Kennie, 315–336. Caithness, U.K.: Whittles.

Shekhar, S., and Vatsavai, R.R. (2008) Object-oriented database management systems. In *The handbook of geographic information science*, ed. J.P. Wilson and A.S. Fotherinham, 111–143. Malden, MA: Blackwell.

Shi, J.J. (2000) Reducing prediction error by transforming input data for neural networks. *Journal of Computing in Civil Engineering* 14: 109–116.

Shi, W., Fisher, P.F., and Goodchild, M.F., eds. (2002) *Spatial data quality*. London: Taylor & Francis.

Shim, J.P., Warkentin, M., Courtney, J.F., Power, D.J., Sharda, R., and Carlsson, C. (2002) Past, present and future of decision support technology. *Decision Support Systems* 33: 111–126.

Sibley, D. (1987) *Spatial applications of exploratory data analysis*. Norwich, U.K.: Geo Books.

Simms, A. (2008) Just 100 months to save the Earth. *The Guardian Weekly* 179 (9): 28–29.

Simon, H.A. (1960) *The new science of management decision*. New York: Harper Brothers.

Sinton, D.F. (1978) The inherent structure of information as a constraint to analysis: mapped thematic data as a case study. In *Harvard papers on geographic information systems*, ed. G. Dutton, 1–17. Harvard University: Graduate School of Design, vol. 7: SINTON.

Skidmore, A.K. (1989) A comparison of techniques for calculating gradient and aspect from gridded digital elevation model. *International Journal of Geographical Information Systems* 3: 323–334.

Skidmore, A.K., ed. (2002) *Environmental modelling with GIS and remote sensing*. London: Taylor & Francis.

Skidmore, A.K. and Turner, B.J. (1992) Map accuracy assessment using line intersect sampling. *Photogrammetric Engineering & Remote Sensing* 58: 1453–1457.

Smith, J.W.F., and Campbell, I.A. (1989) Error in polygon overlay processing of geomorphic data. *Earth Surface Processes and Landforms* 14: 703–717.

Smith, M.B. (1993) A GIS-based distributed parameter hydrologic model for urban areas. *Hydrological Processes* 7: 45–61.

Smith, M.B., and Brilly, M. (1992) Automated grid element ordering for GIS-based overland flow modelling. *Photogrammetric Engineering & Remote Sensing* 58: 579–585.

Snow, J. (1855) *On the mode of communication of cholera*. London: John Churchill.

Soeters, R., and van Westen, C.J. (1996) Slope instability recognition, analysis and zonation. In *Landslides: Investigation and mitigation*, ed. A.K. Turner and R.L. Schuster, 129–177. Washington D.C.: Transport Research Board, National Research Council, Special Report 247.

Sokal, R., and Sneath, P. (1963) *Principles of numerical taxonomy*. San Francisco, CA: Freeman.

Spiekermann, K., and Wegener, M. (2000) Freedom from the tyranny of zones: Towards new GIS-based spatial models. In *patial models and GIS: New potential and new models*, ed. S. Fotheringham and M. Wegener, 45–61. London: Taylor & Francis.

Srinivasan, R., and Engel, B.A. (1991) Effect of slope prediction methods on slope and erosion estimates. *Applied Engineering in Agriculture* 7: 779–783.

StatSoft (1998) *Statistica neural networks*. Tulsa: OK: StatSoft Inc.

Steinitz, C., Parker, P., and Laurie, J. (1976) Hand-drawn overlays: Their history and prospective uses. *Landscape Architecture* 66: 444–455.

Stern, N. (2006) *Stern review: The economics of climate change*. London: HM Treasury (available at: www.hm-treasury.gov.uk/stern-review_report.htm).

Stevens, S.S. (1946) On the theory of scales of measurement. *Science* 103: 677–680.

Stocks, C.E, and Wise, S. (2000) The role of GIS in environmental modelling. *Geographical & Environmental Modelling* 4: 219–235.

Story, M., and Congalton, R.G. (1986) Accuracy assessment: A user's perspective. *Photogrammetric Engineering & Remote Sensing* 52: 397–399.

Sui, A.Z., and Maggio, R.C. (1999) Integrating GIS with hydrological modelling: Practices, problems and prospects. *Computers, Environment and Urban Systems* 23: 33–51.

Swiss Reinsurance Company (1993) Natural catastrophes and major losses in 1992. *Sigma* 2/93.

Taber, C.S., and Timpone, R.J. (1996) *Computational modeling*. Thousand Oaks, CA: Sage.

Tansel, B. (2005) Natural and manmade disasters: Accepting and managing risks. *Safety Science* 20: 91–99.

Tarboton, D.G. (1997) A new method for the determination of flow directions and upslope areas in grid digital elevation models. *Water Resources Research* 33: 309–319.

Thapa, K., and Burtch, R.C. (1990) Issues in data collection in GIS/LIS. *Proceedings of the ACSM-ASPRS Annual Convention*, Denver, CO, 271–283.

Theissen, A.H. (1911) Precipitation averages for large areas. *Monthly Weather Review* 39: 1082–1084.

Therival, R., Glasson, J., and Chadwick, A (2005) *Introduction to environmental impact assessment*, London: Taylor & Francis.

Thornes, J.B., and Brunsden, D. (1977) *Geomorphology and time*. London: Methuen.

Tickell, C. (1993) The human species: A suicidal success? *Geographical Journal* 59: 219–226.

Tobler, W.R. (1970) A computer movie simulating urban growth in the Detroit region. *Economic Geography* 46 (Suppl.) 234–240.

Tokar, A.S., and Johnson, P.A. (1999) Rainfall-runoff modelling using artificial neural networks. *Journal of Hydrologic Engineering* 4: 232–239.

Tomlin, C.D. (1990) *Geographic information systems and cartographic modeling*. Englewood Cliffs, NJ: Prentice Hall.

Tomlinson, R.F. (1984) Geographic information systems—A new frontier. *The Operational Geographer* 5: 31–35.

Townsend, N.R., and Bartlett, J.M. (1992) Formulation of basin management plans for the northern New Territories, Hong Kong. *Proceedings of Conference on Flood Hydraulics*, British Hydraulic Research Group, Florence, Italy, 39–48.

Trudel, B.K., Belore, B.J., Jessiman, B.J., and Ross, S.L. (1987) Development of a dispersant-use decision-making system for oil spills in the U.S. Gulf of Mexico. *Spill Technology Newsletter* 12: 101–110.

Tsou, M., and Buttenfield, B. (2002) A dynamic architecture for distributed geographic information services. *Transactions in GIS* 6: 355–381.

Tsui, P.H.Y., and Brimicombe, A.J. (1997) The hierarchical tessellation model and its use in spatial analysis. *Transactions in GIS* 3: 267–279.

Tufte, E.R. (1983) *The visual display of quantitative information*. Cheshire, CT: Graphics Press.

Turk, G. (1979) GT Index: A measure of success of prediction. *Remote Sensing of the Environment* 8: 65–75.

Turner, A.K., and Schuster, R.L., eds. (1996) *Landslides: Investigation and mitigation.* Washington, D.C.: Transport Research Board, National Research Council, Special Report 247.

Tversky, A., and Kahneman, K, (1974) Judgment under uncertainty: Heuristics and biases. *Science* 185: 1124–1131.

U.S. Bureau of Census (1970) *The DIME geocoding system.* Report No. 4, Census Use Study Washington D.C.: U.S. Department of Commerce.

United Nations (1990) *Urban flood loss and mitigation.* New York: United Nations Publication E.91.II.F.6.

United Nations (1992) *Earth summit agenda 21: The United Nations Programme of Action from Rio.* New York: United Nations Publication E.93.1.11.

United Nations Disaster Relief Organization (1991) *Mitigating natural disasters—Phenomena, effects and options.* New York: United Nations Publication E.90.III.M.1.

Unwin, D. (1995) Geographical information systems and the problem of 'error and uncertainty.' *Progress in Human Geography* 19: 549–558.

Unwin, D. (1996) GIS, spatial analysis and spatial statistics. *Progress in Human Geography* 20: 540–551.

Urbis Travers Morgan Ltd. (1992) *Environmental route assessment study for the Black Point 400kV power transmission system.* Hong Kong.

van Beurden, A.U.C.J., and Douven, W.J.A.M. (1999) Aggregation issues of spatial information and environmental research. *International Journal of Geographical Information Science* 13: 513–527.

van den Brink, A. (1999) Sustainable development and land use planning. In *Land reform and sustainable development,* ed. R. Dixon-Gough, 61–68. Aldershot, U.K.: Ashgate.

van der Meer, F., Schmidt, K.S., Bakker, W., and Bijker, W. (2002) New environmental remote sensing systems. In *Environmental modelling with GIS and remote sensing,* ed. A.K. Skidmore, 26–51. London: Taylor & Francis.

van Deursen, W.P.A., and Wesseling, C.G. (1995) *PCRaster.* Utrecht, The Netherlands: Department of Physical Geography, Utrecht University.

van Gaans, P., and Burrough, P.A. (1993) The use of fuzzy logic and continuous classification in GIS applications: A review. *Proceedings of the EGIS'93,* Genoa, Italy, 1025–1034.

van Niel, K.P., and Lees, B.G. (2000) The case for a data model to reconcile geographical and ecological paradigms. In *Proceedings of the Fourth International Conference/ Workshop on Integrating GIS and Environmental Modeling,* Banff, Alberta, Canada. National Center for Geographic Information and Analysis (NCGIA).

van Westen, C.J. (1993) *Application of geographical information systems to landslide hazard zonation.* Publication No. 15, Enschede, The Netherlands: ITC: International Institute for Aerospace Survey and Earth Sciences.

van Zuidam, R.A. (1985) *Aerial photo-interpretation in terrain analysis and geomorphological mapping.* The Hague, The Netherlands: Smits.

Varekamp, C, Skidmore, A.K., and Burrough, P.A. (1996) Using public domain geostatistical and GIS software for spatial interpolation. *Photogrammetric Engineering & Remote Sensing* 62: 845–854.

Varnes, D.J. (1974) *The logic of geological maps with reference to their interpretation and use for engineering purposes.* Washington, D.C.: U.S. Geological Survey Professional Paper 837.

Varnes, D.J. (1984) *Landslide hazard zonation: A review of principles and practice.* Paris: United Nations Educational, Scientific, and Cultural Organization (UNESCO).

Velleman, P.F., and Wilkinson, L. (1993) Nominal, ordinal, interval and ratio types are misleading. *The American Statistician* 47: 65–72.

Veregin, H. (1989a) Error modelling for the map overlay operation. In *The accuracy of spatial databases*, ed. M.F. Goodchild, and S. Gopal, 3–18. London: Taylor & Francis.

Veregin, H. (1989b) A review of error models for vector to raster conversion. *The Operational Geographer* 7 (1): 11–15.

Veregin, H. (1994) Integration of simulating modeling and error propagation for the buffer operation in GIS. *Photogrammetric Engineering & Remote Sensing* 60: 427–435.

Villa, F., Athanasiadis, I.N., and Rizzoli, A.E. (2009) Modelling with knowledge: A review of emerging semantic approaches to environmental modelling. *Environmental Modelling & Software* 24: 577–587.

Voss, R.F. (1988) Fractals in nature: From characterisation to simulation. In *The science of fractal images*, ed. H.O. Peitgen and D. Saupe, 21–70. New York: Springer-Verlag.

Wachowicz, M. (1999) *Object-oriented design for temporal GIS.* London: Taylor & Francis.

Wagtendonk, A.J., and de Jeu, A.M. (2007) Sensible field computing: Evaluating the use of mobile GIS methods in scientific fieldwork. *Photogrammetric Engineering & Remote Sensing* 73: 651–662.

Walford, N. (1995) *Geographical data analysis.* Chichester, U.K.: John Wiley & Sons.

Walsh, S.J., Lightfoot, D.R., and Buttler, D.R. (1987) Recognition and assessment of error in geographic information systems. *Photogrammetric Engineering & Remote Sensing* 53: 1423–1430.

Wang, F., and Hall, G.B. (1996) Fuzzy representation of geographical boundaries in GIS. *International Journal of Geographic Information Systems* 10: 573–590.

Wang, S.Q., and Unwin, D.J. (1992) Modelling landside distribution on loess soils in China: An investigation. *International Journal of Geographic Information Systems* 6: 391–405.

Weiner, D., and Harris, T.M. (2008) Participatory geographic information systems. In *The handbook of geographic information science*, ed. J.P. Wilson and A.S. Fotherinham, 466–480. Malden, MA: Blackwell.

Welford, R. (1995) *Environmental strategy and sustainable development.* London: Routledge.

Wesseling, C.G., Karssenberg, D-J., Burrough, P.A., and van Deursen, W.P.A. (1996) Integrating dynamic environmental models and GIS: the development of a dynamic modelling language. *Transactions in GIS* 1: 40–48.

Whittemore, R. (2001) Editorial: Is the time right for consensus on model calibration guidance? *Journal of Environmental Engineering* 127: 95–96.

Wiig, K. (1990) *Expert systems: A manager's guide.* Geneva, Switzerland: International Labour Office.

Wilkinson, P.L., Brooks, S.M., and Anderson, M.G. (2000) Design and application of an automated non-circular slip surface search within a combined hydrology and stability model (CHASM). *Hydrological Processes* 14: 2003–2017.

Wilson, J.P., and Fotherinham, A.S., eds. (2008) *The handbook of geographic information science.* Malden, MA: Blackwell.

Wisner, B., Baikie, P., Cannon T., and Davis, I. (2004) *At risk: Natural hazards, people's vulnerability and disasters.* London: Routledge.

Wood, E.F., Sivapalan, M., and Beven, K.J. (1990) Similarity and scale in catchment storm response. *Reviews of Geophysics* 28: 1–18.

Wood, E.F., Sivapalan, M., Beven, K.J., and Band, L. (1988) Effect of spatial variability and scale with implications to hydrologic modeling. *Journal of Hydrology* 102: 29–47.

Wood, J. (2002) *Java programming for spatial sciences.* London: Taylor & Francis.

Wooldridge, M. (1997) Agent-based software engineering. *IEEE Proceedings of Software Engineering* 114: 26–37.

Worboys, M., Hearnshaw, H., and Maguire, D. (1990) Object-oriented data modelling for spatial databases. *International Journal of Geographical Information Systems* 4: 369–383.

Wright, D.J., Goodchild, M.F., and Proctore, J.D. (1997a) GIS: Tool or science. *Annals of the Association of American Geographers* 87: 346–362.

Wright, D.J., Goodchild, M.F., and Proctore, J.D. (1997b) Reply: Still hoping to turn that theoretical corner. *Annals of the Association of American Geographers* 87: 373.

Zadeh, L.A. (1965) Fuzzy sets. *Information & Control* 8: 338–353.

Zadeh, L.A. (1972) A fuzzy-set-theoretical interpretation of linguistic hedges. *Journal of Cybernetics* 2 (3): 4–34.

Zevenbergen, L.W., and Thorne, C.R. (1987) Quantitative analysis of land surface topography. *Earth Surface Processes and Landforms* 12: 47–56.

Zhang, G., and Tulip, J. (1990) An algorithm for the avoidance of sliver polygons in spatial overlay. *4th International Symposium on Spatial Data Handling*, Zurich, Switzerland, 141–150.

Zhang, W., and Montgomery, D.R. (1994) Digital elevation model grid size, landscape representation and hydrologic simulations. *Water Resources Research* 30: 1019–1028.

Index

A

Printed in the United States
by Baker & Taylor Publisher Services